Dopants and Defects in Semiconductors

Second Edition

Dopants and Defects in Semiconductors

Second Edition

Matthew D. McCluskey
Washington State University, Pullman, Washington, USA

Eugene E. Haller
UC Berkeley and Lawrence Berkeley National Laboratory,
Berkeley, California, USA

CRC Press
Taylor & Francis Group
Boca Raton London New York

CRC Press is an imprint of the
Taylor & Francis Group, an **informa** business

CRC Press
Taylor & Francis Group
6000 Broken Sound Parkway NW, Suite 300
Boca Raton, FL 33487-2742

© 2018 by Taylor & Francis Group, LLC
CRC Press is an imprint of Taylor & Francis Group, an Informa business

No claim to original U.S. Government works

International Standard Book Number-13: 978-1-138-03519-5 (Hardback)

This book contains information obtained from authentic and highly regarded sources. Reasonable efforts have been made to publish reliable data and information, but the author and publisher cannot assume responsibility for the validity of all materials or the consequences of their use. The authors and publishers have attempted to trace the copyright holders of all material reproduced in this publication and apologize to copyright holders if permission to publish in this form has not been obtained. If any copyright material has not been acknowledged please write and let us know so we may rectify in any future reprint.

Except as permitted under U.S. Copyright Law, no part of this book may be reprinted, reproduced, transmitted, or utilized in any form by any electronic, mechanical, or other means, now known or hereafter invented, including photocopying, microfilming, and recording, or in any information storage or retrieval system, without written permission from the publishers.

For permission to photocopy or use material electronically from this work, please access www.copyright.com (http://www.copyright.com/) or contact the Copyright Clearance Center, Inc. (CCC), 222 Rosewood Drive, Danvers, MA 01923, 978-750-8400. CCC is a not-for-profit organization that provides licenses and registration for a variety of users. For organizations that have been granted a photocopy license by the CCC, a separate system of payment has been arranged.

Trademark Notice: Product or corporate names may be trademarks or registered trademarks, and are used only for identification and explanation without intent to infringe.

Library of Congress Cataloging-in-Publication Data
Names: McCluskey, Matthew D., author. | Haller, Eugene E., author. Title: Dopants and defects in semiconductors / Matthew D. McCluskey, Eugene E. Haller. Description: Second edition. | Boca Raton, FL : CRC Press, Taylor & Francis Group, [2018] | Includes bibliographical references and index. Identifiers: LCCN 2017042997| ISBN 9781138035195 (hardback ; alk. paper) | ISBN 113803519X (hardback ; alk. paper) Subjects: LCSH: Semiconductor doping. | Semiconductors--Defects. Classification: LCC TK7871.85 .M27 2018 | DDC 660/.2977--dc23 LC record available at https://lccn.loc.gov/2017042997

Visit the Taylor & Francis Web site at
http://www.taylorandfrancis.com

and the CRC Press Web site at
http://www.crcpress.com

Printed and bound in Great Britain by
TJ International Ltd, Padstow, Cornwall

To Jill McCluskey and Marianne Haller

Contents

Preface to the Second Edition xi
Preface to the First Edition xiii
Authors xv
Abbreviations xvii
List of Elements by Symbol xxi

Chapter 1 Semiconductor Basics 1

    1.1   Historical Overview 1
    1.2   Cubic Crystals 3
    1.3   Other Crystals 6
    1.4   Phonons and the Brillouin Zone 8
    1.5   The Band Gap 10
    1.6   Band Theory 11
    1.7   Electrons and Holes 16
    1.8   Doping 19
    1.9   Optical Properties 21
    1.10   Electronic Transport 23
    1.11   Examples of Semiconductors 24
    Problems 25
    References 26

Chapter 2 Defect Classifications 29

    2.1   Basic Definitions 29
    2.2   Energy Levels 30
    2.3   Examples of Native Defects 33
    2.4   Examples of Nonhydrogenic Impurities 35
    2.5   Hydrogen 38
    2.6   Defect Symmetry 39
    2.7   Dislocations 42
    Problems 46
    References 47

Chapter 3 Interfaces and Devices 49

    3.1   Ideal Metal-Semiconductor Junctions 49
    3.2   Real Metal-Semiconductor Junctions 51
    3.3   Depletion Width 55
    3.4   The *p-n* Junction 56
    3.5   Applications of *p-n* Junctions 58
    3.6   The Metal-Oxide-Semiconductor Junction 59
    3.7   The Charge-Coupled Device 61
    3.8   Light-Emitting Devices 62

vii

viii   Contents

	3.9	The 2D Electron Gas	63
		Problems	64
		References	65

Chapter 4   Crystal Growth and Doping 67

	4.1	Bulk Crystal Growth	67
	4.2	Dopant Incorporation During Bulk Crystal Growth	70
	4.3	Thin Film Growth	74
	4.4	Liquid Phase Epitaxy	75
	4.5	Chemical Vapor Deposition	77
	4.6	Molecular Beam Epitaxy	79
	4.7	Alloying	81
	4.8	Doping by Diffusion	83
	4.9	Ion Implantation	85
	4.10	Annealing and Dopant Activation	89
	4.11	Neutron Transmutation	92
		Problems	94
		References	94

Chapter 5   Electronic Properties 97

	5.1	Hydrogenic Model	97
	5.2	Wave Function Symmetry	102
	5.3	Donor and Acceptor Wave Functions	106
	5.4	Deep Levels	108
	5.5	Carrier Concentrations as a Function of Temperature	113
	5.6	Freeze-Out Curves	116
	5.7	Scattering Processes	119
	5.8	Hopping and Impurity Band Conduction	122
	5.9	Spintronics	125
		Problems	126
		References	127

Chapter 6   Vibrational Properties 129

	6.1	Phonons	129
	6.2	Defect Vibrational Modes	133
	6.3	Infrared Absorption	138
	6.4	Interactions and Lifetimes	140
	6.5	Raman Scattering	142
	6.6	Wave Functions and Symmetry	144
	6.7	Oxygen in Silicon and Germanium	147
	6.8	Impurity Vibrational Modes in GaAs	151
	6.9	Hydrogen Vibrational Modes	155
		Problems	157
		References	158

Chapter 7   Optical Properties 161

	7.1	Free-Carrier Absorption and Reflection	161
	7.2	Lattice Vibrations	164

Contents **ix**

7.3	Dipole Transitions	168
7.4	Band-Gap Absorption	169
7.5	Carrier Dynamics	173
7.6	Exciton and Donor–Acceptor Emission	178
7.7	Isoelectronic Impurities	182
7.8	Lattice Relaxation	184
7.9	Transition Metals	187
Problems		189
References		190

Chapter 8 Thermal Properties 193

8.1	Defect Formation	193
8.2	Charge State	195
8.3	Chemical Potential	197
8.4	Diffusion	199
8.5	Microscopic Mechanisms of Diffusion	204
8.6	Self-Diffusion	207
8.7	Dopant Diffusion	210
8.8	Quantum-Well Intermixing	214
Problems		218
References		219

Chapter 9 Electrical Measurements 221

9.1	Resistivity and Conductivity	222
9.2	Methods of Measuring Resistivity	223
9.3	Hall Effect	228
9.4	Capacitance–Voltage Profiling	231
9.5	Carrier Emission and Capture	234
9.6	Deep-Level Transient Spectroscopy	235
9.7	Minority Carriers and Deep-Level Transient Spectroscopy	238
9.8	Minority Carrier Lifetime	240
9.9	Thermoelectric Effect	241
Problems		242
References		243

Chapter 10 Optical Spectroscopy 247

10.1	Absorption	248
10.2	Emission	250
10.3	Raman Spectroscopy	252
10.4	Fourier Transform Infrared Spectroscopy	255
10.5	Photoconductivity	258
10.6	Time-Resolved Techniques	260
10.7	Applied Stress	262
10.8	Electron Paramagnetic Resonance	265
10.9	Optically Detected Magnetic Resonance	269
10.10	Electron Nuclear Double Resonance	270
Problems		273
References		274

x   Contents

Chapter 11 Particle-Beam Methods 277
      11.1   Rutherford Backscattering Spectrometry 277
      11.2   Ion Range 281
      11.3   Secondary Ion Mass Spectrometry 285
      11.4   X-Ray Emission 287
      11.5   X-Ray Absorption 288
      11.6   Photoelectric Effect 290
      11.7   Electron Beams 292
      11.8   Positron Annihilation 294
      11.9   Muons 296
      11.10 Perturbed Angular Correlation Spectroscopy 298
      11.11 Nuclear Reactions 301
      Problems 301
      References 302

Chapter 12 Microscopy and Structural Characterization 305
      12.1   Optical Microscopy 305
      12.2   Scanning Electron Microscopy 310
      12.3   Cathodoluminescence 314
      12.4   Electron Beam Induced Current Microscopy 316
      12.5   Diffraction 318
      12.6   Transmission Electron Microscopy 319
      12.7   Scanning Probe Microscopy 322
      Problems 326
      References 327

Appendices 329

Physical Constants 337

Index 339

Preface to the Second Edition

In this second edition, we have added a new chapter (Chapter 3) on interfaces and devices. This chapter provides a discussion of "area" defects and key semiconductor applications. We also added problems to the end of each chapter. The majority of these exercises have been tested on actual students! Appendices were added to provide background on mathematical methods and quantum mechanics. Finally, the text has been improved, especially in the early chapters, to provide a smoother introduction to the material.

We would like to acknowledge numerous students for fruitful discussions and feedback. Jesse Huso, Slade Jokela, Chris Pansegrau, Violet Poole, Jacob Ritter, and Muad Saleh provided valuable feedback on this second edition. We would also like to especially thank executive editor Lu Han for guiding this project to completion.

Preface to the First Edition

The origin of this book was a graduate course at University of California, Berkeley, "Semiconductor Materials," taught by Eugene Haller from 1981 to 2008 and taken by Matthew McCluskey in 1993. The course and the accompanying lecture notes covered a broad range of topics relating to the growth, characterization, and processing of semiconductors. They filled a crucial gap between solid-state physics and more specialized courses. Many students described the course as the highlight of their academic experience.

This book aims to present readers with a comprehensive overview of dopants and defects in semiconductors. The need for such a text, at the graduate or advanced undergraduate level, has been apparent for a long time. Books and edited volumes on this topic have previously had a narrow focus, targeting subspecialties such as characterization techniques or state-of-the-art theoretical methods. By treating dopants and defects in semiconductors as a unified subject, we hope that this book will help define the field and prepare students for work in technologically important areas.

The contents are divided into three sections. The first covers introductory concepts, from basic semiconductor theory to defect classifications, crystal growth, and doping. The second section looks at electrical, vibrational, optical, and thermal properties. The final section turns to characterization approaches: measurement of electrical properties, optical spectroscopy, particle-beam methods, and microscopy.

We would like to acknowledge numerous students for fruitful discussions and feedback. We thank Kathy Boucart, Karen Bustillo, Slade Jokela, Marie Mayer, Marianne Tarun, and Samuel Teklemichael for their efforts on this book. While there are too many valued colleagues to list here, those who have influenced our research and this book include Joel Ager III, Jeffery Beeman, Leah Bergman, Hartmut Bracht, Manuel Cardona, Marvin Cohen, Susan Dexheimer, Oscar Dubon, Leo Falicov, William Hansen, Noble Johnson, Hans Queisser, Anant Ramdas, Alfred Seeger, Michael Stavola, Chris Van de Walle, Wladek Walukiewicz, Kin Man Yu, and Shengbai Zhang. Administrative assistance by David Hom and Margaret Ragsdale is gratefully acknowledged. We would also like to thank senior editor Lu Han, project editor Linda Leggio, and production coordinator Jennifer Ahringer for guiding this project to completion.

Finally, we would like to thank our spouses, Jill McCluskey and Marianne Haller, to whom this book is dedicated. For many years they listened patiently to the results of our research endeavors, making us feel comfortable as researchers and as teachers. There would be no book without Jill and Marianne.

Corrections or comments are welcomed and may be sent by email to: mattmcc@alum.mit.edu.

Authors

Matthew D. McCluskey is a professor in the Department of Physics and Astronomy and Materials Science Program at Washington State University (WSU), Pullman, Washington. He earned a physics PhD from the University of California (UC), Berkeley, in 1997, and was a post-doctoral researcher at the Xerox Palo Alto Research Center (PARC) (California) from 1997 to 1998. Dr. McCluskey joined WSU as an assistant professor in 1998. His research interests include defects in semiconductors, materials under high pressure, shock compression of semiconductors, and vibrational spectroscopy.

Eugene E. Haller is a professor emeritus at the University of California, Berkeley, and a member of the National Academy of Engineering. He earned a PhD in solid state and applied physics from the University of Basel, Switzerland, in 1967. Dr. Haller joined the Lawrence Berkeley National Laboratory (California) as a staff scientist in 1973. In 1980, he was appointed associate professor in the Department of Materials Science Engineering, UC, Berkeley. His major research areas include semiconductor growth, characterization, and processing; far-infrared detectors; isotopically controlled semiconductors; and semiconductor nanocrystals.

Abbreviations

2DEG	Two-dimensional electron gas
AB	Antibonding
AFM	Atomic force microscopy
ALD	Atomic layer deposition
BC	Bond center
bcc	Body-centered cubic
BGO	Bismuth germanium oxide
CB	Conduction band
CBE	Chemical beam epitaxy (same as MOMBE)
CBM	Conduction-band minimum
CCD	Charge-coupled device
CL	Cathodoluminescence
CMOS	Complementary metal oxide semiconductor
C-V	Capacitance-voltage
CVD	Chemical vapor deposition
cw	Continuous wave
CZ	Czochralski
DAC	Diamond-anvil cell
DAP	Donor-acceptor pair
DFT	Density functional theory
DLTS	Deep-level transient spectroscopy
DM	Dimethyl
EBIC	Electron beam induced current
EBSD	Electron backscattering diffraction
EDS (or EDX)	Energy dispersive x-ray spectroscopy
EELS	Electron energy loss spectroscopy
EFG	Electric field gradient
EL	Electroluminescence
EM	Effective mass
ENDOR	Electron nuclear double resonance
EPR	Electron paramagnetic resonance
EXAFS	Extended x-ray absorption fine structure
fcc	Face-centered cubic
FFT	Fast Fourier transform
FTIR	Fourier transform infrared
FWHM	Full width at half maximum
FZ	Floating zone
hcp	Hexagonal close packed
HRTEM	High-resolution transmission electron microscopy
IR	Infrared
I-V	Current-voltage
LA	Longitudinal acoustical

xviii Abbreviations

LCAO	Linear combination of atomic orbitals
LDA	Local density approximation
LEC	Liquid encapsulated Czochralski
LED	Light-emitting diode
LEED	Low-energy electron diffraction
LO	Longitudinal optical
LPE	Liquid phase epitaxy
LSS	Lindhard, Scharff, and Schiøtt
LVM	Local vibrational mode
MBE	Molecular beam epitaxy
MCT	Mercury cadmium telluride
MIGS	Metal-induced gap states
MIPS	Multiband Imaging Photometer
MOCVD	Metalorganic chemical vapor deposition
MOMBE	Metalorganic molecular beam epitaxy
MOS	Metal-oxide-semiconductor
MOSFET	Metal-oxide-semiconductor field effect transistor
MQW	Multiple quantum well
μSR	Muon spin rotation (or resonance)
NA	Numerical aperture
NAA	Neutron activation analysis
NMR	Nuclear magnetic resonance
NRA	Nuclear reaction analysis
NSOM	Near-field scanning optical microscopy
NTD	Neutron transmutation doping
ODMR	Optically detected magnetic resonance
OMVPE	Organometallic vapor phase epitaxy (same as MOCVD)
PACS	Perturbed angular correlation spectroscopy
PC	Photoconductivity
PHA	Pulse height analyzer
PIXE	Particle-induced x-ray emission
PL	Photoluminescence
PLD	Pulsed laser deposition
PLE	Photoluminescence excitation
PMMA	Poly(methyl methacrylate)
PMT	Photomultiplier tube
PTIS	Photothermal ionization spectroscopy
RBS	Rutherford backscattering spectrometry
RE	Rare earth
RF	Radio frequency
RHEED	Reflection high-energy electron diffraction
RTA	Rapid thermal annealing
SAD	Selected area diffraction
sc	Simple cubic
SEM	Scanning electron microscopy
SI	Semi-insulating
SIMS	Secondary ion mass spectrometry
SPE	Solid phase epitaxy
SPM	Scanning probe microscopy
SRH	Shockley–Read–Hall

STM	Scanning tunneling microscopy
TA	Transverse acoustical
TE	Triethyl
TEES	Thermoelectric effect spectroscopy
TEM	Transmission electron microscopy
TM	Trimethyl
TO	Transverse optical
TSC	Thermally stimulated current
UPS	Ultraviolet photoelectron spectroscopy
UV	Ultraviolet
VB	Valence band
VBM	Valence-band maximum
VPE	Vapor phase epitaxy
XANES	X-ray absorption near edge spectroscopy
XPS	X-ray photoelectron spectroscopy
XRD	X-ray diffraction
YAG	Yttrium aluminum garnet

List of Elements by Symbol

Ac, Actinium
Ag, Silver
Al, Aluminum
Am, Americium
Ar, Argon
As, Arsenic
At, Astatine
Au, Gold
B, Boron
Ba, Barium
Be, Beryllium
Bi, Bismuth
Bk, Berkelium
Br, Bromine
C, Carbon
Ca, Calcium
Cd, Cadmium
Ce, Cerium
Cf, Californium
Cl, Chlorine
Cm, Curium
Co, Cobalt
Cr, Chromium
Cs, Cesium
Cu, Copper
Dy, Dysprosium
Er, Erbium
Es, Einsteinium
Eu, Europium
F, Fluorine
Fe, Iron
Fm, Fermium
Fr, Francium
Ga, Gallium
Gd, Gadolinium

Ge, Germanium
H, Hydrogen
He, Helium
Hf, Hafnium
Hg, Mercury
Ho, Holmium
I, Iodine
In, Indium
Ir, Iridium
K, Potassium
Kr, Krypton
La, Lanthanum
Li, Lithium
Lr, Lawrencium
Lu, Lutetium
Md, Mendelevium
Mg, Magnesium
Mn, Manganese
Mo, Molybdenum
N, Nitrogen
Na, Sodium
Nb, Niobium
Nd, Neodymium
Ne, Neon
Ni, Nickel
No, Nobelium
Np, Neptunium
O, Oxygen
Os, Osmium
P, Phosphorus
Pa, Protactinium
Pb, Lead
Pd, Palladium
Pm, Promethium
Po, Polonium

Pr, Praseodymium
Pt, Platinum
Pu, Plutonium
Ra, Radium
Rb, Rubidium
Re, Rhenium
Rh, Rhodium
Rn, Radon
Ru, Ruthenium
S, Sulfur
Sb, Antimony
Sc, Scandium
Se, Selenium
Si, Silicon
Sm, Samarium
Sn, Tin
Sr, Strontium
Ta, Tantalum
Tb, Terbium
Tc, Technetium
Te, Tellurium
Th, Thorium
Ti, Titanium
Tl, Thallium
Tm, Thulium
U, Uranium
V, Vanadium
W, Tungsten
Xe, Xenon
Y, Yttrium
Yb, Ytterbium
Zn, Zinc
Zr, Zirconium

xxii   List of Elements by Symbol

Periodic Table of the Elements

Outer electron configurations are shown for the neutral atom in its ground state.

IA	IIA	IIIB	IVB	VB	VIB	VIIB	VIIIA	VIIIA	VIIIA	IB	IIB	IIIA	IVA	VA	VIA	VIIA	VIIIA
1 **H** $1s$ 1.008																	2 **He** $1s^2$ 4.003
3 **Li** $2s$ 6.94	4 **Be** $2s^2$ 9.012											5 **B** $2s^2 2p$ 10.81	6 **C** $2s^2 2p^2$ 12.01	7 **N** $2s^2 2p^3$ 14.01	8 **O** $2s^2 2p^4$ 16.00	9 **F** $2s^2 2p^5$ 19.00	10 **Ne** $2s^2 2p^6$ 20.18
11 **Na** $3s$ 22.99	12 **Mg** $3s^2$ 24.31											13 **Al** $3s^2 3p$ 26.98	14 **Si** $3s^2 3p^2$ 28.09	15 **P** $3s^2 3p^3$ 30.97	16 **S** $3s^2 3p^4$ 32.06	17 **Cl** $3s^2 3p^5$ 35.45	18 **Ar** $3s^2 3p^6$ 39.95
19 **K** $4s$ 39.10	20 **Ca** $4s^2$ 40.08	21 **Sc** $3d4s^2$ 44.96	22 **Ti** $3d^2 4s^2$ 47.88	23 **V** $3d^3 4s^2$ 50.94	24 **Cr** $3d^5 4s$ 52.00	25 **Mn** $3d^5 4s^2$ 54.94	26 **Fe** $3d^6 4s^2$ 55.85	27 **Co** $3d^7 4s^2$ 58.93	28 **Ni** $3d^8 4s^2$ 58.69	29 **Cu** $3d^{10} 4s$ 63.55	30 **Zn** $3d^{10} 4s^2$ 65.39	31 **Ga** $4s^2 4p$ 69.72	32 **Ge** $4s^2 4p^2$ 72.64	33 **As** $4s^2 4p^3$ 74.92	34 **Se** $4s^2 4p^4$ 78.96	35 **Br** $4s^2 4p^5$ 79.90	36 **Kr** $4s^2 4p^6$ 83.80
37 **Rb** $5s$ 85.47	38 **Sr** $5s^2$ 87.62	39 **Y** $4d5s^2$ 88.91	40 **Zr** $4d^2 5s^2$ 91.22	41 **Nb** $4d^4 5s$ 92.91	42 **Mo** $4d^5 5s$ 95.96	43 **Tc** $4d^5 5s^2$ (98)	44 **Ru** $4d^7 5s$ 101.1	45 **Rh** $4d^8 5s$ 102.9	46 **Pd** $4d^{10}$ 106.4	47 **Ag** $4d^{10} 5s$ 107.9	48 **Cd** $4d^{10} 5s^2$ 112.4	49 **In** $5s^2 5p$ 114.8	50 **Sn** $5s^2 5p^2$ 118.7	51 **Sb** $5s^2 5p^3$ 121.8	52 **Te** $5s^2 5p^4$ 127.6	53 **I** $5s^2 5p^5$ 126.9	54 **Xe** $5s^2 5p^6$ 131.3
55 **Cs** $6s$ 132.9	56 **Ba** $6s^2$ 137.3	*	72 **Hf** $5d^2 6s^2$ 178.5	73 **Ta** $5d^3 6s^2$ 180.9	74 **W** $5d^4 6s^2$ 183.9	75 **Re** $5d^5 6s^2$ 186.2	76 **Os** $5d^6 6s^2$ 190.2	77 **Ir** $5d^7 6s^2$ 192.2	78 **Pt** $5d^9 6s$ 195.1	79 **Au** $5d^{10} 6s$ 197.0	80 **Hg** $5d^{10} 6s^2$ 200.5	81 **Tl** $6s^2 6p$ 204.4	82 **Pb** $6s^2 6p^2$ 207.2	83 **Bi** $6s^2 6p^3$ 209.0	84 **Po** $6s^2 6p^4$ (209)	85 **At** $6s^2 6p^5$ (210)	86 **Rn** $6s^2 6p^6$ (222)
87 **Fr** $7s$ (223)	88 **Ra** $7s^2$ (226)	**															

*	57 **La** $5d6s^2$ 138.9	58 **Ce** $4f5d6s^2$ 140.1	59 **Pr** $4f^3 6s^2$ 140.9	60 **Nd** $4f^4 6s^2$ 144.2	61 **Pm** $4f^5 6s^2$ (145)	62 **Sm** $4f^6 6s^2$ 150.4	63 **Eu** $4f^7 6s^2$ 152.0	64 **Gd** $4f^7 5d6s^2$ 157.2	65 **Tb** $4f^9 6s^2$ 158.9	66 **Dy** $4f^{10} 6s^2$ 162.5	67 **Ho** $4f^{11} 6s^2$ 164.9	68 **Er** $4f^{12} 6s^2$ 167.3	69 **Tm** $4f^{13} 6s^2$ 168.9	70 **Yb** $4f^{14} 6s^2$ 173.0	71 **Lu** $4f^{14} 5d6s^2$ 175.0
**	89 **Ac** $6d7s^2$ (227)	90 **Th** $6d^2 7s^2$ 232.0	91 **Pa** $5f^2 6d7s^2$ 231.0	92 **U** $5f^3 6d7s^2$ 238.0	93 **Np** $5f^4 6d7s^2$ (237)	94 **Pu** $5f^6 7s^2$ (244)	95 **Am** $5f^7 7s^2$ (243)	96 **Cm** $5f^7 6d7s^2$ (247)	97 **Bk** $5f^9 7s^2$ (247)	98 **Cf** $5f^{10} 7s^2$ (251)	99 **Es** $5f^{11} 7s^2$ (252)	100 **Fm** $5f^{12} 7s^2$ (257)	101 **Md** $5f^{13} 7s^2$ (258)	102 **No** $5f^{14} 7s^2$ (259)	103 **Lr** $5f^{14} 6d7s^2$ (262)

Semiconductor Basics

1

In this chapter, the fundamentals of semiconductor physics are presented. After a brief historical overview, we discuss examples of crystal structures that are relevant to semiconductors. We then discuss the most important properties of semiconductors: phonons, band structure, electronic transport, and optical properties. This is followed by a section on specific semiconductor materials.

1.1 HISTORICAL OVERVIEW

In the early 1870s, Ferdinand Braun, a high school teacher in Leipzig, connected metal wires to natural semiconducting minerals such as galena (PbS). To his surprise, he found that the current flow was not proportional to the applied bias (Braun, 1875). Ohm's law, so successful for metals, did not hold for these metal–mineral contacts. Braun's discovery eventually led to the solid-state rectifier. Although the underlying physical principles were not understood, wireless telegraphy and radio engineers soon used "crystal detectors" for the demodulation of amplitude-modulated radio-frequency signals. Braun played a key role in the development of early radio science and technology and also invented the TV picture tube, which in German is called the *Braunsche Röhre*. He was awarded the Nobel prize in 1909 jointly with Guglielmo Marconi.

A variety of materials were used as rectifiers. PbS was used in crystal radios. Other early semiconductors used as crystal detectors and AC rectifiers included copper oxide (CuO), selenium, and pyrite (FeS). Because uncontrolled impurities and defects led to widely variable crystal properties, however, the crystal detectors were perceived as an unreliable technology. Some famous physicists expressed doubts regarding the very existence of semiconductors. They called the unpredictable phenomena the "physics of dirt."

Starting in the 1920s, a more "modern" device, the vacuum tube, began permeating electronics and kept its dominant position until the 1960s. There was, however, one area of electronics where tubes could not perform. The urgent need for sensitive, ultra high frequency rectifiers for radar reception during World War II led to the development of silicon and germanium point contact rectifier diodes (Torrey and Whitmer, 1948; Seitz, 1995).

In addition to current rectification, semiconductors exhibited unusual temperature-dependent behavior. In contrast to metals,

2   1.1 Historical Overview

when a pure semiconductor is warmed up, its resistance drops. In semiconductors, electrons fill a band of energy states (Wilson, 1931), now called the valence band. Due to the Pauli exclusion principle, the electrons in these filled states could not respond to an electric field. Only electrons that were thermally excited into a higher band (the conduction band) could conduct electricity. Building on these insights, Walter Schottky developed a theory explaining rectification at metal-semiconductor junctions, over a half century after Braun's discovery (Schottky, 1938).

Driven by the need for rectifiers in radar systems, researchers attempted to improve the quality of semiconductor materials. At Bell Labs, Russel Ohl and Jack Scaff purified silicon by melting and recrystallizing ingots (Riordan and Hoddeson, 1997). They noticed there were two types of materials, now called *n*-type and *p*-type. In an *n*-type semiconductor, the mobile charge carriers are electrons. In *p*-type material, the carriers are missing electrons, or "holes." In 1939, a technician cut a crystal such that it contained a *p-n* junction, inadvertently making the world's first solar cell.

Silicon was difficult to purify because its high melting point (1415°C) caused impurities from the furnace to contaminate the material. Germanium, the first elemental semiconductor discovered in 1886 by Clemens Winkler, had a lower melting point (936°C) and could be made exceedingly pure. Members of the Purdue University physics department, led by Karl Lark-Horovitz, developed and characterized pure and deliberately doped germanium. By the early 1940s, it was understood why small quantities of certain elements made germanium electron conducting (*n* type), whereas other elements made germanium hole conducting (*p* type).

At Bell Labs, researchers found that a gold point contact on *n*-type germanium depleted a near-surface region of electrons. When they applied a positive bias to the gold contact, holes were injected into the semiconductor. John Bardeen and Walter Brattain then made a device with two closely spaced point contacts. When holes were injected by the first contact, the current flow through the second contact increased dramatically, due to the infusion of charge carriers (Bardeen and Brattain, 1948). This current amplifier, called the point contact transistor, was demonstrated to Bell Labs management and several scientists on December 24, 1947. A piece of Purdue germanium served as the semiconductor base. While this device never made it into mass commercial production, it nonetheless provided a powerful demonstration that changed the world. There was a time *before* the transistor, and there is a time *after* the invention of the transistor.

Only two months after the invention of the point contact transistor, William Shockley of Bell Labs designed the bipolar transistor, an all solid-state three-layer *n-p-n* device. The delicate point contacts were eliminated. By applying a voltage between the *p*-type layer (called the base) and one of the *n* layers (the emitter), one could control the electron flow from the emitter to the other *n* layer (the collector). To make this device, Gordon Teal and Morgan Sparks grew single-crystal germanium using the Czochralski technique (Czochralski, 1918). This method uses a small crystalline seed for the growth of a large crystal from the molten germanium. They introduced dopants at specific locations along the crystal to form the *n* and *p* layers. In 1950, they succeeded in making an *n-p-n* transistor (Shockley et al., 1951). Further developments in zone refining, invented by William Pfann, resulted in germanium with even higher purity, improving device performance. In 1954, Bell Labs researchers used diffusion of dopant atoms into ultra-pure germanium to form *p-n* junctions. The first integrated circuit, designed and fabricated by Jack Kilby at Texas Instruments in 1958, was also made from germanium.

In the early 1950s, scientists began a renewed effort to develop silicon as an electronic material. Despite its high melting point, silicon has two major advantages over germanium. First, its band gap is larger, so that at room temperature, very few electrons are thermally excited from the valence band to the conduction band. This makes the performance of silicon devices quite temperature independent. Second, it readily forms a tough insulating native oxide (SiO_2) surface layer. Gordon Teal moved to Texas Instruments and made an *n-p-n* transistor from silicon in

1954. In 1958, Jean Hoerni, working at Fairchild Electronics in California, used the SiO_2 layer on silicon as a mask for fabricating planar diodes and transistors. His colleague Robert Noyce used oxide masking to form the first integrated circuit on silicon, around the same time as Kilby.

In 1960, John Atalla's group at Bell Labs oxidized the silicon surface to passivate dangling bonds and reduce the surface states that trap electrons. An aluminum gate contact was deposited on the oxide, forming a metal-oxide-semiconductor (MOS) field-effect transistor (Sah, 1988). In this device, the voltage applied to the gate regulates the flow of current through the silicon. MOS transistors form the basis for modern electronics. Today, hundreds of millions of MOS transistors are packed on silicon wafers, forming data processors and memories. Gordon Moore's famous law states that the number of transistors on a chip doubles every 18 to 24 months, while the size of an MOS transistor is cut in half. This "law" reigned for over 40 years.

Semiconductors have also had a major impact in the area of solid-state lighting (Schubert, 2006). Once again, this application has its historical roots in crystal rectifiers. In 1907, a radio engineer named Henry Joseph Round reported that applying a voltage to silicon carbide (SiC) caused it to glow (Round, 1907). However, the group-IV semiconductors SiC, silicon, and germanium are inefficient light emitters. In contrast, compound semiconductors made from group-III and group-V atoms are much more efficient. Heinrich Welker at Siemens began the growth of bulk gallium arsenide (GaAs) in 1954. In 1962, various researchers reported infrared (IR) light-emitting diodes and laser diodes made from GaAs. Red emission was achieved by adding phosphorus to form GaAsP alloys (Holonyak and Bevacqua, 1962). After these successes, it took a long time to develop efficient green and blue LEDs. Gallium nitride (GaN) emerged as the preferred material, with the first current-injection blue LEDs being reported in the early 1990s (Akasaki, 2006). Shuji Nakamura of Nichia Chemical Industries Corporation, Japan, led the development of blue and green InGaN LEDs and achieved the first current-injection blue laser (Nakamura and Fasol, 1997). Akasaki, Amano, and Nakamura won the Nobel Prize in Physics for 2014 for the invention of blue LEDs.

As with the work in the 20th century, future advances will depend on a fundamental understanding of semiconductors. The following sections present a synopsis of their structural, electronic, and optical properties.

1.2 CUBIC CRYSTALS

A crystal is a regular, repeating pattern of atoms that continues indefinitely. The repeating unit is referred to as a cell. A perfect crystal is created by filling space with an infinite number of cells. (In reality, of course, crystals are finite and have surfaces.) Many technologically important semiconductor crystals are described by a conventional *cubic cell*. The edges of the cube have a length a, the *lattice constant*. Examples of three cubic cells are shown in Figure 1.1: simple cubic (sc), body-centered cubic (bcc), and face-centered cubic (fcc). Of these, the fcc structure is most relevant to semiconductors.

The following nomenclature is used to describe location, direction, and planes:

- *Location.* We designate the location of an atom in the cubic cell by the fractional coordinates (x,y,z), where the values are in terms of the lattice constant(s). The sc cell has only one atom, at the origin, $(0,0,0)$. The other seven atoms (Figure 1.1) belong to adjacent cells. The bcc structure has two atoms per cell, at $(0,0,0)$ and $(½,½,½)$. The fcc cell has four atoms, at the origin and face centers: $(0,0,0)$; $(½,½,0)$; $(½,0,½)$; $(0,½,½)$.
- *Direction.* Directions are defined as pointing from the origin of the cube and are expressed by $[xyz]$, with a negative value indicated by a bar; e.g., $[1\bar{1}0]$. We often refer to a

1.2 Cubic Crystals

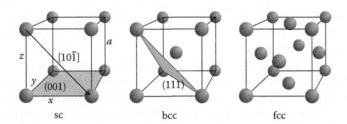

FIGURE 1.1 Simple cubic (sc), body-centered cubic (bcc), and face-centered cubic (fcc) crystal structures. The (001) and (111) planes, and [10$\bar{1}$] direction, are illustrated.

class of directions that are equivalent by symmetry. For example, the "<111> directions" include [111], [1$\bar{1}$1], [$\bar{1}$$\bar{1}$1], etc.
- *Planes.* The atomic plane that is perpendicular to the [hkl] direction is denoted (hkl), where h, k, and l are *Miller indices*. For example, the (001) plane is the x–y plane and is perpendicular to the z axis. Examples of planes are shown in Figure 1.1. Classes of planes are denoted with curly brackets; e.g., "the {111} planes."

The elemental semiconductors diamond (carbon), silicon, and germanium have the *diamond* crystal structure (Figure 1.2). In this structure, atoms reside on the fcc sites. In addition, each atom has a second atom associated with it, displaced by (¼,¼,¼). In all, the diamond cell has eight atoms, located at the following coordinates:

(1.1)
$$(0,0,0) \quad (½,½,0) \quad (½,0,½) \quad (0,½,½)$$
$$(¼,¼,¼) \quad (¾,¾,¼) \quad (¾,¼,¾) \quad (¼,¾,¾)$$

The distance between neighboring atoms is

(1.2)
$$R = \sqrt{(a/4)^2 + (a/4)^2 + (a/4)^2} = \frac{\sqrt{3}}{4}a$$

Because the diamond structure is so important, let's look at it from a couple of different perspectives. First, note that each atom forms a bond with four neighbors. The bonds are oriented along <111> directions. Along the <110> directions, the atoms follow a "zig-zag," shown in Figure 1.3. From the lengths of the sides of the zig-zag, one can see that the angle θ is given by

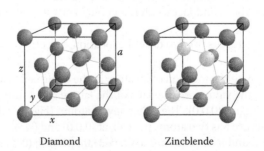

Diamond   Zincblende

FIGURE 1.2 Diamond and zincblende crystal structures.

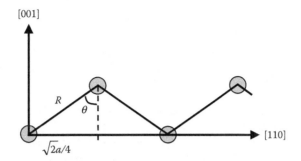

FIGURE 1.3 Atoms along the [110] "zig zag" in the diamond structure.

(1.3) $$\sin\theta = \frac{\sqrt{2}a/4}{R} = \sqrt{\frac{2}{3}}$$

yielding $\theta = 54.74°$. The bond angle is $2\theta = 109.5°$, which is characteristic for *tetrahedral* bonding. The valence electrons of each atom form tetrahedral bonds with the four nearest neighbors.

Another approach is to examine the crystal projection on the (001) plane (Figure 1.4). Atoms that intersect the plane are shown as solid circles. Atoms above and below the plane are shown by different symbols. The result is that the atoms are arranged in a simple square grid, with each atom bonded to four neighbors. This is basically what you see when you look at a model of the diamond structure along a <100> direction. The <100> projection is often used to illustrate a crystal such as silicon. One should remember, however, that the atoms are not all in the same plane.

The group-IV elements have four valence electrons that form bonds with four neighbors, completing the Lewis octet. III-V semiconductors have one atom from the group-III and group-V column of the periodic table. As in the case of silicon, each pair of atoms contributes a total of eight electrons to the covalent bonds, satisfying the Lewis octet. Examples of III-V semiconductors include GaAs, InP, and GaSb. Semiconductors can also be made from group-II and group-VI atoms. II-VI semiconductors include ZnSe and CdTe.

Many compound semiconductors form the *zincblende* structure, also called the sphalerite structure. Like the diamond structure, the zincblende structure has atoms on the fcc sites. Each atom has a second atom associated with it, displaced by (¼,¼,¼). Unlike the diamond structure, this second atom is not the same as the first. In GaAs, for example, the atoms in the cubic cell are located at the following coordinates:

(1.4) Ga atoms: $(0,0,0)$ $(½,½,0)$ $(½,0,½)$ $(0,½,½)$
As atoms: $(¼,¼,¼)$ $(¾,¾,¼)$ $(¾,¼,¾)$ $(¼,¾,¾)$

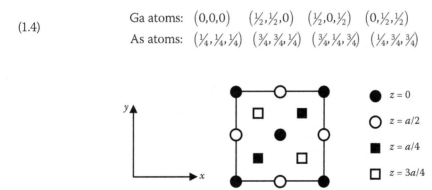

FIGURE 1.4 Atoms in a diamond crystal projected onto the (001) plane.

Atoms in the diamond structure (Equation 1.1) have these same coordinates, but the atoms are all identical.

1.3 OTHER CRYSTALS

Many crystals in nature have the hexagonal close packed (hcp) structure (Figure 1.5), described by the lattice vectors $\mathbf{a}_1$, $\mathbf{a}_2$, and $\mathbf{a}_3$. The $\mathbf{a}_1$ and $\mathbf{a}_2$ vectors are in the *basal plane*, have a length a, and span an angle of 120°. The $\mathbf{a}_3$ vector is perpendicular to the basal plane and has a length c. As before, we use fractional coordinates, where (x,y,z) denotes the position $x\mathbf{a}_1 + y\mathbf{a}_2 + z\mathbf{a}_3$. The hcp cell has two atoms, one at the origin and one at (⅔,⅓,½).

Semiconductors such as CdS, GaN, and ZnO have the hexagonal *wurtzite* structure (Figure 1.5). In the wurtzite cell, one atom resides on the hcp site while the second atom is a distance uc above the first. Wurtzite crystals have four atoms per cell, with the following coordinates for the case of GaN:

$$(1.5) \quad \begin{aligned} \text{Ga atoms:} & \quad (0,0,0) \quad (\tfrac{2}{3},\tfrac{1}{3},\tfrac{1}{2}) \\ \text{N atoms:} & \quad (0,0,u) \quad (\tfrac{2}{3},\tfrac{1}{3},u+\tfrac{1}{2}) \end{aligned}$$

Like the diamond and zincblende structures, atoms in the wurtzite structure bond to four neighbors; i.e., they are fourfold coordinated. The u parameter is different for each crystal. For the special case $u = 3/8$ and $c/a = (8/3)^{1/2}$, an atom has tetrahedral bonding with its nearest neighbors, identical to that of the zincblende structure. In that case, the distance between neighboring atoms is $3/8\,c$.

The c axis in the wurtzite structure is analogous to the [111] direction in the zincblende structure. To capture this similarity, one often introduces an additional Miller index. Directions are then denoted $[x\,y\,-(x+y)\,z]$. With this nomenclature, the c axis is [0001], and the a axis is [10$\bar{1}$0]. The a axis points from an atom to its neighbor in the basal plane. In the zincblende structure, the [10$\bar{1}$] axis performs a similar function, pointing from an atom to its neighbor in the (111) plane.

We can compare the zincblende and wurtzite structures further by looking at the *stacking sequence* along the [111] or c axis, respectively. As shown in Figure 1.6, a layer is denoted as A, B, or C, depending on the positions of the atoms. The zincblende stacking sequence is $ABCABC\ldots$, whereas wurtzite is $ABAB\ldots$ SiC is an example of a crystal with many different structures, or *polytypes*. The only cubic polytype, denoted 3C, is zincblende. The hexagonal polytypes are denoted nH, where n is the periodicity of the stacking sequence. The most common hexagonal polytypes are 2H (wurtzite), 4H ($ABACABAC\ldots$), and 6H ($ABCACBABCACB\ldots$).

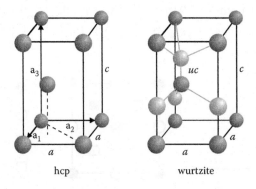

FIGURE 1.5 The hexagonal close packed (hcp) and wurtzite crystal structure.

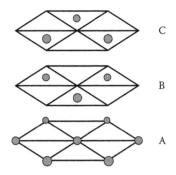

FIGURE 1.6 Stacking sequence for the diamond or zincblende crystal structure, along the [111] direction. Each atom has a second atom associated with it, not shown. The wurtzite structure has an *ABAB* ... stacking sequence along the *c* direction.

Unlike cubic crystals, hexagonal crystals are anisotropic. In GaN, the "Ga-face" denotes the *c*-plane surface where each Ga atom is bonded to three N atoms and has one dangling bond. The "N-face" indicates the opposite surface. Similar notation applies to ZnO (Zn-face, O-face) and other wurtzite crystals. The *c* axis defines the optic axis for the crystal. Due to anisotropy, wurtzite crystals are birefringent, having different indices of refraction for light polarized parallel and perpendicular to the *c* axis.

The diamond, zincblende, and wurtzite structures account for the majority of technologically important semiconductors. We conclude this section with two more structures that are of interest to the field of semiconductors. The *rocksalt* (NaCl) structure is described by a face-centered cubic cell, with cations at the corners and the face centers of the cube (Figure 1.7). Each cation has an anion associated with it, displaced by (½,½,½). PbS is a famous example of a semiconductor with this structure. Although Pb is in the group-IV column of the periodic table, it contributes only two electrons to the bond. The atoms are located at

(1.6)
$$\begin{aligned} \text{Pb atoms:} &\quad (0,0,0) \quad (\tfrac{1}{2},\tfrac{1}{2},0) \quad (\tfrac{1}{2},0,\tfrac{1}{2}) \quad (0,\tfrac{1}{2},\tfrac{1}{2}) \\ \text{S atoms:} &\quad (\tfrac{1}{2},\tfrac{1}{2},\tfrac{1}{2}) \quad (0,0,\tfrac{1}{2}) \quad (0,\tfrac{1}{2},0) \quad (\tfrac{1}{2},0,0) \end{aligned}$$

This structure is also important because many semiconductors, such as ZnO, transform into the rocksalt structure at high pressures.

The *rutile* structure is found for oxide semiconductors such as TiO_2 and SnO_2, which are important in dye-sensitized solar cells and for gas-sensing applications. The titanium (or tin) is in the 4+ oxidation state (i.e., a cation contributes four electrons to the bonds). The structure is described by a tetragonal unit cell, which is a right rectangular solid with two edges of length *a* and a shorter edge of length *c* (Figure 1.8). Titanium atoms reside on the corners and at the center

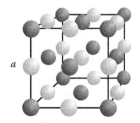

FIGURE 1.7 Rocksalt (NaCl) crystal structure.

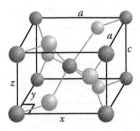

FIGURE 1.8 Rutile crystal structure.

of the cell. For each titanium atom in an x–y plane, there are two oxygen atoms. The titanium atoms are sixfold coordinated, while the oxygen atoms are threefold coordinated. The fractional coordinates for the atoms are as follows:

(1.7)
$$\text{Ti atoms: } (0,0,0) \; (½,½,½)$$
$$\text{O atoms: } (u,u,0) \; (-u,-u,0) \; (½+u,½-u,½) \; (½-u,½+u,½)$$

Another common form of TiO$_2$ is anatase (Horn et al., 1972).

1.4 PHONONS AND THE BRILLOUIN ZONE

Our experience with the macroscopic world shows that solids display, within limits, elastic properties. Hooke's law tells us that the relative dimensional changes are reversible and proportional to the applied forces. A familiar and simple case is the reversible stretching Δl of a steel wire of length l by a force F, such as that exerted by a weight suspended from a wire. The strain, $\varepsilon = \Delta l/l$, is proportional to the stress, $\sigma = F/$(cross-sectional area):

(1.8)
$$\varepsilon = \sigma \cdot \frac{1}{E}$$

where E is a materials constant called Young's modulus.

We can relate the macroscopic law to changes at the atomic level. Solids exist as a result of bonds formed between atoms. In an ionic crystal, such as table salt (NaCl), the cation (Na) gives its outermost electron to the anion (Cl). The oppositely charged ions attract each other via the Coulomb force, with no electron density between them. In a covalent bond, there is a high electron density between two atoms which attracts the two nuclei. Diamond, silicon, and germanium are covalent semiconductors. In general, the bonding types of most crystals lie between these two extremes.

Crystal structures describe the equilibrium positions of the atoms. At room temperature, the atoms oscillate around their equilibrium positions in collective motions, or *lattice waves*. Classically, a lattice wave can have an arbitrary amount of energy. The amplitude is a tunable constant. However, quantum mechanics tells us that this is not really true. The energy of a wave cannot be any arbitrary value; instead, it must be quantized. Ignoring the zero-point energy, the energy of a vibrational mode of frequency ω is

(1.9)
$$E = n\hbar\omega$$

where n is a positive integer, and $\hbar$ is Planck's constant. A *phonon* ("particle of sound") is one quantum ($n = 1$) of a lattice wave.

In addition to energy, phonons have *crystal momentum* $\hbar \mathbf{K}$, where $\mathbf{K}$ is the wavevector, which points in the direction of propagation. The magnitude is $K = 2\pi/\lambda$, where λ is the wavelength. *Acoustical phonons*, or sound waves, have a sound speed given by

$$v_s = \omega/K \tag{1.10}$$

The energy of a single acoustical phonon is therefore $E = \hbar v_s K$. Another type of lattice wave, the "optical" phonon, is discussed in Chapter 6. While phonons carry crystal momentum, they do not carry real mechanical momentum. Like a vibrating molecule, only relative coordinates are involved in the motion. Nonetheless, phonons are treated as quasiparticles that can collide with other particles like electrons. In these collision processes, energy and crystal momentum are conserved.

Because a crystal is composed of discrete atoms, it cannot support phonons with arbitrarily short wavelengths. The *first Brillouin zone* defines all the **k**-points we need to describe lattice waves in the crystal. To illustrate this, we consider a one-dimensional (1D) linear chain of atoms with interatomic distance a. For the linear chain, the first Brillouin zone is defined as the set of points $-\pi/a < K \leq \pi/a$. K points that are positive or negative indicate waves that travel in the $+x$ or $-x$ direction, respectively.

An acoustical phonon dispersion curve (ω versus K) in the first Brillouin zone is shown in Figure 1.9. Near $K = 0$, the curve is linear, per Equation 1.10. Notice how the graph is symmetric about $K = 0$. This means that waves with K or $-K$ have the same frequency ω. There is no meaningful physical difference between a wave that travels "forward" or "backward." Because of this, we sometimes plot the dispersion relation for positive K points only.

At some snapshot in time, the lattice wave is given by $\sin(Kx)$. The value of this sine wave indicates how much an atom is displaced from equilibrium. For a longitudinal wave in 1D, positive or negative values mean displacements in the $+x$ or $-x$ directions, respectively. Consider two specific K points. First, let $K = -\pi/2a$. This gives a sine wave, shown by the solid line in Figure 1.10, with a wavelength $\lambda = 4a$. Next, consider a wave with $K' = 3\pi/2a$, which is outside the first Brillouin zone. Its wavelength is $\lambda' = 4a/3$, shown by the dashed line. The values K and K' give different functions. However, the functions are the same wherever there happens to be an atom. Therefore, K and K' describe the same physical situation. For the 1D case, we can always bring a K-point into the first Brillouin zone by adding an appropriate multiple of $2\pi/a$.

In three-dimensional (3D) crystals, the first Brillouin zone is a 3D solid that contains a set of **k**-points. The first Brillouin zone for the diamond and zincblende structures is defined by a

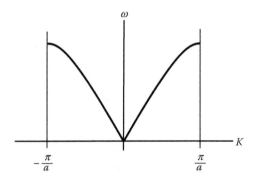

FIGURE 1.9 Acoustical-phonon dispersion relation for a linear chain.

10   1.5 The Band Gap

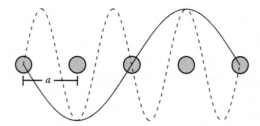

FIGURE 1.10   Examples of two lattice waves in a linear chain. The solid and dashed lines are for K points inside and outside the first Brillouin zone, respectively.

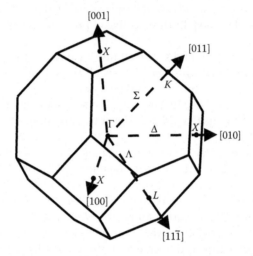

FIGURE 1.11   First Brillouin zone for diamond and zincblende crystals.

truncated octahedron, shown in Figure 1.11. This solid has eight planes that enclose an octahedron, with six planes that truncate the corners, for a total of 14 sides. All **k**-points that are inside the truncated octahedron are considered to be in the first Brillouin zone. Important points and directions are denoted by capital Greek letters. The point $\mathbf{k} = 0$ is called the Γ point. The <100>, <110>, and <111> directions (in **k** space) are denoted Δ, Σ, and Λ, respectively. Lines drawn along these directions from the Γ point will intersect the Brillouin zone edge at X (<100>), K (<110>), and L (<111>).

1.5 THE BAND GAP

The band gap is a defining feature of semiconductors. Figure 1.12a shows a ball-and-stick model of silicon in a <100> projection (Section 1.2). Each covalent bond contains a spin-up and spin-down electron. If thermal or optical energy is given to the crystal, an electron may be liberated from its bond, leaving behind a missing electron or *hole*. The electron is free to move around the crystal. The minimum amount of energy needed to excite an electron from a bond to a conducting state is the *band gap*.

Figure 1.12b shows the energy-level picture. Electrons occupy a band of states called the *valence band*. These electrons reside in the covalent bonds. The *conduction band* is a set of unoccupied states above the valence band, separated by the band gap (E_g). Electrons in the conduction band are free to move and conduct electricity. In a pure semiconductor at 0 K, the valence band is filled

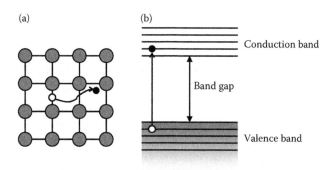

FIGURE 1.12 (a) Ball-and-stick model of an electron (solid circle) and hole (open circle). (b) Simplified energy band diagram.

with electrons, while the conduction band is empty. If we raise the temperature, electrons will be thermally excited, leading to electrons in the conduction band and holes in the valence band.

Electrons can also be promoted to the conduction band by electromagnetic radiation. A photon of energy E_g or greater may be absorbed by the semiconductor, leading an electron–hole pair. A semiconductor therefore absorbs photons of energy $h\nu \geq E_g$. Silicon, for example, has a band gap of 1.1 eV and absorbs visible and UV light. Gallium nitride (GaN), in contrast, has a band gap of 3.4 eV, which is in the UV. This is why silicon appears black (opaque) while GaN is transparent to visible light.

Semiconductor band gaps span a wide range, from IR to UV (Table 1.1). In general, as one goes down columns in the periodic table, bonds get weaker and the band gap decreases. For example, diamond (C), Si, and Ge have band gaps of 5.5, 1.1, and 0.7 eV, respectively. The band gaps of the III-V compound semiconductors GaN, GaP, and GaAs are 3.4, 2.3, and 1.4 eV. The variety of band gaps for compound semiconductors, in particular, have been beneficial for light-emitting applications.

1.6 BAND THEORY

In this section, we approach the idea of energy bands from two opposite directions: the linear combination of atomic orbitals (LCAO) and the free electron approximation.

The LCAO, or tight-binding, approach considers electrons that are bound to specific nuclei. As an example, consider a H_2 molecule. Two hydrogen atoms that are far apart have electrons in the 1s ground state. As the hydrogen atoms approach, the electron clouds begin to overlap. Let ϕ be the 1s wave function for an individual atom. The wave function for the H_2 molecule is approximated by a linear combination of the individual 1s orbitals:

$$(1.11) \qquad \phi_{\pm}(\mathbf{r}) = \tfrac{1}{\sqrt{2}}\left[\phi(\mathbf{r}-\mathbf{r}_1) \pm \phi(\mathbf{r}-\mathbf{r}_2)\right]$$

where $\mathbf{r}_1$ and $\mathbf{r}_2$ are the positions of the nuclei. ϕ_+ is the bonding state, which has a high electron density between the nuclei. ϕ_- is the antibonding state; due to the minus sign, it has a low electron density between the nuclei. The bonding state has a lower energy than the antibonding state.

Now, consider two silicon atoms. The 3s and 3p orbitals of neighboring atoms will overlap, creating bonding and antibonding states (Yu and Cardona, 1996). Because the atoms have a total of eight valence electrons, the s and p bonding states are filled, whereas the antibonding

TABLE 1.1   Properties of Important Semiconductors

Type	Name		Crystal Structure	Lattice Constant (Å)	Band Gap (eV)	
					300 K	0 K
Elemental	C	Carbon (diamond)	D	3.56683	5.47 (I)	5.48
	Si	Silicon	D	5.43095	1.12 (I)	1.17
	Ge	Germanium	D	5.65776	0.66 (I)	0.74
	Sn	Grey tin	D	6.48920	0	
IV-IV	SiC	Silicon carbide, 3C	Z	4.3596	2.36 (I)	2.416
		Silicon carbide, 2H	W	3.073 (a)	3.3 (I)	
				5.048 (c)		
		Silicon carbide, 4H		3.073 (a)	3.3 (I)	
				10.053 (c)		
		Silicon carbide, 6H		3.086 (a)	2.996 (I)	3.03
				15.117 (c)		
III-V	AlN	Aluminum nitride	W	3.112 (a)	6.14 (D)	6.25
				4.982 (c)		
	AlP	Aluminum phosphide	Z	5.4635	2.45 (I)	
	AlAs	Aluminum arsenide	Z	5.6605	2.17 (I)	
	AlSb	Aluminum antimonide	Z	6.1355	1.58 (I)	1.68
	BN	Boron nitride	Z	3.6150	6.4 (I)	
	BP	Boron phosphide	Z	4.5380	2.0 (I)	
	GaN	Gallium nitride	W	3.189 (a)	3.43 (D)	3.51
				5.185 (c)		
	GaP	Gallium phosphide	Z	5.4512	2.26 (I)	2.34
	GaAs	Gallium arsenide	Z	5.6533	1.42 (D)	1.52
	GaSb	Gallium antimonide	Z	6.0959	0.72 (D)	0.86
	InN	Indium nitride	W	3.533 (a)	0.64 (D)	0.69
				5.693 (c)		
	InP	Indium phosphide	Z	5.8686	1.35 (D)	1.42
	InAs	Indium arsenide	Z	6.0584	0.36 (D)	0.42
	InSb	Indium antimonide	Z	6.4794	0.17 (D)	0.23
II-VI	CdS	Cadmium sulfide	Z	5.8320	2.42 (D)	2.56
	CdS	Cadmium sulfide	W	4.16 (a)	2.49 (D)	2.59
				6.756 (c)		
	CdSe	Cadmium selenide	W	4.30 (a)	1.74 (D)	1.85
				7.02 (c)		
	CdTe	Cadmium telluride	Z	6.482	1.43 (D)	1.48
	ZnO	Zinc oxide	W	3.24 (a)	3.4 (D)	3.437
				5.21 (c)		
	ZnS	Zinc sulfide	Z	5.430	3.68 (D)	3.84
	ZnS	Zinc sulfide	W	3.82 (a)	3.80 (D)	3.91
				6.26 (c)		
	ZnSe	Zinc selenide	Z	5.669	2.71 (D)	2.82
	ZnTe	Zinc telluride	Z	6.1034	2.24 (D)	2.394
IV-VI	PbS	Lead sulfide	N	5.9362	0.41 (D)	0.286
	PbTe	Lead telluride	N	6.4620	0.31 (D)	0.19

(Continued)

TABLE 1.1 (*Continued*)   Properties of Important Semiconductors

Type	Name		Crystal Structure	Lattice Constant (Å)	Band Gap (eV) 300 K	0 K
Oxides	SnO$_2$	Tin dioxide	R	4.737 (*a*)	2.7 (I)	
				3.186 (*c*)		
	TiO$_2$	Titanium dioxide	R	4.5931 (*a*)	3.05 (I)	
				2.9586 (*c*)		

Source: After Böer, K.W. 1990. *Survey of Semiconductor Physics.* New York: Van Nostrand Reinhold; Sze, S.M. 1981. *Physics of Semiconductor Devices*, 2nd edn. New York: John Wiley & Sons. See also Madelung, O., ed. 1996. *Semiconductors—Basic Data*, 2nd edn. Berlin: Springer.

Note: The crystal structures are D = diamond, N = NaCl (rocksalt), R = rutile, W = wurtzite, and Z = zincblende. The SiC 4H and 6H structures are discussed in Section 1.3. Lattice constants are for room temperature. The band gap is D = direct or I = indirect.

states are empty (Figure 1.13). If we have a large number of silicon atoms, then each state will broaden into a band of many levels. The filled low-energy *p* states are the valence band. The unfilled high-energy *p* states are the conduction band. The minimum energy difference between the valence band and conduction band is the band gap, E_g. If an electron acquires an amount of energy greater than E_g, then it can be promoted from the valence band to the conduction band, where it is free to roam around the crystal and conduct electricity. In effect, it has been liberated from its bonding orbital.

The LCAO approach also works for compound semiconductors. Consider a Ga and As atom in a GaAs crystal. As, the more electronegative atom, has energy levels that are lower than those in Ga. The Ga and As orbitals overlap and form the conduction band and valence band (Figure 1.13). The valence band consists of *p* states, and the conduction band consists of *s* states. This is the situation for most III-V and II-VI semiconductors, as well as germanium. In the case of GaAs, due to the energy difference between the Ga and As states, the valence band is "As-like" and the conduction band is "Ga-like." In other words, the wave functions for electrons in the valence (conduction) band are primarily composed of As (Ga) wave functions.

The *nearly free electron* approach considers a crystal to be an empty box, into which we place electrons. We ignore the Coulomb interactions between electrons. After we have put in the

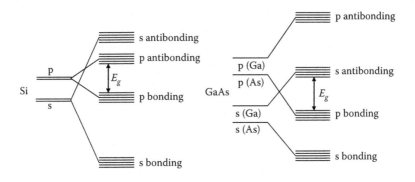

FIGURE 1.13 Formation of bands in silicon and GaAs, from the linear combination of atomic orbitals (LCAO) approach. (After Yu, P.Y. and Cardona, M., *Fundamentals of Semiconductors*, Berlin, Springer-Verlag, 1996.)

1.6 Band Theory

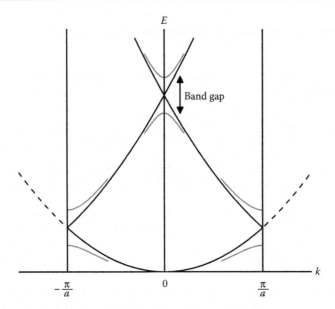

FIGURE 1.14 Electron bands in a 1D crystal, plotted in the reduced zone scheme. The black lines are for free electrons and the gray lines are for electrons perturbed by a periodic potential.

appropriate number of electrons, we slowly turn on the potential due to the atoms in the crystal. From quantum mechanics, a free electron is represented by a plane wave:

$$\psi \sim e^{i\mathbf{k}\cdot\mathbf{r}} \tag{1.12}$$

where $\mathbf{p} = \hbar\mathbf{k}$ is the momentum. The kinetic energy is given by

$$E = \frac{p^2}{2m} = \frac{\hbar^2 k^2}{2m} \tag{1.13}$$

The plot of E versus k is just a parabola. Let's consider a 1D example, where identical atoms are separated by a distance a. The first Brillouin zone is $-\pi/a < k \leq \pi/a$. As discussed in Section 1.4, we can always plot quantities in the first Brillouin zone by adding or subtracting multiples of $2\pi/a$. This way of plotting the band structure is called the *reduced zone scheme* (Figure 1.14).

In the reduced zone scheme, the points at $k = -2\pi/a$ and $k = 2\pi/a$ are actually plotted at $k = 0$. There are therefore two degenerate (same energy) wave functions at the zone center, $e^{i(2\pi x/a)}$ and $e^{-i(2\pi x/a)}$. Using the Euler identities (Appendix A), we can write linear combinations of these functions:

$$\begin{aligned}\psi_1 &\sim e^{i(2\pi x/a)} + e^{-i(2\pi x/a)} \sim \cos(2\pi x/a) \\ \psi_2 &\sim e^{i(2\pi x/a)} - e^{-i(2\pi x/a)} \sim \sin(2\pi x/a)\end{aligned} \tag{1.14}$$

To get the electron density, we square these wave functions. Now, we imagine slowly turning on the potential due to the nuclei. ψ_1 has a high electron density near the positive nuclei, so its energy is low due to Coulomb attraction (Figure 1.15). ψ_2 has a low electron density near the

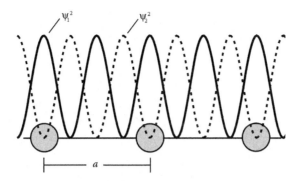

FIGURE 1.15 Electron densities for two wavefunctions, ψ_1 and ψ_2, in a 1D crystal. The energies of the two wavefunctions are plotted at $k = 0$ in the reduced zone scheme (see Figure 1.14).

nuclei, so its energy is high. This interaction lifts the degeneracy, creating a band gap. A similar calculation shows that gaps also occur at the zone edges. In general, band gaps result from a periodic potential, as in a crystal.

Band structure can be depicted by plotting E versus k. We can also depict the band structure more simply, ignoring k, and representing energies as horizontal lines (Figure 1.12b). Owing to Pauli exclusion, only two electrons, one spin up and one spin down, may occupy a particular **k**-point. At zero temperature, the electrons occupy the lowest possible states. In a pure semiconductor at 0 K, electrons completely fill the valence band, leaving the conduction band empty.

Sophisticated theoretical methods have been developed to calculate band structures. Some calculated energy bands are shown in Figure 1.16. At first glance, these band structures look rich and complicated. In all these examples, three valence band maxima are located at the Γ point of the Brillouin zone, and two of these are degenerate. We can understand the splitting of the valence band by recalling that the valence band is composed of p orbitals, which have an angular momentum $l = 1$. Because the electron spin is 1/2, the total angular momentum is $j = 3/2$ or $1/2$. Due to spin-orbit coupling, the $j = 3/2$ band has a higher energy than the $j = 1/2$ band. The shape of the $j = 3/2$ band is discussed in Section 1.7.

Conduction-band minima appear at L, Γ, and X (Figure 1.17). The magnitude of the three energy gaps $E(L)$, $E(\Gamma)$, and $E(X)$ depends on the chemical composition of each semiconductor. GaAs, for example, has the conduction band minimum and the valence band maximum both located at Γ, the center of the Brillouin zone. This is called a *direct* band gap. For silicon and germanium, the conduction band minima do not occur at Γ, but at nonzero values of k (at L for

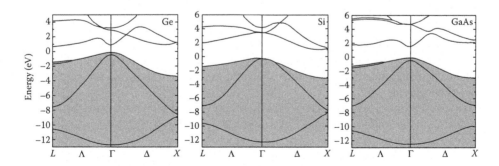

FIGURE 1.16 Calculated band structures for germanium, silicon, and GaAs. (After Chelikowsky, J.R. and M.L. Cohen. 1976. *Phys. Rev. B*, 14, 556–582.)

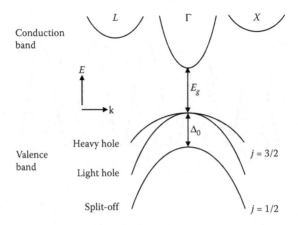

FIGURE 1.17 Schematic band structure for diamond and zincblende semiconductors. In wurtzite semiconductors, the valence band degeneracy at $k = 0$ (Γ) is lifted. (After Böer, K.W. 1990. *Survey of Semiconductor Physics*. New York: Van Nostrand Reinhold.)

germanium and near X for silicon). These are called *indirect* gaps. The ramifications of this classification are discussed in Section 1.9.

The conduction band minima of many semiconductors have been analyzed in great detail. Researchers have also explored energy shifts of the various minima with pressure (Goñi and Syassen, 1998). The shift of an energy level versus strain is governed by *deformation potentials* (Bardeen and Shockley, 1950). Under hydrostatic pressure, the minima at the Γ and L points shift upwards in energy, with the shift being larger for the Γ band, while the X-band minimum shifts downward. These shifts alter the band gap of each semiconductor in a unique way. Silicon shows a constant reduction of the indirect X band gap with applied pressure. In germanium, an initial gradual increase of the indirect L band gap intersects the decreasing indirect X band gap at 3.5 GPa. In GaAs, the upwards shift of the Γ band intersects the X band at ~4 GPa, and the material undergoes a direct-to-indirect transition.

The band parameters also depend on temperature, due to electron–phonon coupling. As temperature increases from 0 K, phonons become thermally populated and the atoms in the crystal oscillate about their equilibrium positions. These displacements modify the band edges via the deformation potentials. In general, the band gap *decreases* as temperature is raised. To model the temperature-dependent shift, empirical formulas are typically used (e.g., Varshni, 1967).

1.7 ELECTRONS AND HOLES

At $T = 0$ K, a pure semiconductor has no electrons in the conduction band and the valence band is completely filled. When an electric field is applied, the valence electrons are unable to respond. The valence band has no empty states into which they can move. The semiconductor is therefore electrically insulating.

When temperature is raised, electrons are thermally excited from the valence band to the conduction band. In the conduction band, electrons are free to move around the crystal and conduct electricity. For a free electron in vacuum, the energy is

$$E = \frac{\hbar^2 k^2}{2m} \tag{1.15}$$

where m is the free electron mass. Now, consider a single electron in the conduction band. Near the minimum energy point, we can approximate the E versus k curve as an isotropic parabola. We write the parabola as

$$(1.16) \qquad E = \frac{\hbar^2 k^2}{2m_e}$$

where m_e is the *electron effective mass* (Kittel, 2005). The effective mass is inversely proportional to the curvature of the band:

$$(1.17) \qquad m_e = \hbar^2 \left(\frac{\partial^2 E}{\partial k^2} \right)^{-1}$$

A band with high curvature, like the conduction band in GaAs, gives rise to a small electron effective mass. Electrons do not travel through a crystal with their free mass but with this modified or effective mass. This is a consequence of the wave nature of the electron interacting with the regularly spaced atoms in the semiconductor crystal.

The effective mass defines how an electron responds to an external force $\mathbf{F}$, such as that produced by an applied voltage. The acceleration $\mathbf{a}$ of the electron is simply governed by Newton's second law:

$$(1.18) \qquad \mathbf{F} = m_e \mathbf{a}$$

Most values of m_e are actually *lower* than the free-electron mass. The reason for this surprising result is that the effective mass accounts for the forces exerted by the atoms in the crystal. These forces help push along the electron. However, if an electron is accelerated too much, it will encounter a part of the band where m_e is high. It is left to the reader's imagination to figure out what happens to the effective mass at the values of $\mathbf{k}$ where the second derivative becomes zero.

Table 1.2 lists the effective masses of electrons at various conduction band minima for several semiconductors. For the minimum at Γ we obtain one value for m_e because the mass is isotropic. For indirect-gap minima along different directions (e.g., X and L), a parallel $m_\parallel$ and a perpendicular mass $m_\perp$ are listed. GaP is an unusual case because it has a double-humped "camel's back" minimum along the Δ direction.

When an electron is excited from the valence band to the conduction band, it leaves behind a missing electron, called a *hole*, in the valence band. We treat the hole as a quasiparticle with all the properties of a real particle: mass, charge, energy, and momentum. Because the hole is the absence of an electron, its energy equals the negative of the electron energy. Therefore, its energy is minimized at the valence band *maximum*. As in the case of electrons, holes have an effective mass m_h, given by

$$(1.19) \qquad m_h = \hbar^2 \left(\frac{\partial^2 E_h}{\partial k^2} \right)^{-1}$$

where E_h is the hole energy. For the valence band, the curvature of E is negative. Because $E_h = -E$, m_h is positive.

TABLE 1.2 Band Gaps (0 K) and Electron Effective Masses for Several Semiconductors

Crystal	Direct Gap (eV)	m_e	Indirect Gap (eV)	$m_{e//}$	$m_{e\perp}$
Si	4.19		1.17 (X)	0.916	0.191
Ge	0.898	0.038	0.74 (L)	1.57	0.081
GaN	3.51	0.2			
GaP	2.895	0.093	2.34 (Δ)	0.91	0.25
GaAs	1.52	0.067	1.82 (L)	1.9	0.075
			2.03 (X)	1.8	0.257
GaSb	0.86	0.041	1.22 (L)	0.95	0.11
			1.72 (X)	1.2	0.25
InN	0.69	0.11			
InP	1.42	0.077	2.19 (L)		
InAs	0.42	0.024	1.53 (L)		
InSb	0.23	0.0136	1.03 (L)		
ZnO	3.44	0.26			
ZnS (cubic)	3.84	0.34			
ZnS (hex)	3.91	0.28			
ZnSe	2.82	0.16	3.96 (L)		
ZnTe	2.394	0.122			
CdS	2.56	0.21			
CdSe	1.85	0.112			
CdTe	1.48	0.096	2.82 (L)		

Source: Böer, K.W. 1990. *Survey of Semiconductor Physics.* New York: Van Nostrand Reinhold.

Note: Effective masses m_e are in units of the free-electron mass. For indirect gaps, // and ⊥ refer to the directions parallel and perpendicular to the zone-edge wave vector (X, L, etc.). For hexagonal (wurtzite) structures, m_e is an average of the mass parallel and perpendicular to the c axis.

Like electrons, holes carry electric charge. Consider a valence band that has a single hole. In the absence of an applied electric field, the hole resides at the valence band maximum, $\mathbf{k} = 0$. Now, suppose an electric field is applied along the x direction. The force on the negatively charged electrons will be in the $-x$ direction. An electron with $k_x > 0$ will be "pushed" into the hole. This results in a hole with a positive k_x value. The hole accelerates in the same direction as the electric field. Therefore, we can think of holes as having a charge of $+|e|$.

The valence band maxima of all diamond and zincblende semiconductors are degenerate and are described by two curved surfaces that touch at the Γ point. Therefore, we must consider two bands, approximated as parabolas. The *heavy holes* belong to the band with the smaller curvature (E_h). The *light holes* define the larger curvature band (E_l). As in the case of electrons, the effective hole masses differ from the free electron mass. Table 1.3 contains a collection of effective light and heavy hole masses for a number of semiconductors.

In general, hole masses are more complicated than those of electrons. Luttinger (1956) showed that the E versus k dependence of the light and heavy holes can be described accurately by

$$(1.20) \qquad E_{l,h} = \frac{\hbar^2}{2m}\left[\gamma_1 k^2 \pm \sqrt{4\gamma_2^2 k^4 + 12\left(\gamma_3^2 - \gamma_2^2\right)\left(k_x^2 k_y^2 + k_y^2 k_z^2 + k_z^2 k_x^2\right)}\right]$$

TABLE 1.3 Luttinger Parameters, Valence-Band Splitting, and Hole Effective Masses for Several Diamond and Zincblende Semiconductors

Crystal	γ_1	γ_2	γ_3	Δ_0 (eV)	m_{hh}	m_{lh}
Si	4.285	0.339	1.446	0.045	0.537	0.153
Ge	13.38	4.28	5.69	0.297	0.284	0.044
GaP	4.05	0.49	1.25	0.08	0.419	0.16
GaAs	6.95	2.25	2.86	0.34	0.51	0.082
GaSb	13.3	4.4	5.7	0.76	0.28	0.05
InP	5.15	0.94	1.62		0.56	0.12
InAs	20.4	8.3	9.1	0.38	0.35	0.026
InSb	3.25	−0.2	0.9	0.85	0.34	0.016
ZnSe	4.3	1.14	1.84	0.403		
ZnTe	3.9	0.83	1.30	0.97		
CdTe	5.3	1.7	2	0.811	0.72	0.13

Source: Böer, K.W. 1990. *Survey of Semiconductor Physics*. New York: Van Nostrand Reinhold.
Note: Heavy and light holes are denoted m_{hh} and m_{lh}, respectively.

where γ_1, γ_2, and γ_3 are called *Luttinger parameters*. In addition, due to spin-orbit coupling, the *split-off* band is below the valence band maximum by an energy Δ_0. In semiconductors with small values of Δ_0, the split-off band has a major influence on the shape of the light and heavy hole bands.

1.8 DOPING

The discussion so far has been limited to pure, or *intrinsic*, semiconductors. The free electrons are intrinsic to the material and do not come from an external source. In reality, whether by design or accident, semiconductors always contain impurities. In silicon, for example, phosphorus is a common dopant. The phosphorus impurity atom occupies a *substitutional* site, replacing a silicon atom. Because phosphorus is in the group-V column of the periodic table, it has five valence electrons, one more than the silicon atom it replaced. At room temperature, the phosphorus atom gives its extra electron to the conduction band. Because phosphorus "donates" a free electron, it is called a *donor*. Semiconductors that contain a preponderance of donor impurities are called *n*-type because the free electrons carry negative charge.

Boron dopants also occupy substitutional sites in silicon. Because boron is in the group-III column of the periodic table, it has three valence electrons, one fewer than silicon. The boron atom readily accepts a fourth electron from the valence band, leaving behind a hole. Boron is therefore an *acceptor* impurity. Semiconductors that have more acceptors than donors are called *p*-type because holes carry positive charge. Semiconductors are called *extrinsic* if their electrical conduction is dominated by donor or acceptor dopants, as opposed to electrons that are excited across the band gap. Heavy *n*- or *p*-type doping is denoted n^+ or p^+.

Compound semiconductors also have substitutional donors and acceptors, but it is important to consider whether the impurity resides on the cation or anion site. In GaAs, tellurium is a group-VI atom that replaces arsenic, which is group V, making tellurium a donor. Zinc is a group-II impurity that replaces a group-III gallium atom and is therefore an acceptor. Group-IV impurities can, in principle, substitute for either gallium or arsenic. In practice, carbon always

1.8 Doping

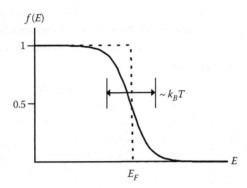

FIGURE 1.18 Fermi–Dirac distribution function, for $T = 0$ (dashed line) and $T > 0$ (solid line).

substitutes for arsenic, making it an acceptor, whereas silicon predominantly substitutes for gallium, making it a donor.

The concentration of electrons in a semiconductor is determined by the *Fermi–Dirac distribution*:

$$(1.21) \qquad f(E) = \frac{1}{e^{(E-E_F)/k_B T} + 1}$$

where E_F is the *Fermi energy*, k_B is the Boltzmann constant, and T is the temperature in Kelvin. If there is a state with an energy E, then $f(E)$ tells us the probability that the state will be occupied by an electron. Figure 1.18 shows plots of $f(E)$ for zero and nonzero temperature. At zero temperature, E_F defines a sharp cutoff point: below E_F, all states are filled; above E_F, all states are empty. For $T > 0$, however, electrons are thermally excited to higher energy states. The function gets smeared out around E_F, with a width of $\sim k_B T$.

Consider the following regimes:

- $f(E_F) = \tfrac{1}{2}$. At the Fermi energy, if there is a state, then there is a 50% chance that it will be occupied.
- For low energies ($E - E_F \ll -k_B T$), $f(E) \approx 1$. Low-energy states are almost totally filled.
- For high energies ($E - E_F \gg k_B T$), the distribution decays exponentially, as shown in the following equation:

$$(1.22) \qquad f(E) \approx e^{(E_F - E)/k_B T}$$

Equation 1.22 is referred to as the "Boltzmann tail." It has the same functional form as that for classical particles in a gas.

When one dopes a semiconductor, one is tuning the Fermi energy. For an intrinsic semiconductor, E_F is near the middle of the gap. The probability that a particular conduction-band state is filled is exceedingly low. The semiconductor therefore has few electrons (or holes) and is *semi-insulating*. For an *n*-type semiconductor, E_F is near the conduction band minimum, such that the Boltzmann tail extends into the band, populating it with free electrons. For a *p*-type semiconductor, E_F is near the valence band maximum, resulting in unfilled states (holes) in the valence band. In Chapter 5, we use the Fermi–Dirac distribution to calculate the concentration of free carriers as a function of temperature.

1.9 OPTICAL PROPERTIES

Absorption of photons can be used to explore the band structure in detail over a wide energy range. Modern spectrometers can handle photons with energies ranging from μeV to many eV and beyond, providing a convenient and nondestructive characterization method. When a photon is absorbed by a semiconductor, energy and momentum must be conserved. Red-orange light, for example, has

$$(1.23) \qquad k_{ph} = \frac{2\pi}{\lambda} \approx \frac{6}{600 \text{ nm}} = 10^7 \text{ m}^{-1}$$

The magnitude of k at the Brillouin zone edge, in contrast, is

$$(1.24) \qquad k \approx \frac{\pi}{a} \approx 10^{10} \text{ m}^{-1}$$

Therefore, on the scale of a band-gap diagram, the photon momentum is negligible. As an approximate description, we can say that *photons cause vertical transitions.*

These vertical transitions result in strong absorption of light. The linear absorption coefficient α is defined by the Beer–Lambert law:

$$(1.25) \qquad I = I_0 e^{-\alpha x}$$

where I_0 is the intensity (or irradiance) of light impinging on a wafer of semiconductor with parallel faces, I is the intensity exiting the semiconductor sample, and x is the thickness of the semiconductor wafer. Reflection effects are assumed to be negligible, a condition that can be closely approximated through efficient antireflection coatings. Solving this equation for α, we find

$$(1.26) \qquad \alpha = \frac{1}{x} \ln(I_0/I)$$

When an electron absorbs a photon, its momentum is unchanged,

$$(1.27) \qquad \Delta \mathbf{k} = 0$$

but its energy is increased:

$$(1.28) \qquad \Delta E = h\nu$$

where $h\nu$ is the photon energy. In a direct-gap semiconductor, photons can excite electrons across the minimum band gap without a change in momentum. This is an efficient process involving only two particles, a photon and an electron. The high efficiency leads to a sharp rise of the absorption coefficient α for energies larger than E_g. For photon energies lower than E_g, the semiconductor is transparent. This is why GaN ($E_g = 3.4$ eV) is transparent to visible light, whereas GaAs ($E_g = 1.4$ eV) is opaque.

For an indirect-gap semiconductor, the conduction band minimum occurs at $\mathbf{k} \neq 0$. To conserve momentum, a photon *and* a phonon are required to excite the electron across the minimum band gap. The phonon energy $\hbar\Omega$ is typically tens of meV, much smaller than the band gap. Therefore, phonons cause nearly *horizontal* transitions (Figure 1.19). In the three-particle process, the electron momentum is changed:

$$(1.29) \qquad \Delta \mathbf{k} = \pm \mathbf{K}$$

1.9 Optical Properties

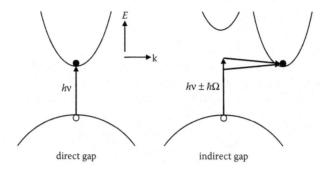

FIGURE 1.19 Direct and indirect optical absorption. The vertical transitions are from photon absorption. The nearly horizontal transitions are from phonon absorption ($+\hbar\Omega$) or phonon creation ($-\hbar\Omega$).

where **K** is the phonon momentum and ± refers to the destruction or creation of a phonon. The electron energy is increased by

$$\Delta E = h\nu \pm \hbar\Omega \tag{1.30}$$

A three-particle process is less likely than the two-particle process discussed above. This is evident from the rise of α at the minimum band-gap energy of 0.66 eV at 300 K to a value of only 10–100 cm^{-1} for germanium (Figure 1.20). A second sharp rise up to 10^4 cm^{-1} occurs at higher energies where direct (phonon-less) transitions become possible. In silicon, the indirect gap is 1.1 eV at room temperature, and photon energies larger than 3 eV are required for direct transitions.

In a band-gap absorption process, a photon creates an electron–hole pair. The inverse of this process is *emission*, where an electron and hole recombine, resulting in the creation of a photon. The same momentum and energy conservation laws that govern absorption also apply to emission. Consider electrons and holes that are excited by a light source with photons above the band gap. The electrons rapidly fall to the conduction band minimum by creating phonons (i.e., heat). Likewise, the holes float up to the valence band maximum. The electrons and holes then recombine, resulting in the emission of photons of energy $\sim E_g$.

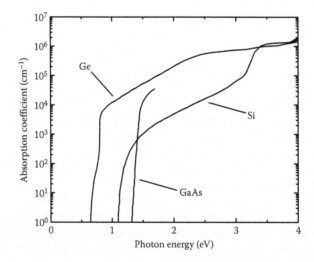

FIGURE 1.20 Band-gap absorption of germanium, silicon, and GaAs, at 300 K. (After Sze, S.M. 1981. *Physics of Semiconductor Devices*, 2nd ed. New York: John Wiley & Sons.)

Chapter 1—Semiconductor Basics   **23**

In an indirect-gap semiconductor, the recombination of an electron and a hole requires a phonon in order to conserve momentum. This three-particle process is less likely than a two-particle process. Therefore, indirect-gap semiconductors such as silicon have low emission efficiencies. This is why practical light emitting devices are not made from silicon, despite years of research efforts. GaAs and GaN, in contrast, are direct-gap semiconductors and are therefore preferred materials for optoelectronic applications. These examples illustrate why one classifies semiconductors as direct or indirect band gap materials.

1.10 ELECTRONIC TRANSPORT

The electrical properties of semiconductors can be tuned over a large range. One of the fundamental properties of intrinsic semiconductors is the strong dependence of the concentration of free electrons (n) and holes (p) on absolute temperature T. As temperature increases, electrons are thermally promoted from the valence band to the conduction band, increasing n and p. The rapid increase of these concentrations leads to a sharp decrease of the intrinsic resistivity with increasing temperature, opposite to the behavior for the resistivity of metals.

Electronic transport encompasses phenomena associated with the motion of *free carriers*, electrons in the conduction band and holes in the valence band. These phenomena include motion in low and high electric fields, motion in combined electric and magnetic fields, and scattering from impurities and phonons. In this section, we will restrict ourselves to the most basic aspects of carrier transport.

In low electric fields ($\mathbf{E}$), the motion of a free carrier of charge e can be described with the effective mass (m^*) and Newtonian mechanics:

$$(1.31) \qquad \mathbf{F} = e\mathbf{E} = m^*\mathbf{a}$$

where the acceleration $\mathbf{a}$ is due to the applied field. Under normal conditions, free carriers do not undergo ballistic transport. Instead, they scatter off phonons and defects such as impurities. After being scattered, let us assume that the free carrier has a random velocity with a mean of zero. Let $\mathbf{v}$ be the velocity component along the direction of $\mathbf{E}$. If the mean time between scattering events is τ, then in a time dt, this velocity should be reduced by a fraction dt/τ. The change in velocity is therefore given by

$$(1.32) \qquad d\mathbf{v} = \mathbf{a}dt - \mathbf{v}dt/\tau$$

Dividing through by dt and using Equation 1.31 yields

$$(1.33) \qquad \frac{d\mathbf{v}}{dt} + \frac{\mathbf{v}}{\tau} = \frac{e\mathbf{E}}{m^*}$$

τ is often referred to as the *momentum relaxation time* (Seeger, 1982).

When the $\mathbf{E}$ field is turned on, the velocity of the free carrier increases until it reaches a steady-state value. The steady-state velocity occurs when $d\mathbf{v}/dt = 0$:

$$\mathbf{v}_d = \mu\mathbf{E}$$
$$(1.34)$$
$$\mu = \frac{e\tau}{m^*}$$

24 1.11 Examples of Semiconductors

where $\mathbf{v}_d$ is the *drift velocity*, and μ is the *mobility*. High mobilities result in high drift velocities for a given electric field. Mobility is inversely proportional to effective mass. For high-speed devices, it is desirable to have low effective masses. This is why GaAs, with an electron effective mass of 0.067 *m*, has been used for high-speed electronics. Hole mobilities are typically lower than electron mobilities, due to the larger effective mass.

At room temperature, the drift velocity is an averaged quantity that is superimposed on the thermal motion of the carriers. For small electric fields, the average gain in kinetic energy is small compared to the thermal energy. The energy gain of the carrier in the electric field can be lost in inelastic collisions. However, for small electric fields, such collisions are infrequent compared to the number of elastic collision events.

The mobility is limited by scattering events. At room temperature, phonons are thermally excited. Phonons are quasiparticles that have momentum and energy. These quasiparticles collide with free carriers, like billiard balls, in a process called *phonon scattering*. As temperature is lowered, the number of phonons decreases so the mobility increases (up to a point). One can visualize phonon scattering by imagining a perfect crystal at zero temperature. Carriers move through spaces in the ordered lattice at high speeds. As the crystal is warmed up, the atoms jiggle and bump into the carriers, reducing their average speed. Another way to approach the problem is to consider that lattice waves modulate the band gap through the deformation potential. This modulation interacts with the electron wave functions, causing them to scatter.

A second source of scattering is due to ionized impurities. Consider an *n*-type semiconductor with N_d donors and N_a acceptors, $N_d > N_a$. Each acceptor will take one electron from a donor, a transaction that results in a negatively ionized acceptor and a positively ionized donor. The acceptors are said to electrically *compensate* the donors. The remaining neutral donors can then donate electrons to the conduction band. These free electrons will scatter off the ionized impurities as well as phonons. *Ionized impurity scattering* is similar to Rutherford scattering of an alpha particle from an atomic nucleus. As the free carrier approaches the ionized center, its trajectory deviates by an angle θ. Heavily compensated semiconductors have many ionized impurities and hence much ionized impurity scattering. Therefore, for applications that require high mobilities, it is necessary to minimize the concentration of compensating impurities. Phonon and ionized impurity scattering are discussed in more detail in Chapter 5.

At sufficiently high velocities, carriers undergo inelastic scattering. The most common such process is the creation of an optical phonon near $\mathbf{k} = 0$. The carrier loses kinetic energy, which goes into lattice vibrations. When the kinetic energy of a carrier reaches the optical phonon energy, inelastic scattering will begin, limiting the velocity. Semiconductors with high optical phonon frequencies, like GaN and SiC, are therefore well suited to high-power applications, because carriers can be accelerated to high kinetic energies before optical phonon scattering takes place. The maximum velocity for an electron (or hole) is called the *saturation velocity*.

1.11 EXAMPLES OF SEMICONDUCTORS

Table 1.1 contains a collection of important properties for a large number of semiconductors. There are several trends that are apparent. Consider the elemental semiconductors diamond, silicon, and germanium. As we go down the group-IV column, the atomic mass increases, and the covalent bonds get weaker and longer. These factors result in (1) larger lattice constants, (2) lower phonon frequencies, (3) lower melting temperature, and (4) smaller band gap. Tin, which is below germanium, is a semiconductor where the conduction band and valence band touch, yielding a "band gap" of 0 eV. Similar trends are observed for the compound semiconductors. SiC, for example, has a band gap that is between diamond and silicon. The band gaps of GaP and AlAs are greater than that of GaAs, due to the smaller atomic numbers of P and Al, respectively.

The elemental semiconductors are covalently bonded. Compound semiconductors, in contrast, all exhibit some degree of ionic bonding. Roughly, the degree of ionicity is related to the difference in the electronegativity of the constituent atoms (Phillips, 1970). III-V semiconductors such as GaAs are slightly ionic, whereas the III-nitride and II-VI semiconductors are more ionic. Ionic crystals are also characterized by high melting points and a tendency to form the rocksalt structure. Most wurtzite and zincblende II-VI semiconductors transform into the rocksalt structure at high pressures (Nelmes and McMahon, 1998).

The wide range of properties allows for great flexibility in choosing semiconductor materials for different applications. In addition to binary semiconductors, one can form ternary alloys such as $Al_xGa_{1-x}As$, where x is the atomic fraction of Al. The crystal structure is zincblende, as in GaAs, except that a Ga site randomly contains either a Ga atom or an Al atom. By increasing the fraction of Al, the band gap increases. In $In_xGa_{1-x}N$ alloys, the band gap spans all the way from the ultraviolet (UV) ($x = 0$, $E_g = 3.4$ eV) to the infrared (IR) ($x = 1$, $E_g = 0.7$ eV). Through alloying, the band gap can be engineered for devices such as light-emitting diodes (LEDs) or laser diodes. By adding a fourth component, one can tune both the band gap and the lattice constant. Red LEDs, for example, utilize quaternary AlGaInP alloys. Detailed band parameters for technologically important III-V semiconductors may be found in Vurgaftman et al. (2001).

Beyond the well-known applications of silicon and germanium, GaAs is used in high-speed switches, due to its high electron mobility. Due to its wide band gap, SiC is used in high-temperature, high-voltage electronics. There also exists a large number of sensor and display applications. Television and oscilloscope screens used "phosphors" such as ZnS and ZnSe, and CdS was used as a light-sensitive resistor. Infrared detectors are made from silicon, InAs, germanium, and $Hg_xCd_{1-x}Te$ (MCT). Radiation detectors, which convert the energy of energetic particles or quanta into electron–hole pairs, include silicon (x-rays), germanium (gamma-rays), and $Cd_xZn_{1-x}Te$.

Oxide materials, historically regarded as insulators, are emerging as potentially important wide-band-gap semiconductors. The conduction band minima in oxide semiconductors are low on an absolute energy scale (i.e., they have a high electron affinity). TiO_2 uses this attribute to capture electrons in dye-sensitized solar cells. The transparent conductive oxide SnO_2:In, or ITO, is used widely in display applications. Another oxide semiconductor is ZnO, which is used as a transparent electrical contact and may find use as a blue/UV light-emitting material. Complex oxides consist of more than two atoms. Strontium titanate ($SrTiO_3$), for example, is a semiconductor with an indirect band gap of 3.2 eV.

PROBLEMS

1.1 A silicon crystal has a density of 2.33 g/cm³. Find the distance between Si nearest neighbors.

1.2 Germanium has a lattice constant $a = 5.66$ Å. What is the density (atoms/cm³)?

1.3 Strontium titanate ($SrTiO_3$) has the *perovskite* structure. In a cubic cell, Sr atoms are on the corners, Ti is in the center, and O atoms are on the face centers.
   a. Sketch the unit cube with the atoms.
   b. To how many O atoms is the Ti atom bonded?
   c. The lattice parameter is $a = 3.9$ Å. Calculate the density (g/cm³).

1.4 In 3D, the free-electron energy is $E = \hbar^2(k_x^2 + k_y^2 + k_z^2)/2\,m$. The first Brillouin zone for a simple cubic cell is:

$$-\pi/a < k_x \leq \pi/a, \quad -\pi/a < k_y \leq \pi/a, \quad -\pi/a < k_z \leq \pi/a$$

To plot in the reduced zone scheme, add multiples of $2\pi/a$ to the k values:

$$E = \frac{\hbar^2}{2m}\left[\left(k_x + \frac{2\pi}{a}n_x\right)^2 + \left(k_y + \frac{2\pi}{a}n_y\right)^2 + \left(k_z + \frac{2\pi}{a}n_z\right)^2\right]$$

Plot E versus k_x along the [100] direction, up to a maximum energy of $5\hbar^2\pi^2/2ma^2$.

1.5 What is the difference between an intrinsic and an extrinsic semiconductor?

1.6 At room temperature, what is the probability that a state 0.1 eV above the Fermi energy will be occupied by an electron?

1.7 Sketch and label semiconductor valence bands for heavy and light holes as well as the split-off band in a semiconductor with a diamond crystal lattice. In which valence band will the effective mass be the largest?

1.8 Suppose a conduction band has an energy given by

$$E = \sqrt{ak^2 + b}$$

Find the electron effective mass at $k = 0$. What if $b = 0$?

1.9 Pure diamond is electrically insulating and transparent at room temperature. Pure InSb is black and conducting at room temperature. Explain why.

1.10 GaAs-based materials are used for IR light emitting diodes (LEDs). Why is no one selling Si LEDs?

REFERENCES

Akasaki, I. 2006. Key inventions in the history of nitride-based blue LED and LD. *J. Cryst. Growth* 300: 2–10.

Bardeen, J. and W.H. Brattain. 1948. The transistor, a semiconductor triode. *Phys. Rev.* 74: 230–231.

Bardeen, J. and W. Shockley. 1950. Deformation potentials and mobilities in non-polar crystals. *Phys. Rev.* 80: 72–80.

Böer, K.W. 1990. *Survey of Semiconductor Physics*. New York: Van Nostrand Reinhold.

Braun, F. 1875. Über die Stromleitung durch Schwefelmetalle. *Annalen der Physik und Chemie* 229: 556–563. Translation in: Sze, S.M. 1991. *Semiconductor Devices: Pioneering Papers*. Singapore: World Scientific.

Chelikowsky, J.R. and M.L. Cohen. 1976. Nonlocal pseudopotential calculations for the electronic structure of eleven diamond and zinc-blende semiconductors. *Phys. Rev. B* 14: 556–582.

Czochralski, J. 1918. Ein neues Verfahren zur Messung der Kristallisationsgeschwindig-heit der Metalle. *Z. Phys. Chemie* 92: 219–221.

Goñi, A.R. and K. Syassen. 1998. Optical properties of semiconductors under pressure. In *High Pressure in Semiconductor Physics I*, eds. T. Suski and W. Paul, pp. 146–244. San Diego: Academic Press.

Holonyak Jr., N. and S.F. Bevacqua. 1962. Coherent (visible) light emission from $Ga(As_{1-x}P_x)$ junctions. *Appl. Phys. Lett.* 1: 82–3.

Horn, M., C.F. Schwerdtfeger, and E.P. Meagher. 1972. Refinement of the structure on anatase at several temperatures. *Zeitschrift für Kristallographie* 136: 273–281.

Kittel, C. 2005. *Introduction to Solid State Physics*, 8th edn. New York: John Wiley & Sons.

Luttinger, T.M. 1956. Quantum theory of cyclotron resonance in semiconductors: General theory. *Phys. Rev.* 102: 1030–1041.

Madelung, O., ed. 1996. *Semiconductors—Basic Data*, 2nd edn. Berlin: Springer.

Nakamura, S. and G. Fasol. 1997. *The Blue Laser Diode*. Berlin: Springer.

Nelmes, R.J. and M.I. McMahon. 1998. Structural transitions in the group IV, III-V and II-VI semiconductors under pressure. In *High Pressure in Semiconductor Physics I*, eds. T. Suski and W. Paul, pp. 146–244. San Diego: Academic Press.

Phillips, J.C. 1970. Bonds and bands in semiconductors. *Science* 169: 1035–1042.

Riordan, M. and L. Hoddeson. 1997. *Crystal Fire: The Birth of the Information Age*. New York: W.W. Norton.

Round, H.J. 1907. A note on carborundum. *Electrical World* 49: 309.

Sah, C.T. 1988. Evolution of the MOS transistor—From conception to VSLI. *Proc. IEEE* 76: 1280–1326.

Schottky, W. 1938. Halbleitertheorie der Sperrschicht. *Naturwissenschaften* 26: 843.

Schubert, E.F. 2006. *Light-Emitting Diodes*, 2nd edn. Cambridge: Cambridge University Press.

Seeger, K. 1982. *Semiconductor Physics*, 2nd edn. Berlin: Springer-Verlag.

Seitz, F. 1995. Research on silicon and germanium in World War II. *Phys. Today* 48(1): 22–27.

Shockley, W., M. Sparks, and G.K. Teal. 1951. P-n junction transistors. *Phys. Rev.* 83: 151–162.

Sze, S.M. 1981. *Physics of Semiconductor Devices*, 2nd edn. New York: John Wiley & Sons.

Torrey, H.C. and C.A. Whitmer. 1948. *Crystal Rectifiers*. New York: McGraw-Hill.

Varshni, Y.P. 1967. Temperature dependence of the energy gap in semiconductors. *Physica* 34: 149–154.

Vurgaftman, I., J.R. Meyer, and L.R. Ram-Mohan. 2001. Band parameters for III-V compound semiconductors and their alloys. *J. Appl. Phys.* 89: 5815–5875.

Wilson, A.H. 1931. The theory of electronic semi-conductors. *Proc. Roy. Soc. London Ser. A* 133: 458–491.

Yu, P.Y. and M. Cardona. 1996. *Fundamentals of Semiconductors*. Berlin: Springer-Verlag.

Defect Classifications

Perfect crystals exist only in theory or in textbooks. In reality, all crystals have imperfections. Some of these imperfections make semiconductors highly interesting from electrical and optical device points of view. These are the electrically active dopant impurities that either replace a crystal host atom (substitutional) or occupy an interstitial site. Other imperfections are undesirable because they can produce energy levels in the band gap that influence device performance in a detrimental way.

Defects are broadly defined as any perturbations in an otherwise perfect crystal. In this chapter, we discuss ways to classify and understand different types of defects in semiconductors. First, they can be classified by their atomic structure. Second, defects are classified by the energy levels that they introduce into the band gap. Examples of defect systems are given in order of increasing dimensionality: point defects (impurities and native defects) are followed by line defects (dislocations); area defects (surfaces and interfaces) are discussed in Chapter 3.

2.1 BASIC DEFINITIONS

In Chapter 1, we discussed common semiconductor crystal structures. Defects are also characterized by their atomic structure. Intrinsic defects, or *native defects*, involve only the atoms that are present or absent in the perfect crystal. The simplest example of a native defect is a missing atom, or vacancy. Extrinsic defects, in contrast, are composed of *impurities*, atoms that are foreign to the crystal. *Dopants* usually refer to impurity atoms that are introduced intentionally. A *complex* involves a combination of two or more impurities or native defects (e.g., the boron–hydrogen complex in silicon).

Four different kinds of native defects are of importance in elemental semiconductors. They are

- The lattice *vacancy* (V_A), where A is a host atom.
- The *interstitial* host atom (A_i), also called a self-interstitial.
- The *Schottky defect*: A host atom that leaves its position and moves to the surface, forming a vacancy.
- The *Frenkel defect*: A host atom that moves to an interstitial position, forming a vacancy–interstitial pair.

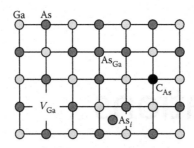

FIGURE 2.1 Defects in GaAs. Gallium vacancy (V_{Ga}), arsenic interstitial (As_i), arsenic antisite (As_{Ga}), and substitutional carbon (C_{As}).

For elemental semiconductors, V_A and A_i are sometimes denoted V and I. In compound semiconductors, we find two kinds of each of these defects (Figure 2.1). In GaAs, for example, a gallium vacancy is denoted V_{Ga} while an arsenic vacancy is V_{As}. In addition, there is a new kind of defect, the *antisite* (e.g., a gallium atom residing on an arsenic site, Ga_{As}, or an arsenic atom sitting on a gallium site, As_{Ga}). Deviations from exact stoichiometry may lead to large concentrations of interstitials and antisites of the excess constituent.

Impurities can reside on various sites. A substitutional site is a site that is normally occupied by an atom in the perfect lattice. For a compound semiconductor, the atomic site is indicated by a subscript. In GaAs, for example, silicon that substitutes for a gallium or an arsenic atom is denoted GaAs:Si_{Ga} or GaAs:Si_{As}, respectively. A site that is not substitutional is called interstitial and is identified by the subscript i. Interstitial copper, for example, is denoted Cu_i.

2.2 ENERGY LEVELS

The replacement of atoms in a semiconductor crystal with substitutional impurities, interstitial impurities, intrinsic defects, complexes between point defects, and dislocations can lead to electron states inside the band gap. The existence of such energy states, also called energy levels, is the major reason why semiconductors have gained enormous scientific and technical importance. Among the hundreds of known energy levels, there are only a few that are used to fabricate electronic devices.

Donor (acceptor) levels that are close to the conduction (valence) band edge are called *shallow levels*. In the group-IV semiconductors silicon and germanium, the elements of the third group (B, Al, Ga, In, Tl) require an extra electron to form a complete Lewis octet when they substitute a host atom. We call these impurities shallow acceptors. Elements of the fifth group (P, As, Sb, Bi) have one valence electron more than required to form a Lewis octet. This extra electron is given up easily to the conduction band, and we call these impurities shallow donors. Shallow donors (acceptors) are also called *hydrogenic* because they can be modeled as an electron (hole) orbiting around an ion, analogous to the hydrogen atom.

For an intrinsic (pure) semiconductor, the band gap is devoid of electronic states. As shown in Figure 2.2, impurities introduce energy levels into the band gap. For a donor impurity, the energy difference between the conduction band minimum and the donor level is the ionization energy, or *donor binding energy*. This is the amount of energy required to excite an electron from the donor atom into the conduction band. In Chapter 5, we discuss models that are used to calculate the binding energy. For a shallow donor, the binding energy is small compared to the band gap. If a donor is sufficiently shallow, then it will readily give up its electron to the conduction band at room temperature.

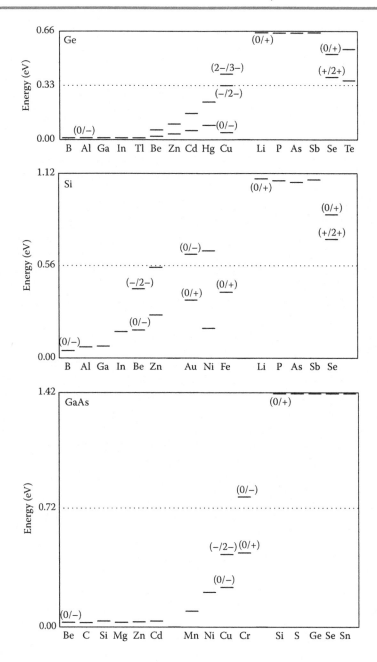

FIGURE 2.2 Energy levels of impurities in germanium, silicon, and GaAs. Band gaps are at room temperature. With the exception of Li, which is interstitial, the levels shown are for the substitutional sites. (From Sze, S.M. 1981. *Physics of Semiconductor Devices*, 2nd edn., p. 42. New York: John Wiley & Sons.)

A donor is neutral when it has an electron bound to it and is positive when the electron is liberated. The donor level is therefore called the (0/+) level, also referred to as E_d. As we discussed in Chapter 1, the Fermi level E_F can be tuned by doping with donor or acceptor impurities. For an intrinsic semiconductor, E_F is near the middle of the band gap. To see the relation between impurity levels and the Fermi level, consider a semiconductor at $T = 0$ K, so that the Fermi–Dirac distribution has a sharp cutoff at E_F. If the Fermi level is above the (0/+) level, then the level is occupied, so the donor is neutral. When the Fermi level is below the (0/+) level, the level is unoccupied, and the donor has a charge of +1.

A similar reasoning applies to acceptors. Here, the ionization energy is the energy difference between the acceptor level and the valence band maximum. The *acceptor binding energy* is the energy required to promote a hole from the acceptor atom into the valence band. Equivalently, it is the minimum amount of energy needed to excite an electron from the valence band to the acceptor level. As with donors, a shallow acceptor has a small binding energy as compared to the band gap. An acceptor level is called the (0/−) level, or E_a. If the Fermi level is above the (0/−) level, then the acceptor has a charge of −1. When the Fermi level is below the (0/−) level, the level is unoccupied so the acceptor is neutral.

In general, semiconductors contain a mix of donors and acceptors. If a semiconductor contains more donors than acceptors, for instance, then all of the acceptors will be negatively ionized. The acceptors are then said to *compensate* the donors. As shown in Figure 2.3, compensation can be viewed in a couple of different ways. In a real-space picture, donors give their electrons to acceptors. This results in positive donor ions and negative acceptor ions. In the energy-level picture, electrons fall from the donor level to the acceptor level, minimizing the energy. Consider a semiconductor with 3×10^{16} cm^{-3} donors and 10^{16} cm^{-3} acceptors. Such a sample would have 10^{16} cm^{-3} positive donors, 10^{16} cm^{-3} negative acceptors, and 2×10^{16} cm^{-3} neutral donors. Only the neutral donors can contribute electrons to the conduction band. The *net* donor concentration is therefore 2×10^{16} cm^{-3}.

The previous examples considered single donors and acceptors, which have only one level in the band gap. Other defects have multiple states. Whereas single donors are modeled as hydrogen atoms, double donors (or acceptors) are modeled as helium atoms (Chapter 5). In germanium, for example, selenium is a double donor. The shallow selenium (0/+) level, 0.14 eV below the conduction-band minimum, gives the first ionization energy (Figure 2.2). The deeper (+/2+) level at 0.28 eV gives the second ionization energy, the energy required to promote a second electron into the conduction band. Substitutional copper in germanium is a triple acceptor. The (0/−) and (−/2−) levels are 0.043 and 0.33 eV above the valence band maximum, respectively, and the (2−/3−) level is 0.26 eV below the conduction band minimum.

Some impurities and defects are *amphoteric*, meaning they have acceptor and donor levels. Consider the case of substitutional gold in silicon (Figure 2.2). It has an acceptor level (0/−) in the upper part of the band gap and a donor level (0/+) in the lower part. What does this mean? If E_F is above the acceptor level, then the gold atom accepts an electron and becomes negatively charged. If E_F is below the donor level, then gold donates an electron and becomes positively charged. When E_F lies between the two levels, the gold impurity is neutral.

In compound semiconductors and their alloys (e.g., GaN, ZnSe, GaAs, Al$_x$Ga$_{1-x}$As), there exist two sublattices. The dopant impurities must be incorporated on the appropriate sublattice in order to act as desired. In III-V semiconductors, one finds that group-II elements occupy the

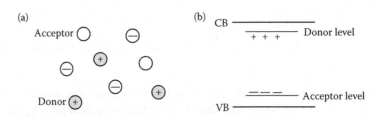

FIGURE 2.3 Compensation in a *p*-type semiconductor at low temperature. (a) Real-space picture. (b) Energy-level picture.

group-III sublattice and act as acceptors, while group-VI elements prefer the group-V sublattice and form donors. Group-IV elements are in some cases amphoteric (i.e., they may occupy group-III sublattice sites and act as donors or group-V sublattice sites and form acceptors).

Deep-level impurities such as transition metals create energy levels that may be far from the band-gap edges. Such impurities do not generate a hydrogenic ground state because the electron (hole) is bound strongly at a short distance from the impurity core. Due to this localization, the averaging of the ground state wave function over a large volume does not occur. For these reasons, deep-level impurity ground state energies strongly depend on the particular details of the potential near the impurity core. There exists no simple model that accurately describes deep levels.

Impurities that generate deep levels in the band gap typically do not fit well into the crystal, generating significant local strains. The localized levels can trap carriers, impeding conduction. Iron and gold are much-studied deep level impurities in silicon. These two impurities diffuse rapidly into silicon crystals, and they dominate electrical properties even at very low concentrations. In the early days of silicon technology, the minority carrier lifetime was controlled by the addition of gold. The first fast-switching diodes were fabricated from such crystals.

Historically, deep levels and shallow levels were always defined according to their proximity to the band edges. More recently, however, localized states have been discovered that happen to lie close to the conduction or valence band. It is therefore useful to define a defect as localized or delocalized. In the literature, "deep" generally refers to a defect wave function that is localized in real space. We will explore this distinction further in Chapter 5.

2.3 EXAMPLES OF NATIVE DEFECTS

Native defects are always present in semiconductors and may introduce deep levels into the band gap. In this section, several illustrative examples are discussed.

An important deep-level defect in GaAs is the arsenic antisite As_{Ga} (Kaminska and Weber, 1993). This deep double donor controls the position of the Fermi level of semi-insulating GaAs used for integrated circuits. The As_{Ga} center has a number of unusual properties that are related, in part, to the change of the relative positions of As_{Ga} and its four As neighbors with a change in charge state. This lattice relaxation is now known to occur for a large number of deep-level centers in various semiconductors.

In extreme cases, the lattice relaxation can be strong enough to allow a center to bind two electrons and assume a lower total energy than binding one electron, despite the fact that two equally charged carriers repel each other. Such defects are called negative-U, where U refers to the Coulomb repulsion term in the Anderson model (Anderson, 1961). One example of a negative-U center is the oxygen vacancy (V_O) in ZnO, which is a deep double donor (McCluskey and Jokela, 2009). An illustration of this vacancy is shown in Figure 2.4. When an oxygen atom is

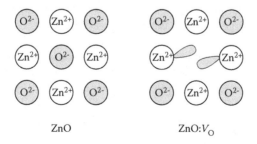

FIGURE 2.4 Ionic picture of an oxygen vacancy in ZnO. The removal of an oxygen atom leaves two dangling bonds.

removed from the lattice, two electrons are left in the vacancy. These electrons are not part of a Zn-O bond, so they are dangling bonds. This is the neutral charge state, V_O^0. If the Fermi level is sufficiently low, then the V_O defect will give up two electrons, becoming V_O^{2+}. The V_O^{1+} is not thermodynamically stable; its energy is always higher than V_O^0 or V_O^{2+}.

In compound semiconductors, anion vacancies are generally donors while cation vacancies are acceptors. For example, ZnO:V_O and GaN:V_N are deep donors while ZnO:V_{Zn} and GaN:V_{Ga} are deep acceptors. These native defects are important as compensating centers. In n-type GaN, the formation of gallium vacancies introduces acceptor levels into the band gap, enabling electrons to reduce their energy by several eV. This reduction in energy encourages the formation of gallium vacancies. As we will see in Chapter 8, the formation energy of electrically active intrinsic defects depends on the position of the Fermi level.

The silicon vacancy is a negative-U center: the V^+ state is higher than that of V^0 or V^{2+}. The (0/2+) energy level is 0.13 eV above the valence band maximum, making it a very deep donor. In p-type silicon, where the Fermi level is below the (0/2+) level, the vacancy can compensate acceptors by providing two electrons. The vacancy can also act as an acceptor (V^- or V^{2-}) but the positions of these levels are not well known (Sueoka et al., 2016).

In silicon, isolated vacancies are unstable at room temperature, but they readily form pairs with other defects or impurities. An E center is a vacancy plus a neighboring donor impurity (Watkins and Corbett, 1964). The formation of E centers is energetically favorable for two reasons. First, the donor electron can lower its energy by occupying a vacancy acceptor level (compensation). Second, the Coulomb attraction between a positively charged donor and a negatively charged vacancy is maximized when they are neighbors. In silicon and germanium, E centers are formed with all the group-V donors. In diamond, vacancies readily form pairs with nitrogen impurities (Davies et al., 1992). In the compound semiconductor CdTe, a cadmium vacancy can pair with a group-VII donor such as Cl_{Te} (Biernacki et al., 1993). Confusingly, in CdTe, this defect pair is called an A-center.

Vacancies also form complexes with hydrogen impurities. The hydrogen atoms attach onto the dangling bonds in a process called *passivation*, discussed further in Section 2.5. By supplying the needed electron, hydrogen passivation removes the energy level associated with the dangling bond from the band gap. Extensive studies of hydrogen in silicon and germanium have shown that vacancies can be decorated with one, two, three, or four hydrogen atoms (Chapter 6). Hydrogen also forms complexes with cation vacancies in compound semiconductors. Hydrogen complexed with gallium vacancies in GaP and GaN, for example, forms strong P-H or N-H bonds. Hydrogen can also occupy anion vacancies and act as shallow donors, as is believed to be the case in ZnO:V_O (Janotti and Van de Walle, 2007) and many other oxides (Kilic and Zunger, 2002).

A pair of neighboring vacancies is called a *divacancy* (Watkins and Corbett, 1965). Unlike the isolated vacancy in silicon, the divacancy is stable at room temperature. A divacancy has six dangling bonds and may be partially passivated by hydrogen atoms. In germanium, the divacancy–hydrogen complex (V_2H) has an acceptor level 72 meV above the valence band maximum (Haller, 1991). The structures of larger vacancy defects—trivacancies, tetravacancies, etc.—are less well-known. Characterizing "missing atoms" is a difficult task. One method, electron paramagnetic resonance (EPR), measures the spin state of defects and is useful in determining the energy levels of vacancies. A second technique is positron annihilation, which measures the annihilation of positrons with electrons in the sample. A fraction of the positrons are trapped by vacancies and enjoy a longer lifetime due to the low electron density. The long-lifetime component of the annihilation gives information about the vacancy. EPR and positron annihilation are discussed in Chapters 10 and 11, respectively.

Interstitials are usually generated through electron irradiation at cryogenic temperatures. At room temperature, as in the case of silicon vacancies, silicon interstitials tend to diffuse rapidly

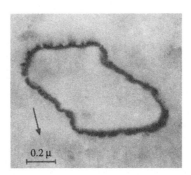

FIGURE 2.5 Transmission electron micrograph of an A-swirl in silicon. The sample surface is (111), and the arrow indicates a <220> direction. (With kind permission from Springer Science+Business Media: *Appl. Phys. A*, Formation and nature of swirl defects in silicon, 8, 1975, 319–331, Föll, H. and B.O. Kolbesen. Copyright 1975, Springer.)

until they reach another defect or the crystal surface. In addition to the high-symmetry interstitial sites discussed in Section 2.6, there is the *split interstitial*, where two atoms share a single lattice site. This configuration is also called an *interstitialcy*.

Intrinsic point defects can agglomerate into larger, extended defects. When interstitials condense, forming an extra section of a lattice plane, we obtain an extrinsic *stacking fault*. As discussed in Section 2.7, these stacking faults are surrounded by partial dislocations. When vacancies condense, they form intrinsic stacking faults (i.e., holes in lattice planes). These defects gave rise to spiral patterns on the silicon wafer, which led to the names A-swirls, B-swirls, and so forth (Föll and Kolbesen, 1975). Figure 2.5 shows a transmission electron micrograph of an A-swirl, which was determined to be an extrinsic stacking fault. The partial dislocations at the boundary of the A-swirl give rise to a dark loop in the picture. In the 1960s and 1970s, the existence of interstitials was still hotly debated (Seeger and Chick, 1968). The discovery of these two-dimensional (2D) defects firmly established the existence of interstitials and vacancies near the melting point (Föll et al., 1981).

Compound semiconductors exhibit extended defects when there is an excess (or deficiency) of one of the components. Such materials are referred to as *nonstoichiometric*. In GaAs grown at low temperatures, for example, excess arsenic atoms precipitate after annealing. These arsenic *precipitates* form depletion regions in the surrounding GaAs material, resulting in high electrical resistivity.

2.4 EXAMPLES OF NONHYDROGENIC IMPURITIES

This section discusses several impurities that are not shallow donors or acceptors. An example of a deep-level defect is the *DX* center, which involves the displacement of a donor-type impurity. The name "*DX* center" was chosen because it was originally believed that the deep levels arose from donors (*D*) complexed with unknown (*X*) impurities. Although it is now known that *DX* centers do not involve additional unknown impurities, for historical reasons the name has remained.

DX centers have been intensively studied in III–V semiconductors (Mooney, 1990). In $Al_xGa_{1-x}As$ alloys with $x > 0.22$, the *DX* center is the lowest-energy state of silicon donors (Nelson, 1977). Silicon donors also become *DX* centers in GaAs under hydrostatic pressures greater than approximately 2 GPa. Chadi and Chang (1988) proposed a model for the negatively charged *DX* center in which the silicon atom is displaced into an interstitial position (Figure 2.6). The Chadi and Chang model was supported by the results of infrared spectroscopy experiments on GaAs:Si under hydrostatic pressure (Wolk et al., 1991). The local vibrational mode frequency

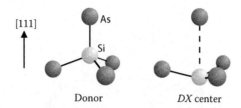

FIGURE 2.6 Chadi and Chang model of the *DX* center in GaAs. The dashed line represents a broken Si-As bond.

of the silicon *DX* center was found to be ~2% lower than that of the substitutional silicon donor due to the softening of the bonds.

Transition metals introduce a range of energy levels into the band gap and sometimes harm device performance. They can occupy substitutional or interstitial sites. When occupying a substitutional site, they are deep-level defects. Interstitial transition metals, in contrast, are usually shallow donors. Iron in silicon, for example, diffuses interstitially and forms complexes with shallow acceptors (Istratov et al., 1999). In the literature, it is customary to label the transition-metal impurity by its valence number, or oxidation state. For example, consider a nickel atom that substitutes for a silicon atom. If the nickel forms bonds with the four silicon neighbors, then its charge state is labeled Ni^{4+}. In the semiconductor terminology used in this book, the Ni^{4+} is considered to be "neutral," because its charge state is the same as the silicon atom that it replaced. Following this convention, Ni^{3+} and Ni^{5+} would be the "−" and "+" charge states. In GaAs, on the other hand, nickel replaces a group-III gallium atom, so that Ni^{3+} is the neutral charge state.

Because the *d* orbitals are localized, the electronic energy levels are relatively unaffected by the positions of the valence-band maxima or conduction-band minima. The (0/−) acceptor levels for transition metals were plotted for GaAs, InP, and GaP, relative to the vacuum level (Ledebo and Ridley, 1982). There are uncertainties in the actual positions of the valence-band maxima relative to vacuum, but the trend is compelling. For each transition metal impurity, the deep acceptor level appears to be constant on an absolute-energy scale (Figure 2.7). Theoretical work (Caldas et al., 1984) indicated that this rule applies for transition metals in a range of semiconductors.

Isoelectronic impurities have the same number of valence electrons as the atoms they replace (e.g., Al$_{Ga}$ in GaAs). If there is no significant lattice strain, then the isoelectronic impurity is electrically neutral. However, some isoelectronic impurities, such as GaP:N, are highly mismatched. Due to the short Ga-N bond as compared to the Ga-P bond, the gallium atoms around the

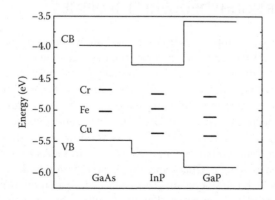

FIGURE 2.7 Acceptor (0/−) levels for Cr, Fe, and Cu in three III-V semiconductors. (After Ledebo, L.-Å. and B.K. Ridley. 1982. *J. Phys. C* 15: L961–L964.)

nitrogen atom experience significant lattice relaxation. The effect of this local strain, along with the high electronegativity of nitrogen, is to attract an electron. The nitrogen acts as an electron trap, with an acceptor level close to the conduction band minimum (Yu and Cardona, 1996). In III-V semiconductors heavily doped with nitrogen, the nitrogen level interacts strongly with the conduction band, affecting the band gap and electron effective mass (Shan et al., 1999). This is a general feature of highly mismatched alloys, in which the components (e.g., GaAs and GaN) have much different bond lengths (Walukiewicz et al., 2000).

In silicon, *carbon* is an isoelectronic impurity that assumes a substitutional position. Because of its small size, it produces considerable local strain. One way to relax this strain is to have a pair of carbon atoms occupy one silicon site, similar to the split interstitial. Although carbon by itself does not produce any energy levels in the gap, it participates in many defect reactions that occur at high temperatures. The *X* center, for example, is a defect pair in silicon with a group-III acceptor on one substitutional site and a carbon on a neighboring substitutional site. The acceptor level is slightly shallower than that of the isolated group-III acceptor. Nitrogen is also found in bulk silicon crystals, with a low solubility of ~10^{15} cm^{-3}, primarily in the form of nitrogen–nitrogen pairs (Stein, 1986).

An impurity that does not introduce energy levels into the gap is *interstitial oxygen*. Oxygen in silicon, a common defect in Czochralski-grown crystals, may be the most intensively studied semiconductor impurity in history (Hrostowski and Kaiser, 1957). Oxygen densities of ~10^{18} cm^{-3} are incorporated in Czochralski-grown silicon crystals, a result of the reduction of the SiO$_2$ crucible by liquid silicon at the growth temperature (1688 K). Control of the oxygen concentration in modern silicon crystals is of paramount importance for device processing.

Interstitial oxygen rests between two neighboring silicon atoms, forming bonds with both neighbors (Pajot, 1994). The Si-Si bond length is too short to accommodate the oxygen impurity, which in turn moves perpendicularly to the bond direction, relieving the local strain (Figure 2.8). A similar phenomenon is observed for interstitial oxygen in germanium. In this off-centered position, the oxygen atom can vibrate at distinct frequencies. In the infrared part of the spectrum, these vibrations lead to absorption peaks that act as quantitative fingerprints for the presence of this impurity. Measurements of these local vibrational modes are discussed in more detail in Chapter 6.

When a Czochralski-grown Si crystal is heated, the O$_i$ impurities diffuse and form clusters called *thermal donors* (Newman, 2000). Thermal donors are denoted TD(1), TD(2), ... TD(N), where N increases with the number of oxygen atoms in the defect. Although the exact structure is still not known, it is believed that the defect consists of a row of O$_i$ atoms along the [110] direction. TD(N) defects are double donors, with the ionization of the first and second electron in the range of 0.04 to 0.07 eV and 0.07 to 0.15 eV, respectively. As N increases, the ionization energies decrease. After sufficient heating, clusters of oxygen impurities grow into SiO$_2$ precipitates. SiO$_2$ precipitates have a beneficial effect in that they attract, or *getter*, transition-metal impurities, which would otherwise become deep defects.

Oxygen has also been shown to combine with intrinsic defects. In silicon, vacancy–oxygen complexes are referred to as *A* centers. They are formed by electron or ion-induced damage of Si:O (Corbett et al., 1961). In *A* centers, the oxygen is almost substitutional, but it relaxes in an off-center direction and binds preferentially to two neighboring silicon atoms (Figure 2.8).

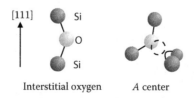

FIGURE 2.8 Interstitial oxygen and *A* center (vacancy–oxygen) defects in silicon.

The other two silicon neighbors form a Si-Si bond. The *A* center has a deep acceptor level 0.18 eV below the conduction band minimum; hence, it acts as an electron trap. When the acceptor level is occupied, the *A* center is negatively charged, leading to a stiffening of the bonds and an increase in the vibrational frequency.

2.5 HYDROGEN

Due to its omnipresence in growth and processing, hydrogen is an especially important impurity. Hydrogen readily forms bonds with host or impurity atoms, introducing or removing electronic states from the band gap. Hydrogen-containing donors and acceptors were first discovered in germanium (Haller, 1978). Subsequent to that observation, hydrogen has been observed to play numerous roles, beneficial as well as detrimental, in elemental and compound semiconductors.

In Section 2.3, we discussed how hydrogen passivates dangling bonds in vacancies. Hydrogen is also important because it passivates electrically active impurities. Sah et al. (1983) observed that hydrogen from water-related molecular species in the oxide layer in metal-oxide-semiconductor (MOS) capacitors diffused into the silicon and passivated the acceptors. Hydrogen neutralizes acceptors by supplying the additional electron needed to complete the Lewis octet (Pearton et al., 1992). The energy level is removed from the band gap, resulting in a decrease in the free hole concentration. Because neutral complexes are formed, the decrease in ionized impurity scattering increases the mobility. Hydrogen passivation has both beneficial and detrimental effects. Hydrogen passivation of deep levels increases the minority carrier lifetime. However, the omnipresence of hydrogen in growth processes can hinder reliable *p*- or *n*-type doping of semiconductors. GaN:Mg layers in laser diodes, for example, must be annealed to break up the Mg-H complexes and activate the Mg acceptors.

The passivation process is shown schematically in Figure 2.9 for the case of Si:B. The hydrogen acts as a donor and compensates the boron acceptor. The proton then feels the Coulomb attraction of the negatively ionized boron. The hydrogen assumes a "bond-centered" orientation, between a silicon and boron, forming a neutral complex (Pankove et al., 1985). In this case, the hydrogen binds more strongly with silicon than boron, so it is not truly bond-centered (Figure 2.10).

In the case of an *n*-type semiconductor such as Si:P, the hydrogen acts as an acceptor. An electron from the donor ionizes the hydrogen, forming H$^-$. The H$^-$ experiences the Coulomb attraction of the positively ionized phosphorus donor. The hydrogen assumes an *antibonding* or "back-bonded" orientation (Figure 2.11), attached to a silicon in a direction opposite to the donor (Johnson et al., 1986). The bond-centered orientation is energetically unfavorable because the electrostatic repulsion of the electrons is too high.

Impurities can also be *partially* passivated. In germanium, for example, beryllium and zinc are double acceptors. When they pair with a single hydrogen atom, they become single acceptors. Copper in germanium is a triple acceptor; when it pairs with two hydrogen atoms, it also is

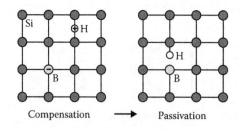

FIGURE 2.9 Hydrogen compensation, followed by passivation, of a boron acceptor in silicon.

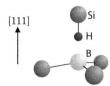

FIGURE 2.10 Boron–hydrogen complex in silicon.

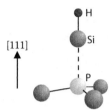

FIGURE 2.11 Phosphorus–hydrogen complex in silicon. The dashed line represents a broken bond.

transformed into a single acceptor. In addition to passivating impurities, hydrogen can also *activate* neutral impurities. In germanium, the electrically inactive impurities silicon, carbon, and oxygen form complexes with hydrogen (Haller, 1991). The Si-H and C-H complexes are shallow acceptors while the O-H complex is a shallow donor. In ZnO, hydrogen can activate isoelectronic impurities such as calcium, turning them into shallow donors (McCluskey and Jokela, 2009).

It is generally accepted that hydrogen is an amphoteric, negative-U impurity. Depending on the position of the Fermi level, isolated hydrogen exists either as H^+ or H^-; the H^0 charge state is thermodynamically unstable. The hydrogen (+/−) level in silicon has been estimated to be 0.4 eV below the conduction-band minimum (Johnson and Van de Walle, 1999). When the Fermi level is below the (+/−) level, H^+ is the stable species. When the Fermi level is above this level, as in *n*-type silicon, then H^- is dominant. The interstitial species is not observed at room temperature, because at that temperature, hydrogen diffuses rapidly until it attaches onto a defect or reaches the crystal surface.

Pairs of hydrogen atoms exist in a couple of different configurations. In silicon, isolated H_2 molecules reside at an interstitial site and are free to rotate (Shi et al., 2005). H_2 molecules are also observed in compound semiconductors (McCluskey and Haller, 1999). Alternatively, hydrogen can form H_2^* defects, in which one hydrogen binds to a bond-centered site and one hydrogen binds to an antibonding site. H_2^* defects form the building blocks for *hydrogen platelets*, extended defects in the (111) plane that can be seen with a transmission electron microscope (Johnson, 1991). The presence of hydrogen atoms causes the separation between (111) atomic planes to increase by 20% to 30%.

Other group-I elements, such as copper and lithium, passivate acceptors in much the same way as hydrogen. They diffuse through a silicon crystal as positive ions and form neutral complexes with the negatively charged acceptors. In the electric field of a reverse-biased *p-n+* junction, lithium drifted into *p*-type silicon or germanium forms almost perfectly passivated material. This process allows the formation of several centimeter-wide depletion layers, which are used as gamma-ray or particle detectors.

2.6 DEFECT SYMMETRY

Point defects can be classified according to their symmetry. A perfect crystal has translational symmetry. Due to the periodicity of the lattice, one can move the crystal by a translation vector

2.6 Defect Symmetry

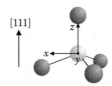

FIGURE 2.12 Substitutional impurity in a diamond or zincblende crystal (T_d symmetry).

and it will look the same. As soon as we introduce a defect, however, the translational symmetry is broken. The defect symmetry is defined by a group of operations, performed with respect to a specific point, that do not change the defect (Falicov, 1989; Cotton, 1990). The most trivial such operation is the *identity* element. The identity element effectively multiplies by one (i.e., it does nothing). Other elements of a point group are reflections, rotations, and inversions.

Several point groups are relevant to defects in semiconductors. An important example is T_d, or tetrahedral, symmetry. Substitutional impurities in diamond or zincblende crystals have T_d symmetry (Figure 2.12). In the figure, x and y are parallel to the $[1\bar{1}0]$ and $[11\bar{2}]$ crystallographic axes, respectively, and z points in the [111] direction. In addition to the identity element, the T_d point group contains the following:

- Threefold *rotations* (120° and 240°) around the four bond axes.
- Twofold rotations (180°) around the three axes that point from the center atom to the midpoint of two neighboring atoms.
- *Reflections* across six planes, defined by the center atom plus two neighboring atoms.
- In addition to rotations and reflections, there are six rotation-plus-reflection operations called *improper rotations*. The rotational axes are orthogonal to the reflection planes. In the T_d group, these involve a ±90° rotation followed by a reflection. The axes of rotation are the same as for the twofold rotations.

Imagine that you have a toy model of this defect. You close your eyes while your friend does something to the model, and then you open your eyes. If your friend performed one of the point-group operations, then the defect would look the same as it did before. For example, if the sample was rotated about the z axis by 120°, its appearance would not change. The same would be true if your friend reflected the atoms across one of the reflection planes.

Another important example is the C_{3v} (trigonal) point group. Figures 2.10 and 2.11 show examples of defects with C_{3v} symmetry: the B-H and P-H complexes in silicon. There are three classes of operations that leave the defect invariant. In addition to the identity element, the C_{3v} point group contains

- Two threefold rotations (120° and 240°) about the z axis.
- Three reflections across vertical planes (a plane contains the top atom and one of the three bottom atoms).

This is a lower symmetry than T_d (i.e., it contains fewer symmetry elements). In diamond and zincblende semiconductors, most impurity–hydrogen pairs have C_{3v} symmetry.

Two groups that have lower symmetry than C_{3v} are C_s and C_1. The C_s group is like C_{3v}, except that one of the atoms in the tetrahedron base is distinct from the other two. An example is the Ca–H complex in wurtzite ZnO (Figure 2.13). In this complex, the O-Ca-O-H series of atoms define a vertical plane (Li et al., 2008). We define x to be perpendicular to this plane. Besides the identity element, C_s has just one operation, reflection across the plane. C_1, the lowest symmetry

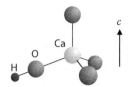

FIGURE 2.13 Calcium–hydrogen complex in ZnO (C_s symmetry).

group, has no symmetry operations besides identity. Any point in an amorphous material has C_1 symmetry.

The previous examples involved impurities. Symmetry is also used to classify native defects such as vacancies. To first approximation, a vacancy in a diamond or zincblende crystal structure has T_d symmetry, just like a substitutional impurity. A silicon vacancy, for example, has four *dangling bonds* due to the removal of a silicon atom. However, this is an overly simplistic approximation. In a real vacancy, the atoms around the vacancy move away from their ideal-crystal positions, or *relax*.

In silicon, V^{2+} has T_d symmetry. For V^+ or V^0, pairs of silicon atoms move toward each other, partially *reconstructing* the dangling bonds (Watkins, 1986). The lowering of the symmetry in this process is called a *Jahn–Teller distortion*. In this case, the distortion lowers the symmetry from T_d to D_{2d} (Figure 2.14). The D_{2d} point group contains the following elements besides the identity:

- A twofold rotation (180°) around the z axis.
- Twofold rotations (180°) around two axes (besides z) that point from the center to the midpoint of two atoms.
- Improper rotations (±90°) around the z axis.
- Reflections about two planes.

Along with substitutional sites, there are several high-symmetry interstitial sites where impurity atoms may reside. Figure 2.15 shows these interstitial sites for silicon. The bond-centered B site is midway between two neighboring silicon atoms, which have a bond length of R. The tetrahedral T site is along the [111] direction, displaced a distance R from a silicon atom. A second, equivalent T site is a distance R from the first T site. The hexagonal H site is midway between the two T sites.

In silicon, the lowest-energy position for the H^+ species is the bond-centered B site. H^-, on the other hand, prefers the T site. In a zincblende crystal such as GaAs, the two T sites are not

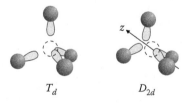

FIGURE 2.14 Model of the silicon vacancy, without relaxation (T_d) and with relaxation (D_{2d}).

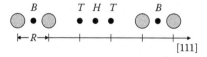

FIGURE 2.15 Interstitial sites in a diamond or zincblende lattice.

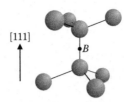

FIGURE 2.16 Bond-centered (*B*) interstitial site in a diamond crystal.

equivalent. They are labeled *T*(Ga) or *T*(As) depending on whether the site is closest to a gallium or arsenic atom.

In silicon and GaAs, the *T* sites have T_d symmetry, the same as that for substitutional impurities. In GaAs, the *H* and *B* sites have C_{3v} symmetry. In silicon, on the other hand, the *H* and *B* sites have a higher symmetry called D_{3d} (Figure 2.16). In addition to identity, threefold rotations, and reflections, the D_{3d} point group has the following operations:

- *Inversion*: if there is an atom at (x,y,z), then there is an identical atom at $(-x,-y,-z)$.
- Improper rotations about the [111] axis by $\pm 60°$.
- Three 180° rotations about axes perpendicular to the [111] axis.

The point groups discussed in this section cover the majority of point defects in technologically relevant semiconductors. The application of group theory to electronic and vibrational defect states is discussed in Chapter 5 and Chapter 6, respectively.

2.7 DISLOCATIONS

The examples presented so far have focused on point defects, with spatial dimensions on the order of a lattice constant. Point defects can, however, agglomerate into larger clusters, forming *extended defects* (Holt and Yacobi, 2007). Examples already discussed are A-swirls, arsenic precipitates, SiO$_2$ precipitates, and hydrogen platelets. Other technologically important defects include (to name a few) micropipes in SiC and nanopipes in GaN, which are columns of empty space along the growth direction, and micron-sized rounded or pyramidal surface features called hillocks. This section focuses on dislocations.

Dislocations are *line defects* that involve a series of atoms that are not in their perfect lattice sites. To illustrate how a dislocation is formed, we consider the textbook example of a simple cubic crystal that experiences a shear stress along a <100> direction. This direction is called a *slip* direction, and {100} is the slip plane. As shown in Figure 2.17, atoms above the slip plane are

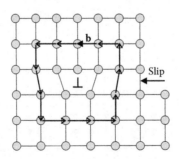

FIGURE 2.17 An edge dislocation. The arrows show a circuit and the Burgers vector **b**. The ⊥ symbol indicates that the dislocation line is perpendicular to the page.

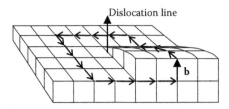

FIGURE 2.18 A screw dislocation. The arrows show a circuit and the Burgers vector **b**. (From Holt, D.B. and B.G. Yacobi. 2007. *Extended Defects in Semiconductors*. New York: Cambridge University Press.)

pushed to the left, while the atoms on the left edge are kept fixed. The resultant strain increases the density of atoms in the upper portion of the crystal. These extra atoms are accommodated by an extra "half-plane." The symbol "⊥" denotes the dislocation, which runs perpendicular to the page. This is an *edge dislocation*, defined by the edge of the terminated atomic plane.

A *Burgers vector* is used to describe the extended defect quantitatively. To obtain the Burgers vector, we define a series of atom-to-atom steps that, in a perfect crystal, form a counterclockwise closed path. In our simple picture of a crystal in Figure 2.17, one could describe such a path as "two atoms left, three atoms down, three atoms right, three atoms up, and one atom left." We draw a circuit around the edge dislocation, following these directions. Because the upper portion of the crystal contains an extra half-plane of atoms, the circuit does not form a closed path. To make a closed path, we must add a vector **b**, which in this case is $[\bar{1}00]$. This is the Burgers vector. (Note that although the lattice is strained around the dislocation, we express the Burgers vector in terms of the unstrained lattice.) For a pure edge dislocation, the Burgers vector is *perpendicular* to the line of the dislocation (Read, 1953).

A second pure dislocation is the *screw dislocation*. As illustrated in Figure 2.18, this dislocation is also formed by shear deformation, but in this case the deformation does not introduce an extra half-plane of atoms. Instead, the crystal is turned into a kind of spiral staircase around the dislocation line. When one draws a circuit around the dislocation line, the end of the path is one atomic plane below the starting point. In contrast to the edge dislocation, the Burgers vector for a pure screw dislocation is *parallel* to the line of the dislocation.

These models of pure screw and edge dislocation were derived for a simple cubic lattice. In the diamond lattice, there are several possible dislocation structures, some of which have been characterized in detail using electron microscopy. The most common line defect in silicon, germanium, and III-V compounds is the *60° dislocation*. Figure 2.19 shows a projection of the diamond $(1\bar{1}0)$ plane. The slip plane is (111) and is perpendicular to the page. To form a dislocation,

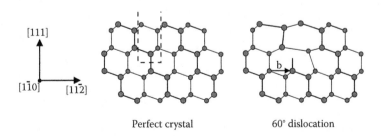

FIGURE 2.19 Projection of a silicon crystal onto the $(1\bar{1}0)$ plane. Larger and smaller circles represent atoms in the $(1\bar{1}0)$ plane or just below it, respectively. The 60° dislocation is formed by removing the slab indicated by the dashed lines. The dislocation line runs along the $[1\bar{1}0]$ direction, out of the page.

a vertical slab of material is removed, shown by the dashed lines. The atoms then form bonds and fill up the removed portion. One can see that this procedure introduces a line of dangling bonds that runs out of the page. Hence, the dislocation line is along the [1$\bar{1}$0] direction. The Burgers vector **b** points along the [10$\bar{1}$] direction, which is 60° to the line of the dislocation. The reader is encouraged to draw a circuit around the dislocation and verify the direction of **b**.

As we discussed in Section 1.3, the diamond and zincblende crystal structures have an *ABCABC*... stacking sequence along the [111] direction. For the 60° dislocation, the Burgers vector lies in a {111} plane. The Burgers vector therefore points from an atom in the *A* layer to another atom in the *A* layer. A *partial dislocation*, in contrast, is described by a Burgers vector that points from an *A* atom to an atom in a different layer (e.g., *A* to *B*).

The 60° dislocation often splits into two partial dislocations, a 30° and 90° partial dislocation. The dissociation of a dislocation into two Shockley partials is described by the following equation for the Burgers vectors:

$$\frac{a}{2}[10\bar{1}] \rightarrow \frac{a}{6}[2\bar{1}\bar{1}] + \frac{a}{6}[11\bar{2}] \tag{2.1}$$

Here, the Burgers vectors for the partial dislocations make angles of 30° and 90°, respectively, with the [1$\bar{1}$0] dislocation lines. These partial dislocations are illustrated schematically in Figure 2.20. To the right lies the 90° partial dislocation, and on the left we see the 30° partial dislocation. This figure shows how the stacking sequence is interrupted by the partial dislocations. The first partial dislocation shifts the *ABCABC*... sequence to *ACABCABC*... The second partial dislocation restores the stacking sequence. Between the two partials, the interruption of the stacking sequence (the *C* plane between the dashed lines) is a stacking fault.

A ball and stick model gives only a rather crude picture of a dislocation. The details of the dislocation core are not revealed. It is quite obvious that in the simple model a large number of broken or dangling bonds exist. The core structure is of utmost interest because the core is the location of the possible electrical activity of a dislocation (Queisser, 1983). Recent studies indicate that many of the neighboring dangling bonds in the core form covalent bonds. This is called dislocation core reconstruction or minimization of the dislocation energy. A reconstructed dislocation shows very little electrical activity, confirming many experimental observations on dislocations in a large variety of semiconductor crystals. Most of the energy levels attributed to dislocations are likely due to point defects near the dislocation core.

Dislocations never end inside a crystal without interacting with other defects. They can end at the crystal surface, form a closed loop, or branch into more dislocations. A low concentration of dislocations can be advantageous because they can absorb excess vacancies or interstitials present in the crystal after solidification from the melt. In this case, it is said that an edge dislocation "climbs." Atoms of the extra half-plane at the dislocation line are removed when they fill arriving vacancies. Conversely, the half-plane can grow by absorbing interstitials. Dislocations also attract

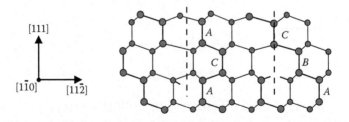

FIGURE 2.20 Projection of a silicon crystal onto the (1$\bar{1}$0) plane. A 30° dislocation is on the left, and a 90° dislocation is on the right. Between the dashed lines, the stacking sequence is altered.

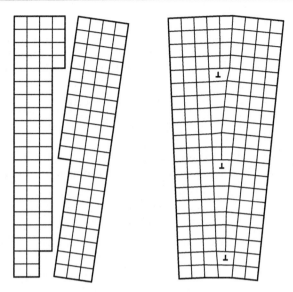

FIGURE 2.21 Low-angle grain boundary in a simple cubic crystal. (From Shockley, W. and W.T. Read. 1949. *Phys. Rev.* 75: 692.)

impurities, forming a *Cottrell atmosphere*. Small impurity atoms (those that form short bonds with the host atoms) tend to incorporate in regions of compressive strain, whereas large impurity atoms are attracted to regions of tensile strain.

The dislocation density in high-quality bulk single crystals is low. Dislocation-free silicon, germanium, and GaAs are commercially available. Other materials, especially those for which the growth technology is in its infancy, suffer from large concentrations of dislocations and other defects. In samples that are not perfect single crystals, *grain boundaries* exist between crystallites of different orientations. Low-angle (<10°) grain boundaries contain dislocations that allow the lattices of adjacent grains to match up (Shockley and Read, 1949). Such dislocations occurring at regular intervals along a grain boundary were first observed in germanium (Vogel et al., 1953). The symbol Σ denotes the period of the dislocation pattern (i.e., for a $\Sigma = 9$ grain boundary, every ninth lattice plane matches up). Figure 2.21 shows a $\Sigma = 7$ grain boundary for a simple cubic lattice.

Dislocations also crop up at interfaces between dissimilar materials. Consider a thin film of InGaN grown on GaN. The InGaN layer is said to be *pseudomorphic* if each atom binds to an atom of the GaN substrate. Because the equilibrium lattice constant of InGaN is larger than that of GaN, the InGaN experiences considerable strain. An illustration of pseudomorphic strain for a simple cubic system is shown in Figure 2.22. If the thickness of the film exceeds a critical value,

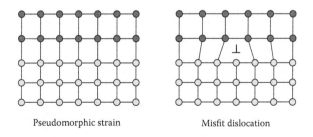

FIGURE 2.22 Pseudomorphic strain and misfit dislocation for a thin film grown on a substrate with a smaller lattice constant.

then the film will relax via the creation of *misfit dislocations*. These dislocations allow the film to expand laterally and hence reduce the strain. Misfit dislocations run parallel to the interface. If the dislocation turns away from the interface and runs toward the surface, then it is called a *threading dislocation*. GaN or InN grown on sapphire substrates has a high density (>10^9 cm^{-2}) of threading dislocations. However, these dislocations are mostly located at boundaries between columnar grains. The columns contain relatively few dislocations.

PROBLEMS

2.1 Indicate whether each of the impurities is a donor, acceptor, or isoelectronic. If it is a donor or acceptor, indicate whether it is a single, double, or triple donor/acceptor.

GaAs:C$_{As}$	GaAs:N$_{As}$	ZnO:Al$_{Zn}$
Si:Se	GaAs:Cu$_{Ga}$	ZnS:Cu$_{Zn}$
Ge:Cu	Si:Zn	GaAs:Si$_{Ga}$

2.2 Silicon is doped with 5×10^{16} cm^{-3} substitutional Zn impurities and 7×10^{16} cm^{-3} substitutional As impurities. Is it *n*-type or *p*-type?

2.3 At room temperature, the Fermi level of a certain semiconductor is 0.025 eV above a single-donor level. Calculate the fraction of donors that are in each charge state.

2.4 A semiconductor has 3×10^{16} cm^{-3} single donors and 8×10^{16} cm^{-3} single acceptors.
   a. At 0 K, how many donors and how many acceptors will be ionized?
   b. At room temperature, is the semiconductor *n*-type or *p*-type?

2.5 A crystal is doped with 5×10^{16} cm^{-3} single donors and 1×10^{16} cm^{-3} single acceptors. The donor binding energy is 40 meV and the acceptor binding energy is 100 meV. At 0 K,
   a. What is the concentration of ionized donors?
   b. What is the concentration of ionized acceptors?
   c. Where is the Fermi level? (Hint: Consider the donor level to be a narrow band of energy levels.)

2.6 What is the difference between compensation and passivation? Give an example of each.

2.7 Sketch simple ball-and-stick models of the following defects and indicate the symmetry:
   a. Carbon acceptor in GaAs.
   b. Zinc–hydrogen complex in GaAs (bond-centered).

2.8 Indicate how the atoms in Figure 2.23 transform under the T_d point-group operations. There are eight threefold rotations, three twofold rotations, six reflections, and six improper rotations (Section 2.6). For example, a threefold rotation about the "1" axis gives

$$(1,2,3,4) \rightarrow (1,4,2,3)$$

which means atom 2 is replaced by atom 4, 3 is replaced by 2, and 4 is replaced by 3.

FIGURE 2.23 Model of a substitutional impurity in a diamond or zincblende crystal (Problem 2.8).

2.9 Sketch a Burgers circuit and indicate the Burgers vector for one of the dislocations in Figure 2.21.

2.10 GaP and GaAs are zincblende semiconductors with lattice constants of 5.45 and 5.65 Å, respectively. A GaP thin film is grown on a (100) oriented GaAs substrate.

    a. Sketch the two extreme cases where the thin film is pseudomorphically strained and fully relaxed by misfit dislocations at the interface.

    b. Estimate the number of misfit dislocations per cm^2 required for the GaP layer to fully relax.

REFERENCES

Anderson, P.W. 1961. Localized magnetic states in metals. *Phys. Rev.* 124: 41–53.

Biernacki, S., U. Scherz, and B.K. Meyer. 1993. Electronic properties of A centers in CdTe: A comparison with experiment. *Phys. Rev. B* 48: 11726–11731.

Caldas, M.J., A. Fazzio, and A. Zunger. 1984. A universal trend in the binding energies of deep impurities in semiconductors. *Appl. Phys. Lett.* 45: 671–673.

Chadi, D.J. and K.J. Chang. 1988. Theory of the atomic and electronic structure of DX centers in GaAs and $Al_xGa_{1-x}As$ alloys. *Phys. Rev. Lett.* 61: 873–876.

Corbett, J.W., G.D. Watkins, R.M. Chrenko, and R.S. McDonald. 1961. Defects in irradiated silicon. II. Infrared absorption of the Si-A center. *Phys. Rev.* 121: 1015–1022.

Cotton, F.A. 1990. *Chemical Applications of Group Theory*, 3rd ed. New York: John Wiley & Sons.

Davies, G., S.C. Lawson, A.T. Collins, A. Mainwood, and S.J. Sharp. 1992. Vacancy-related centers in diamond. *Phys. Rev. B* 46: 13157–13170.

Falicov, L.M. 1989. *Group Theory and Its Physical Applications*, Midway Reprint ed. Chicago: University of Chicago Press.

Föll, H. and B.O. Kolbesen. 1975. Formation and nature of swirl defects in silicon. *Appl. Phys. A* 8: 319–331.

Föll, H., U. Gösele, and H.O. Kolbesen. 1981. Microdefects in silicon and their relation to point defects. *J. Cryst. Growth* 52: 907–916.

Haller, E.E. 1978. Isotope shifts in the ground state of shallow, hydrogenic centers in pure germanium. *Phys. Rev. Lett.* 40: 584–586.

Haller, E.E. 1991. Hydrogen-related phenomena in crystalline germanium. In *Semiconductors and Semimetals*, vol. 34, eds. J.I. Pankove and N.M. Johnson, pp. 113–137. San Diego: Academic Press.

Holt, D.B. and B.G. Yacobi. 2007. *Extended Defects in Semiconductors*. New York: Cambridge University Press.

Hrostowski, H.J. and R.H. Kaiser. 1957. Infrared absorption of oxygen in silicon. *Phys. Rev.* 107: 966–972.

Istratov, A.A. H. Hieslmair, and E.R. Weber. 1999. Iron and its complexes with silicon. *Appl. Phys. A* 69: 13–44.

Janotti, A. and C.G. Van de Walle. 2007. Hydrogen multicentre bonds. *Nature Materials* 6: 44–47.

Johnson, N.M. 1991. Neutralization of donor dopants and formation of hydrogen-induced defects in *n*-type silicon. In *Semiconductors and Semimetals*, vol. 34, eds. J.I. Pankove and N.M. Johnson, pp. 113–137. San Diego, CA: Academic Press.

Johnson, N.M., C. Herring, and D.J. Chadi. 1986. Interstitial hydrogen and neutralization of shallow-donor impurities in single-crystal silicon. *Phys. Rev. Lett.* 56: 769–772.

Johnson, N.M. and C.G. Van de Walle. 1999. Isolated monatomic hydrogen in silicon. In *Semiconductors and Semimetals*, vol. 61, ed. N.H. Nickel, pp. 113–137. San Diego, CA: Academic Press.

Kaminska, M. and E.R. Weber. 1993. EL2 defect in GaAs. In *Semiconductors and Semimetals*, vol. 38, ed. E. Weber, pp. 59–89. San Diego, CA: Academic Press.

Kilic, C. and A. Zunger. 2002. *N*-type doping of oxides by hydrogen. *Appl. Phys. Lett.* 81: 73–75.

Ledebo, L.-Å. and B.K. Ridley. 1982. On the position of energy levels related to transition-metal impurities in III-V semiconductors. *J. Phys. C* 15: L961–L964.

Li, X.-B., S. Limpijumnong, W.Q. Tian, H.-B. Sun, and S.B. Zhang. 2008. Hydrogen in ZnO revisited: Bond center versus antibonding site. *Phys. Rev. B* 78: 113203 (4 pages).

McCluskey, M.D. and E.E. Haller. 1999. Hydrogen in III-V and II-VI semiconductors. In *Semiconductors and Semimetals*, vol. 61, ed. N.H. Nickel, pp. 373–440. San Diego, CA: Academic Press.

48 References

McCluskey, M.D. and S.J. Jokela. 2009. Defects in ZnO. *J. Appl. Phys.* 106: 071101 (13 pages).

Mooney, P.M. 1990. Deep donor levels (DX centers) in III-V semiconductors. *J. Appl. Phys.* 67: R1–R26.

Nelson, R.J. 1977. Long-lifetime photoconductivity effect in n-type GaAlAs. *Appl. Phys. Lett.* 31: 351–353.

Newman, R.C. 2000. Oxygen diffusion and precipitation in Czochralski silicon. *J. Phys. Condens. Matter* 12: R335–R365.

Pajot, B. 1994. Some atomic configurations of oxygen. In *Semiconductors and Semimetals*, vol. 42, ed. F. Shimura, pp. 191–249. New York: Academic Press.

Pankove, J.I., P.J. Zanzucchi, C.W. Magee, and G. Lucovsky. 1985. Hydrogen localization near boron in silicon. *Appl. Phys. Lett.* 46: 421–423.

Pearton, S.J., J.W. Corbett, and M. Stavola. 1992. *Hydrogen in Crystalline Semiconductors*. Berlin: Springer-Verlag.

Queisser, H.J. 1983. Electrical properties of dislocations and boundaries in semiconductors. In *MRS Symp. Proc. Vol. 14, Defects in Semiconductors II*, eds. S. Mahajan and J.W. Corbett, pp. 323–341. New York: North-Holland.

Read, W.T. 1953. *Dislocations in Crystals*. New York: McGraw-Hill.

Sah, C.T., J.Y.C. Sun, and J.J. Tzou. 1983. Deactivation of the boron acceptor in silicon by hydrogen. *Appl. Phys. Lett.* 43: 204–206.

Seeger, A. and K.P. Chick. 1968. Diffusion mechanisms and point defects and silicon and germanium. *Phys. Stat. Solid.* 29: 455–542.

Shan, W., W. Walukiewicz, J.W. Ager et al. 1999. Band anticrossing in GaInNAs alloys. *Phys. Rev. Lett.* 82: 1221–1224.

Shi, G.A., M. Stavola, W. Beall Fowler, and E.E. Chen. 2005. Rotational-vibrational transitions of interstitial HD in Si. *Phys. Rev. B* 72: 085207 (6 pages).

Shockley, W. and W.T. Read. 1949. Quantitative predictions from dislocation models of crystal grain boundaries. *Phys. Rev.* 75: 692.

Stein, H.J. 1986. Nitrogen in crystalline Si. In *MRS Symp. Proc. Vol. 59, Oxygen, Carbon, Hydrogen and Nitrogen in Silicon*, eds. J.C. Mikkelson, Jr. et al., pp. 523–535. New York: North-Holland.

Sueoka, K., E. Kamiyama, P. Śpiewak, and J. Vanhellemont. 2016. Review–properties of intrinsic point defects in Si and Ge assessed by density functional theory. *ECS Journal of Solid State Science and Technology* 5: P3176–P3195.

Sze, S.M. 1981. *Physics of Semiconductor Devices*, 2nd edn., p. 42. New York: John Wiley & Sons.

Vogel, F.L., W.G. Pfann, H.E. Corey, and E.E. Thomas. 1953. Observations of dislocations in lineage boundaries in germanium. *Phys. Rev.* 90: 489–490.

Walukiewicz, W., W. Shan, K.M. Yu et al. 2000. Interaction of localized electronic states with the conduction band: Band anticrossing in II-VI semiconductor ternaries. *Phys. Rev. Lett.* 85: 1552–1555.

Watkins, G.D. 1986. The lattice vacancy in silicon. In *Deep Centers in Semiconductors*, ed. S.T. Pantelides, pp. 147–183. New York: Gordon & Breach.

Watkins, G.D. and J.W. Corbett. 1964. Defects in irradiated silicon: Electron paramagnetic resonance and electron-nuclear double resonance of the Si-E center. *Phys. Rev.* 134: A1359–A1377.

Watkins, G.D. and J.W. Corbett. 1965. Defects in irradiated silicon: Electron paramagnetic resonance of the divacancy. *Phys. Rev.* 138: A543–A555.

Wolk, J.A., M.B. Kruger, J.N. Heyman, W. Walukiewicz, R. Jeanloz, and E.E. Haller. 1991. Local-vibrational-mode spectroscopy of DX centers in Si-doped GaAs under hydrostatic pressure. *Phys. Rev. Lett.* 66: 774–777.

Yu, P.Y. and M. Cardona. 1996. *Fundamentals of Semiconductors*. Berlin: Springer-Verlag.

Interfaces and Devices

3

Perfect crystals have translational symmetry. A diamond or zinc-blende semiconductor, for example, can be displaced a distance a along a <100> direction without changing its appearance. However, the introduction of a defect destroys translational symmetry. The previous chapter discussed point and line defects. We now turn our attention to two-dimensional "area" defects that occur at interfaces (Mönch, 2001). Even a perfectly clean, cleaved surface is a kind of defect, since it interrupts the periodicity of the lattice. This chapter focuses on metal-semiconductor, *p-n*, and metal-oxide-semiconductor junctions, followed by examples of important devices that are enabled by interfaces.

3.1 IDEAL METAL-SEMICONDUCTOR JUNCTIONS

Metal-semiconductor junctions are an important class of interfaces (Brillson, 1983). A semiconductor device requires a number of electrical contacts to provide electrical power for operation or transfer signals to and from the device (Henisch, 1984). These contacts are made with a variety of metals. A great amount of effort has been spent on the understanding of the physics of metal-semiconductor and metal-oxide-semiconductor junctions, and on the fabrication of reliable, stable contacts that have nearly ideal properties.

The ideal metal-semiconductor junction can be understood using simple band theory (Sharma, 1984). In a metal, electrons occupy energy states up to the Fermi level, E_F. The energy difference between E_F and the vacuum is called the *work function*, $e\Phi_M$. Similarly, the semiconductor work function $e\Phi_S$ is the energy difference between the vacuum level and the semiconductor Fermi level. The *electron affinity* $e\chi$ is the energy difference between the vacuum level and the conduction-band minimum.

Consider the case of an *n*-type semiconductor and a metal with a large work function such that $e\Phi_M > e\Phi_S$ (Figure 3.1a). When the metal and semiconductor are brought into contact with each other, electrons move from the semiconductor to the metal to minimize their energy. They leave behind positively charged donors. These donor ions constitute positive space charge known as the *depletion*

3.1 Ideal Metal-Semiconductor Junctions

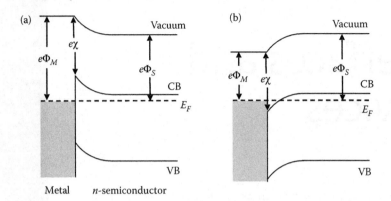

FIGURE 3.1 Ideal metal-semiconductor junctions. The valence band (VB) maximum, conduction-band (CB) minimum, and vacuum level are shown. (a) $e\Phi_M > e\Phi_S$ results in a Schottky barrier with a height $e(\Phi_M-\chi)$. (b) $e\Phi_M < e\Phi_S$ results in an Ohmic contact (no barrier).

layer. The pile-up of electrons at the interface creates a *Schottky barrier* that prevents more electrons from crossing over.

The metal and semiconductor are in diffusive equilibrium. In the absence of applied bias, the Fermi level is constant across the junction. In the semiconductor, far from the junction, E_F is close to the conduction-band minimum. As one approaches the metal, the semiconductor bands bend upward. At the junction, the difference between the conduction-band minimum and metal work function is given by $e(\Phi_M-\chi)$. An energy barrier

$$e\Phi_{Bn} = e(\Phi_M - \chi) \tag{3.1}$$

exists for electrons attempting to travel from the metal to the semiconductor. (The subscript n refers to an n-type semiconductor). This barrier, the Schottky barrier, cannot be changed by the application of a bias across the junction. The barrier against electron travel from the semiconductor to the metal is $e(\Phi_M-\Phi_S)$, which *can* be changed with an applied bias.

Suppose the metal is electrically grounded. If we apply a negative voltage to the semiconductor, the barrier height is reduced, so current can flow (Figure 3.2). In this *forward bias* condition, electrons are given the "lift" they need to make it across the junction. If we apply a positive voltage

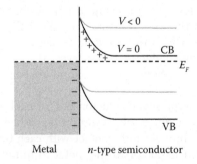

FIGURE 3.2 Schottky barrier formed at an ideal metal–semiconductor interface. CB is the conduction band and VB is the valence band. An applied forward bias ($V < 0$) causes electrons to flow from the semiconductor to the metal.

to the semiconductor, however, very few electrons can surmount the Schottky barrier. This is called *reverse bias*.

The Schottky barrier blocks current flow when the metal is biased negatively relative to the semiconductor (reverse bias), but electric current flows toward the metal for a positive bias applied to the metal (forward bias). For an ideal Schottky contact, the current density is given by

$$(3.2) \qquad J = J_0(e^{eV/k_BT} - 1)$$

where V is the externally applied bias. The constant J_0 depends on the particular theory that applies to the case in question. Schottky developed the diffusion theory, which assumes that the Schottky barrier is much larger than k_BT. The diffusion of electrons is driven by a combination of concentration gradient and electric field. For moderately doped, high-mobility semiconductors (germanium, silicon, or GaAs), thermionic emission describes the constant J_0 more accurately. The thermionic emission model also assumes that the Schottky barrier is larger than k_BT but that a small fraction of the electrons have enough energy to surmount the barrier and generate current (Sze and Ng, 2007).

Now, consider a metal with a low work function such that $e\Phi_M < e\Phi_S$ (Figure 3.1b). We again assume an n-type semiconductor. To minimize their energy, electrons diffuse from the metal to the semiconductor. Because the metal has positive charge, the bands bend downward as one goes from the semiconductor to the metal. Near the interface, the conduction band dips below E_F. This means that a reservoir of free electrons has accumulated in the semiconductor. The junction will not act as a rectifier—instead, electrons can travel across the interface, unimpeded, for either bias condition. The contact is Ohmic.

As Ferdinand Braun discovered, metal-semiconductor junctions may be formed to obtain a rectifier that passes current with little hindrance for one polarity but blocks current flow for the opposite polarity. High-power and high-frequency rectifiers use this metal-semiconductor junction configuration. When Ohmic contacts are required, the rectifying properties must be avoided or circumvented. This is achieved in one of two ways: either (1) a metal is found which forms a junction with the particular semiconductor offering little resistance to current flow in both directions or (2) the near-surface region of the semiconductor is very heavily doped, leading to efficient transfer of charges via tunneling processes.

3.2 REAL METAL-SEMICONDUCTOR JUNCTIONS

How realistic is the model of the ideal Schottky barrier? For a moment, recall the conditions we considered when developing simple band theory. We assumed an infinitely extended semiconductor with lateral translational symmetry. Surfaces with broken bonds, oxidized surfaces, or reconstructed surfaces were never part of the picture. But the simple Schottky barrier model assumes that both the metal and the semiconductor energy bands can be described by bulk properties right to the very junction between the two.

While this simple picture applies rather well to wide-band-gap and ionic semiconductors, it does not describe the situation accurately for metals on silicon, germanium, or GaAs. A large number of early Schottky barrier investigations showed that the measured barrier was not Φ_M–Φ_S. For covalent semiconductors such as silicon, the metal work function did not seem to have any effect on the barrier height. This behavior is quantified by the interface index S (Kurtin et al., 1969):

$$(3.3) \qquad S = \frac{d\Phi_{Bn}}{d\Phi_M}$$

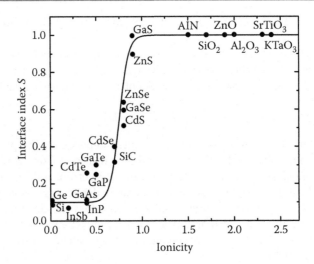

FIGURE 3.3 Interface index versus ionicity (electronegativity difference) for various semiconductors and insulators. The solid line is a guide to the eye. (After Kurtin, S., T.C. McGill, and C.A. Mead. 1969. *Phys. Rev. Lett.* 22: 1433–1436.)

Figure 3.3 shows a compilation of the slope S for numerous semiconductors and oxides as a function of the Pauling electronegativity difference (Pauling, 1988) of the elements forming the semiconductor. The strongly ionic materials (large electronegativity difference) experience the full metal work function in the barrier height ($S \approx 1$). Covalent semiconductors such as silicon and germanium, in contrast, have small values of S.

Various modifications of the simple Schottky barrier model have been proposed to explain this behavior. Most of these extensions of the simple model are based on electronic band-gap states residing at the metal–semiconductor interface. These states are either derived from surface states (Bardeen, 1947) or are induced in the semiconductor by the nearby metal (Louie and Cohen, 1975). Metallurgical reactions between the metal and silicon (silicide formation) have also been considered. We will first discuss the surface/interface state picture and then review silicide formation.

Terminating a semiconductor surface abruptly leads to a variety of possible structures that depend, in part, on the ambient conditions (Brillson, 2010). In an ultra-high vacuum, a cleaved surface will reconstruct in regular atomic patterns, depending on the crystalline orientation. The surface reconstruction and its imperfections form energy states that are different from those of the bulk. Intrinsic surface states from a perfect, clean surface usually lie outside the semiconductor band gap and do not play a dominant role in semiconductor–metal interfaces.

Metal-induced gap states (MIGS) involve metal wavefunctions with energies that lie inside the semiconductor band gap (Louie and Cohen, 1976). MIGS decay exponentially into the semiconductor, similar to a free-electron wavefunction tunneling into a potential barrier. Calculations are in good agreement with the phenomenological dependence of S on ionicity (Louie et al., 1977). The main point is that the presence of a metal introduces new states into the semiconductor gap, near the interface. Other sources of gap states include impurities, native defects, and structural imperfections such as step edges.

Donor states in the upper part of the gap give their electrons to acceptor states in the lower part of the gap, filling the states up to the Fermi level. These interface states "pin" the Fermi level at an energy position that is independent of the metal work function. In the case of a silicon surface, the density of states peaks at $\sim 1/3$ E_g above the valence band maximum, pinning the Fermi level at that energy. The Fermi levels in the silicon bulk and at the surface must equilibrate, which leads to bending of the bands. For n-type silicon, the resulting Schottky barrier is $\sim 2/3$ E_g (Figure 3.4).

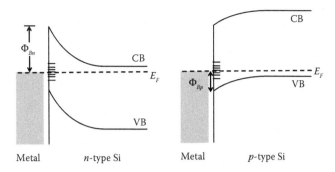

FIGURE 3.4 Real metal–silicon interface. Surface states, indicated by horizontal lines, pin the Fermi level at ∼1/3 E_g above the valence band (VB) maximum. This results in a Schottky barrier of $\Phi_{Bn} \sim 2/3\ E_g$ and $\Phi_{Bp} \sim 1/3\ E_g$ for n- and p-type silicon, respectively.

For p-type silicon, the Schottky barrier is only ∼1/3 E_g. Similarly, n-type germanium has a larger Schottky barrier than p-type germanium.

A simple experimental method takes advantage of this difference in Schottky barrier heights to determine carrier type (electrons or holes). A metal tip touches an n- or p-type crystal, and a deposited large-area contact serves as the "Ohmic" contact. The metal tip and the semiconductor form a Schottky junction. The I-V characteristics for n- or p-type crystals are quite different, due to the difference in Schottky barrier heights. For silicon and germanium, n-type doping leads to a "good" (nearly ideal) diode, while p-type doping leads to a "bad" diode (Ruge, 1975). The *metal tip rectifier* is a qualitative technique to determine carrier type. It is easy to set up and serves as a simple and quick test for the type of conduction.

Other covalent semiconductors also exhibit Fermi-level pinning. In compound semiconductors, native donors and acceptors may form at the interface, compensating each other electrically (Walukiewicz, 1989). While in GaAs pinning occurs between 1/3 E_g to 1/2 E_g above the valence band maximum, it is close to the valence band edge in GaSb and near the conduction band edge in InP.

An extensive summary of measured Schottky barrier heights for many semiconductor/metal combinations is presented in Table 3.1. The metals platinum, palladium, and gold consistently form the highest barriers on n-type semiconductors. The barrier heights for p-type semiconductors can be estimated by subtracting the n-type barrier from the band gap.

From a practical point of view, Schottky barrier contacts formed by metal evaporation are more an art than a science. Especially highly blocking barriers may be difficult to form, and recipes for success abound. One of the authors (EEH) recalls the long days and nights spent forming pure gold barriers on large area planar ultra-pure germanium for gamma-ray detection. When things worked well, devices with tens of cm³ depletion volume would exhibit reverse leakage currents in the pA range at 77 K. At other times, weeks would pass without one working device.

Because of reproducibility and reliability problems, Schottky barriers in the past have found limited applications in electronic devices. This limitation has been largely overcome for silicon by the introduction of *silicides*, compounds or eutectoids formed between a large number of metals (foremost transition metals) and silicon. To form a silicide, a thin metal film is deposited on clean, oxygen-free silicon. A diffusion-driven solid-phase reaction can be initiated at the silicon–metal interface at very moderate temperatures. In general, the formation of the silicide proceeds into the silicon and after a very short time, the original interface is completely consumed. The new silicide–silicon interface is free from the surface state pinning effects described earlier. Ultimate perfection can be achieved with cobalt silicide, which grows epitaxially on the silicon substrate.

An interesting linear relationship between the silicide–silicon barrier height and the heat of formation of the silicide ΔH_f was first reported by Andrews and Phillips (1975). This finding

TABLE 3.1 Schottky Barrier Heights (V) at 300 K for n-Type Semiconductors

Metal	$e\Phi_M$	Si	GaAs	Ge	SiC	GaP	GaSb	InP	ZnS	ZnSe	ZnO	CdS	CdTe
E_g		1.12	1.42	0.66	3.0	2.26	0.72	1.35	3.68	2.71	3.4	2.42	1.43
Ag	4.6	0.83	1.03		0.54	1.2	0.45	0.54	1.81	1.21		0.56	0.8
Al	4.1	0.81	0.93	0.48	1.3	1.06	0.6	0.5	0.8	0.75	0.68		0.76
Au	5.3	0.83	1.05	0.59	1.4	1.3	0.61	0.52	2.2	1.51	0.65	0.78	0.86
Bi	4.3		0.9			0.2				1.14			0.78
Ca	2.9	0.4	0.56										
Co	5.0	0.81	0.86	0.5	1.4								
Cr	4.5	0.60	0.82		1.2	1.18		0.45					
Cu	4.8	0.8	1.08	0.5	1.3	1.2	0.47	0.42	1.75	1.1	0.45	0.5	0.82
Fe	4.7	0.98	0.84	0.42						1.11			0.78
Hf	3.9	0.58	0.82			1.84							
In	4.1		0.83	0.64			0.6		1.5	0.91	0.3		0.69
Ir	5.3	0.77	0.91	0.42									
Mg	3.7	0.6	0.66			1.04	0.3		0.82	0.49			
Mo	4.7	0.69	1.04		1.3	1.13							
Ni	5.2	0.74	0.91	0.49	1.4	1.27		0.32				0.45	0.83
Os	5.9	0.7			0.4							0.53	
Pb	4.3	0.79	0.91	0.38						1.15		0.59	0.68
Pd	5.4	0.8	0.93		1.2		0.6	0.41	1.87		0.68	0.62	0.86
Pt	5.5	0.9	0.98		1.7	1.45			1.84	1.4	0.75	1.1	0.89
Rh	5.0	0.72	0.90	0.4									
Ru	4.7	0.76	0.87	0.38									
Sb	4.6		0.86				0.42			1.34			0.76
Sn	4.4		0.82					0.35					
Ta	4.4		0.85					1.1			0.3		
Ti	4.3	0.6	0.84		1.1	1.12						0.84	
W	4.8	0.66	0.8	0.48									

Source: After Sze, S.M. and K.K. Ng. 2007. *Physics of Semiconductor Devices*, 3rd ed. Hoboken, NJ: John Wiley & Sons; work functions are from Lide, D.R., ed., *CRC Handbook on Chemistry and Physics*, 89th ed, CRC Press, Boca Raton, FL, 2008.

Note: Band gaps (E_g) and metal work functions ($e\Phi_M$) are given in eV. Where there is a spread in values, the average is listed.

showed that chemical bond formation at the interface plays a decisive role in the resulting Schottky barrier height. An improved picture was offered by Ottaviani et al. (1980) who showed that the silicide barrier height depends linearly on the eutectic (melting) temperature over a broad range of silicides. Tu et al. (1981) extended this plot to include rare earth silicides, which form the lowest barriers on n-type silicon (Figure 3.5).

Silicide-silicon technology has rapidly matured and has led to applications in various kinds of electronic devices. Silicides are preferred over more simple metal-silicon contacts because they are highly stable, in many cases to over 800°C, especially the disilicides (MSi_2). The interfaces exhibit excellent planarity and reproducible electrical barrier properties. The fundamental understanding and theory of the barrier formation have, however, not progressed to the point where first-principles calculations can predict barrier heights or formation temperatures based on fundamental properties of silicon and the metal in question.

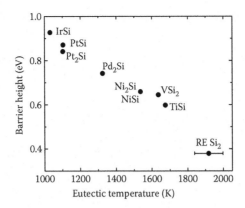

FIGURE 3.5 Schottky barrier heights versus eutectic (melting) temperature for various metal silicides. The rare earth (RE) silicides have the lowest barriers. (After Tu, K.N., R.D. Thompson, and B.Y. Tsaur. 1981. *Appl. Phys. Lett.* 38: 626–629.)

3.3 DEPLETION WIDTH

Consider the example of a metal Schottky barrier on an *n*-type semiconductor (Figure 3.6). Work function differences and surface state pinning lead to a potential barrier, which forms a depletion region of width W in the semiconductor. The net ionized donor concentration in this depletion region constitutes a positive space charge. The Poisson equation quantitatively relates the potential Φ and space charge concentration:

(3.4) $$\frac{d^2\Phi}{dx^2} = -\frac{|e|N}{\varepsilon\varepsilon_0}$$

where $N = N_d^+ - N_a^-$. Two integration steps lead to

(3.5) $$\Phi(x) = -(x-W)^2 \frac{|e|N}{2\varepsilon\varepsilon_0} \quad (0 < x < W)$$

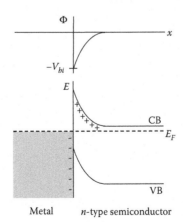

FIGURE 3.6 Schottky junction. Going from the metal to the semiconductor, the potential rises from $-V_{bi}$ to 0, where V_{bi} is the built-in voltage.

56   3.4 The p-n Junction

where the potential at $x = W$ has been set to zero. In the absence of an applied voltage, the total potential difference across the depletion layer corresponds to the Schottky barrier height. This is the built-in voltage V_{bi}. When an external voltage V_r is applied to the semiconductor, the potential difference across the depletion width is $V_{bi} + V_r$, yielding

$$(3.6) \qquad V_{bi} + V_r = W^2 \frac{|e|N}{2\varepsilon\varepsilon_0}$$

Equation 3.6 can be solved for the depletion width,

$$(3.7) \qquad W = \sqrt{\frac{2\varepsilon\varepsilon_0}{|e|N}(V_{bi} + V_r)}$$

From this equation, one can see that reverse bias (positive V_r) increases the depletion width. The amount of charge in the depletion layer is

$$(3.8) \qquad Q = |e|NAW$$

where A is the area of the metal contact. The metal contains a charge $-Q$. In effect, we have a capacitor. For a depletion layer, we define a differential capacitance,

$$(3.9) \qquad C = \frac{dQ}{dV_r}$$

This is the incremental change in charge Q that results from a change in the voltage. Equations 3.7 through 3.9 yield

$$(3.10) \qquad C = \varepsilon\varepsilon_0 \frac{A}{W}$$

In summary, as the reverse bias is increased,

- The depletion width W increases
- The space charge Q increases
- The capacitance C decreases

These qualitative relations also hold for p-n junctions. The depletion layer capacitance can be exploited to characterize electrically active defects, discussed in Chapter 9.

3.4 THE p-n JUNCTION

As discussed in Section 3.1, metal-semiconductor (Schottky) junctions provide current rectification. Similar rectifying behavior occurs in a p-n junction (Figure 3.7). Imagine that we bring an n- and p-type semiconductor together. When they are brought into contact, electrons diffuse from the n-type side to the p-type side. These free electrons leave behind positively charged donors. When they reach the p side, they "fill" the holes. The p side then has negatively charged acceptors.

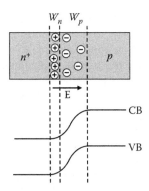

FIGURE 3.7 A p-n^+ junction. W_n and W_p are the depletion widths on the n and p side, respectively. The bending of the conduction-band minimum and valence-band maximum are shown.

The positive donors and negative acceptors produce an electric field, forming a depletion region. In this region, there is space charge due to ionized impurities but very few free electrons or holes. The depletion region is therefore resistive.

As with the Schottky barrier, we can also describe a p-n junction from an energetic point of view. An isolated n-type semiconductor has its Fermi level near the conduction band. A p-type semiconductor has its Fermi level near the valence band. When they are brought in contact, E_F is constant across the junction. The negative space charge on the p side repels electrons. Therefore, the electron energy levels rise as one approaches the p side. As shown in Figure 3.8, the energy rise is approximately equal to the band gap E_g.

If a positive voltage (forward bias) is applied to the p side, then the barrier will be reduced and current will flow. If a negative voltage (reverse bias) is applied, then the barrier will increase and only a small leakage current will flow. The current is given by

$$(3.11) \qquad I = I_0(e^{eV/nk_BT} - 1)$$

where I_0 is the leakage current, V is the applied bias, and n is the "ideality factor," equal to unity for a perfect p-n junction.

The depletion region has a concentration N_d positively charged donors on the n side and N_a negatively charged acceptors on the p side. From charge neutrality,

$$(3.12) \qquad N_a W_p = N_d W_n$$

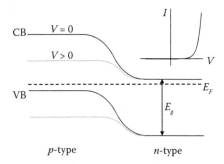

FIGURE 3.8 p-n junction and current-voltage (I-V) plot.

where W_p and W_n are the depletion widths on the p and n sides, respectively (Figure 3.7). The opposite charges create a built-in electric field that bends the bands. Consider the case of an n^+-p junction, in which $N_d \gg N_a$. Given that condition, Equation 3.12 tells us that $W_n \ll W_p$ (i.e., the depletion region is almost entirely in the p-type semiconductor). Under forward bias, free electrons flow from the n^+ side to the p side, where they become minority carriers. The current-versus-voltage plot is given by Equation 3.11. Under reverse bias, the depletion width W_p widens, and only a small leakage current flows.

If the reverse bias is too large, the semiconductor will undergo electrical *breakdown* and a large current will flow. There are three main causes of breakdown in p-n junctions, all of which involve generating electrons and holes in the depletion region (Sze and Ng, 2007). *Avalanche* breakdown, or impact ionization, occurs when carriers flowing through the depletion region generate electron–hole pairs. If the electric field is high (>10^4 V/cm in silicon), the bands tilt steeply and *tunneling* occurs. In that process, an electron tunnels directly from the valence band to the conduction band a short distance away. Finally, semiconductors with low band gaps at room temperature (e.g., germanium) may exhibit *thermal instability*. Resistive heating causes the depletion region temperature to rise, generating electrons and holes.

3.5 APPLICATIONS OF *p-n* JUNCTIONS

p-n diodes are important as *photovoltaic detectors* and *solar cells* (Bube, 1998). A photon that is absorbed in the depletion region will create an electron and a hole (Figure 3.9). The electron will descend to the conduction-band minimum via phonon creation (i.e., the generation of heat). The hole will rise to the valence band maximum by the same mechanism. Then, the electron and hole will be swept in opposite directions by the built-in electric field. The electron and hole will be accelerated to energies of E_e and E_h, respectively, where $E_e + E_h \sim E_g$. Therefore, an absorbed photon of energy $h\nu$ produces electrical energy of $\sim E_g$.

For a solar cell, this results in some inherent inefficiencies. In order to be absorbed, $h\nu$ must be greater than E_g. This means that $h\nu - E_g$ of energy is wasted as heat. Given the spectrum of sunlight, the best possible efficiency for a single-junction device is the *Shockley–Queisser limit*, 31%. One way to circumvent this limit is to fabricate a multi-junction device. For example, the top layer could absorb UV, the middle layer could absorb visible, and the bottom layer could absorb IR. Multijunction solar cells have been reported with efficiencies over 40% (Leite et al., 2013).

The logical extension of the *p-n* junction is the *n-p-n* transistor, also called the *bipolar transistor*. In this device, two *n*-type regions are separated by a thin *p*-type layer. The *p*-type layer is called the "base." A small voltage applied to the base, V_B, acts as a valve to regulate current flow from one *n* layer to the other. Consider the arrangement shown in Figure 3.10a, where one *n* layer is grounded and the other has an applied voltage ($V_C \sim 10$ V). These layers are called the "emitter" and "collector" (of electrons), respectively. Because the second *p-n* junction (base-collector) is

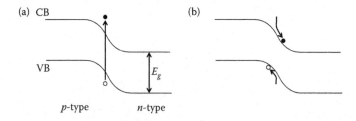

FIGURE 3.9 (a) Absorption of light in a *p-n* junction. (b) The electron and hole lose energy via phonon creation and are swept in opposite directions by the built-in electric field.

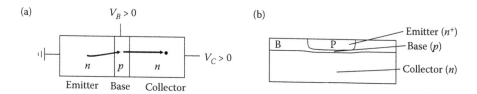

FIGURE 3.10 (a) Schematic diagram of a bipolar n-p-n transistor. The flow of electrons is shown; the corresponding collector current is in the opposite direction. (b) Illustration of the doping layers in a typical n-p-n transistor.

under reverse bias, if $V_B = 0$, then no current will flow. If $V_B > 0$, then the first p-n junction will be under forward bias. Electrons will flow into the thin p layer. Most of these electrons will then be swept into the collector, resulting in collector current. The bipolar transistor is usually modeled as a current amplifier because a small base current produces a large collector current (Barnaal, 1982).

Figure 3.10b shows the doping layers in a typical n-p-n bipolar transistor. To fabricate this device, one starts with a lightly n-type silicon wafer. An acceptor dopant such as boron is diffused in, forming a p-type layer. Finally, a high concentration of donors (e.g., phosphorus) is diffused or implanted into the top layer. The doping conditions are adjusted to optimize the thickness of the p-type base layer for the desired application. Methods for doping, including diffusion and implantation, are discussed in Chapter 4.

3.6 THE METAL-OXIDE-SEMICONDUCTOR JUNCTION

From a practical applications point of view, the metal-oxide-semiconductor (MOS) structure is by far the most important materials configuration for modern silicon electronics. The MOS field effect transistor (MOSFET) is the most abundantly manufactured electronic device. What are the reasons for its phenomenal performance and reliability? Basically, it is the exceptionally low interface state density that can be obtained at a properly formed silicon–silica (SiO_2) interface. This special property of silicon and its oxide is unique among semiconductors and at least as important for the success of silicon technology as are the convenient band gap of silicon, its excellent free carrier mobility, high doping capability, abundance, and nontoxicity.

In an n-channel MOSFET device, two n-type regions, the "source" and "drain," are separated by a p-type layer (Figure 3.11). The p layer is capped with an oxide (SiO_2) and a metal contact called the "gate." The insulator provides a barrier to current flow. If the gate is at ground, then we have two back-to-back p-n junctions. Since one of those junctions is under reverse bias, current cannot flow from the source to the drain. When a positive gate voltage is applied, a thin conducting channel of electrons is formed at the interface. This channel allows electrons to flow from source to drain. The gate therefore acts as a current valve.

A band diagram for an ideal, defect-free structure is shown in Figure 3.12. Let us assume that the semiconductor is grounded and consider what happens when a voltage V is applied to the metal. For $V < 0$, the electric field in Figure 3.12 points to the left. This field pushes holes to

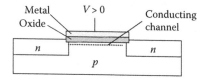

FIGURE 3.11 Schematic diagram of a MOSFET.

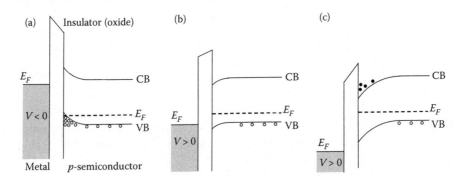

FIGURE 3.12 Ideal metal-oxide-semiconductor (MOS) structure: (a) accumulation, (b) depletion, (c) inversion. The actual barrier height of the oxide is larger than what is shown.

the insulator–semiconductor interface, resulting in the *accumulation* of excess holes. We can also understand this process by examining the semiconductor bands that bend upward near the interface. Because there is no current flow, the system is in equilibrium and the semiconductor Fermi level is constant. Therefore, near the interface, the Fermi level is close to the valence-band maximum. This means that there is a high concentration of holes at that location.

For $V > 0$, we look at two cases. First, we consider a small positive voltage. This results in an electric field that pushes holes away from the interface, resulting in depletion. The semiconductor bands bend downward near the interface. Second, we consider a large positive voltage. In this case, the bands bend downward such that the Fermi level near the interface is in the upper part of the band gap. This condition, called *inversion*, attracts minority electrons to the interface and results in an *n*-type conducting channel. This *n*-type channel is responsible for the MOSFET operation. Even though the potential on the gate electrode leads to all of these free-carrier concentration changes in the semiconductor, no appreciable exchange of free carriers between the gate electrode and the semiconductor takes place.

A major issue for this MOS structure is the reduction of electrically active states at the oxide–semiconductor interface. These states can pin the Fermi level and prevent the inversion condition from taking place. For the example of a *p*-type semiconductor, the surface states can trap the minority electrons. One electrically active defect is called the P_b defect, which involves a silicon dangling bond and can be detected by EPR (Brower, 1983). Hydrogen is often used to passivate dangling bonds at the SiO_2/Si interface, thereby improving the performance of the MOSFET device.

Figure 3.13 shows a micrograph of a small section of a Si–SiO_2 interface imaged at atomic resolution with high-resolution transmission electron microscopy. The crystalline silicon lattice terminates sharply within one plane. No defects are visible in this terminating plane. This

FIGURE 3.13 Transmission electron microscopy image of a metal-oxide-semiconductor (MOS) structure. The crystalline Si abruptly ends with a (100) plane. The extremely thin oxide displays a mottled contrast because the atoms are randomly distributed in the SiO_2 glass. The Al metallization on top of the thin oxide shows lattice fringes, indicating its polycrystalline nature. (Courtesy of R. Jamison and R. Gronsky.)

structural perfection results in surface state densities as low as 10^8 cm^{-2} V^{-1} and lower. With such a low surface density, it is rather easy to have an electric field originating at a metal film on the oxide layer (the gate) penetrate through the Si–SiO$_2$ interface into the underlying silicon. The field inside the crystalline silicon will control the space charge density and the free carrier concentration in this near-surface region.

The gate oxide is a good insulator. The gate SiO$_2$ thickness has shrunk to dimensions (<20 Å) that allow effective electron tunneling between the gate metal and the semiconductor. In addition to interfering with normal device operation, the kinetic energy from tunneling electrons can cause defects in the dielectric (Sze and Ng, 2007). When these defects reach a critical value, current paths exist that short out the insulator. This process, described by percolation theory, leads to dielectric breakdown. To address these problems, hafnium-based dielectrics and other materials have emerged as alternatives to SiO$_2$ (Schlom et al., 2008).

3.7 THE CHARGE-COUPLED DEVICE

Charge-coupled device (CCD) detector arrays are important for applications such as digital cameras and spectroscopy. Because of its widespread impact on imaging, the invention of the CCD was recognized by the Nobel Prize in Physics in 2009. The basic structure is shown in Figure 3.14 for the case of a surface-channel CCD array (Sze and Ng, 2007). A single CCD element, or pixel, consists of a semitransparent gate electrode and insulator on a p-type semiconductor substrate. Most commercial CCD arrays are based on MOS technology, where the insulator is SiO$_2$ and the substrate is silicon.

During operation, a positive gate pulse V_G is applied. This causes the bands to bend downward and a depletion layer is formed. When a photon is absorbed in the semiconductor, an electron–hole pair is generated. The electron is attracted to the semiconductor–insulator interface. Over the integration time, more electrons accumulate at the interface. The light intensity at a particular CCD pixel is inferred from the amount of charge thus collected.

After the integration time, the charges must be read out. An example of this is shown for the case of an interline transfer readout architecture (Figure 3.15). Charge is transferred from

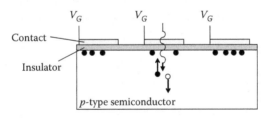

FIGURE 3.14 Schematic diagram of a CCD array.

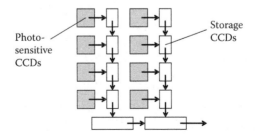

FIGURE 3.15 Interline transfer readout mechanism for a CCD array.

the photo-sensitive CCD pixels to storage CCDs that are not light sensitive. The vertical storage CCDs then transfer the charges to the horizontal storage CCDs, also called the "output register." In this step-by-step procedure, the charge of each pixel is recorded.

Charge is transferred from one CCD to another by cleverly changing the V_G values (Boyle and Smith, 1971). Consider a charged CCD with a positive V_G, next to an empty one with $V_G = 0$. First, the empty CCD's V_G is raised, attracting electrons. Then, the other CCD's V_G is decreased to zero, causing the rest of the electrons to leave. After this cycle, all the electrons have been transferred from one CCD to the other.

There are several variations on this basic design concept. One drawback of the design shown here is that the gates absorb some of the light. For applications that require high sensitivity, it is advantageous to have the light absorbed by the back surface (bottom surface in Figure 3.14). In that case, the back surface must be thinned so that electrons and holes are created within the depletion region.

3.8 LIGHT-EMITTING DEVICES

In addition to their obvious importance to computing, semiconductors are important as materials for *light-emitting diodes* (LEDs). A Schottky contact or a *p-n* junction can emit light, albeit inefficiently (Schubert, 2006). Consider a *p-n* junction under forward bias. Holes travel from the *p* side to the *n* side, while electrons travel in the opposite direction. Occasionally, an electron falls from the conduction band to fill a hole in the valence band, resulting in the emission of a photon. This process, electron-hole *recombination*, is the basis for solid-state lighting.

The inefficiency of a *p-n* junction as a light-emission device stems from the fact that most electrons and holes travel past each other without recombining. To address that problem, researchers use sophisticated growth techniques to fabricate layers of different materials, called *heterostructures*. Consider the InGaN/GaN heterostructure shown in Figure 3.16 (for simplicity, we omit band bending). This structure is an InGaN layer sandwiched by an *n*-type and *p*-type GaN layer. The InGaN layer, which has a smaller band gap than GaN, forms a *quantum well*. Under forward bias, electrons from the *n*-type layer, and holes from the *p*-type layer, become trapped in the quantum well. The electrons and holes recombine and emit photons with energy close to the InGaN band gap. In practice, an LED structure has multiple quantum wells, to give electrons and holes several chances to recombine.

Whereas an LED operates by spontaneous emission, a *laser diode* produces coherent light by stimulated emission. To fabricate a laser diode, one needs to confine the emitted light. This is accomplished by adding layers that have a lower index of refraction *n* than the active region. For example, in blue lasers, AlGaN layers are included on either side of the InGaN quantum wells (Figure 3.17). These layers confine the emitted radiation to a thin layer. A second requirement is that photons need to reflect back and forth in a Fabry–Perot cavity, to induce stimulated

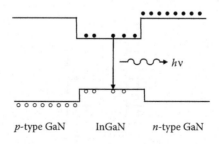

FIGURE 3.16 Band diagram of a blue LED.

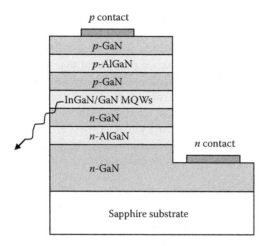

FIGURE 3.17 Blue diode laser. Light is emitted from the multiple quantum well (MQW) region. (After Johnson, N.M., A.V. Nurmikko, and S.P. DenBaars. 2000. *Phys. Today* 53(10): 31–36.)

emission. Such a cavity can be produced simply by etching the material to form a rectangular shape. The difference in n between the semiconductor and air produces high reflectivity, which can be enhanced by depositing reflective material. The laser beam then exits through the lower-reflectivity edge.

3.9 THE 2D ELECTRON GAS

In general, superlattices and individual quantum wells (e.g., a thin layer of GaAs between AlGaAs) exhibit a host of novel optical and electronic properties that are studied extensively worldwide. Every compact disc device (CD, DVD, or Blu-Ray) contains quantum wells. The accomplishment of this remarkable breakthrough required advances in semiconductor growth techniques, discussed in Chapter 4.

A fascinating example of a heterostructure is n-type $Al_xGa_{1-x}As$ ($x \sim 0.3$) grown on lightly p-type GaAs (Figure 3.18). The conduction-band and valence-band offsets are given by the alloy composition x. Electrons from $Al_xGa_{1-x}As$ migrate to the GaAs side in order to minimize their energy. This causes the $Al_xGa_{1-x}As$ bands to bend upward as one approaches the interface. As discussed in Section 3.1, in the absence of an applied bias, the Fermi level is constant. Far from the interface, E_F is near the valence band maximum for GaAs and the conduction-band minimum for $Al_xGa_{1-x}As$.

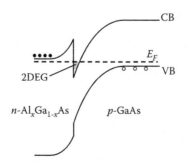

FIGURE 3.18 Band diagram of a 2DEG heterostructure.

64   3.9 The 2D Electron Gas

From the band diagram, we see that the conduction band of GaAs drops below the Fermi level near the heterointerface. This results in a high density of electrons confined to a very narrow region ($\leq$100 Å). Because of the small width along one dimension, the electron states become quantized and form a two-dimensional electron gas (2DEG) (Davies, 1998). 2DEGs reach low-temperature mobility values that far exceed electron mobilities in any other GaAs structures. There are mainly two reasons for this increase. Most importantly, the electrons in the 2DEG are spatially separated from the donors that are the source of these electrons. The mobility limiting ionized impurity scattering is eliminated in this 2DEG structure. Secondly, the heterointerface can be made atomically flat with advanced growth techniques (Chapter 4), eliminating a second source of scattering for the electrons moving along this interface.

In summary, the properties of semiconductors have been harnessed to produce a range of important electronic and optoelectronic devices. Dopants and defects play key roles in determining the properties and performance of semiconductor materials.

PROBLEMS

3.1   The electron affinity of a p-type semiconductor is 4.2 eV. Its band gap is 1.0 eV. The work function of a metal is 4.6 eV. Sketch the energy bands for the metal-semiconductor junction, assuming an ideal interface. Indicate the Schottky barrier height.

3.2   Estimate the depletion width for a real Schottky junction between a metal and p-type silicon (10^{15} cm^{-3} net acceptors), with no applied bias. The dielectric constant of silicon is $\varepsilon = 11.7$.

3.3   Schottky and p-n junctions both rectify current. In which type of junction is minority carrier transport more important?

3.4   Consider an n^+-p junction in germanium, in which $N_d \gg N_a$. Calculate the depletion width for $N_a = 7.7 \times 10^{12}$ cm^{-3} at 77 K and no applied bias. Use a dielectric constant $\varepsilon = 16$ and band gap 0.7 eV. (Because essentially all the depletion occurs on the p side, you can use the equation for a Schottky junction to find the depletion width.)

3.5   A 405-nm LED shines on an ideal silicon photodiode (no applied bias). 6.25×10^{12} photons/second are absorbed by the depletion region. The photodiode is connected to a 100 Ohm resistor. Assume each absorbed photon generates one electron of current (i.e., we are approximating this as a short circuit).
   a.   What is the voltage drop across the resistor?
   b.   How much heat (W) is generated by the photodiode?

3.6   Complementary metal oxide semiconductor (CMOS) circuits use pairs of n-type and p-type MOSFETs. The MOSFET shown in Figure 3.11 is referred to as n-type.
   a.   Sketch a p-type MOSFET and indicate the sign of the gate voltage.
   b.   For a p-type MOSFET, sketch the band diagram given the inversion condition.

3.7   Sketch a CCD array for an n-type semiconductor and indicate the sign of the gate voltages.

3.8   What is a major difference between an LED and a laser-diode structure?

3.9   Sketch the band diagram for an interface between two p-type semiconductors with band gaps of 2 eV and 1 eV.
   a.   Assume there is no valence-band offset (the conduction-band offset is 1 eV).
   b.   Assume the conduction-band and valence-band offsets are both 0.5 eV.

3.10 Sketch a band diagram for a two-dimensional hole gas, forming at the interface of p-type AlAs ($E_g = 2.2$ eV) and n-type GaAs ($E_g = 1.42$ eV).

REFERENCES

Andrews, J.M. and J.C. Phillips. 1975. Chemical bonding and structure of metal-semiconductor interfaces. *Phys. Rev. Lett.* 35: 56–59.

Bardeen, J. 1947. Surface states and rectification at a metal semi-conductor contact. *Phys. Rev.* 71: 717–727.

Barnaal, D. 1982. *Analog Electronics for Scientific Application*, p. A292. Prospect Heights, IL: Waveland Press.

Boyle, W.S. and G.E. Smith. 1971. Charge-coupled devices–a new approach to MIS device structures. *IEEE Spectrum* 8(7): 18–27.

Brillson, L.J. 1983. Advances in understanding metal-semiconductor interfaces by surface science techniques. *J. Phys. Chem. Solids* 44: 703–733.

Brillson, L.J. 2010. *Surfaces and Interfaces of Electronic Materials*. Weinheim, Germany: Wiley.

Brower, K.L. 1983. ^{29}Si hyperfine structure of unpaired spins at the Si/SiO_2 interface. *Appl. Phys. Lett.* 43: 1111–1113.

Bube, R.H. 1998. *Photovoltaic materials*. London: Imperial College Press.

Davies, J.H. 1998. *The Physics of Low-dimensional Semiconductors: An Introduction*. Cambridge: Cambridge University Press.

Henisch, H.K. 1984. *Semiconductor Contacts*. Oxford: Clarendon Press.

Johnson, N.M., A.V. Nurmikko, and S.P. DenBaars. 2000. Blue laser diodes. *Phys. Today* 53(10): 31–36.

Kurtin, S., T.C. McGill, and C.A. Mead. 1969. Fundamental transition in the electronic nature of solids. *Phys. Rev. Lett.* 22: 1433–1436.

Leite, M.S., R.L. Woo, J.N. Munday et al. 2013. Towards an optimized all lattice-matched InAlAs/InGaAsP/InGaAs multijunction solar cell with efficiency >50%. *Appl. Phys. Lett.* 102: 033901 (5 pages).

Lide, D.R., ed. 2008. *CRC Handbook on Chemistry and Physics*, 89th ed. Boca Raton, FL: CRC Press.

Louie, S.G. and M.L. Cohen. 1975. Self-consistent pseudopotential calculation for a metal-semiconductor interface. *Phys. Rev. Lett.* 35: 866–869.

Louie, S.G. and M.L. Cohen. 1976. Electronic structure of a metal-semiconductor interface. *Phys. Rev. B* 13: 2461–2469.

Louie, S.G., J.R. Chelikowsky, and M.L. Cohen. 1977. Ionicity and the theory of Schottky barriers. *Phys. Rev. B* 15: 2154–2162.

Mönch, W. 2001. *Semiconductor Surfaces and Interfaces*, 3rd edn. Berlin: Springer.

Ottaviani, G., K.N. Tu, and J.W. Mayer. 1980. Interfacial reaction and Schottky barrier in metal-silicon systems. *Phys. Rev. Lett.* 44: 284–287.

Pauling, L. 1988. *General Chemistry*. New York: Dover, p. 182.

Ruge, I. 1975. *Halbleiter Technologie*. Berlin: Springer-Verlag.

Schlom, D.G., S. Guha, and S. Datta. 2008. Gate oxides beyond SiO_2. *MRS Bulletin* 33: 1017–1025.

Schubert, F. 2006. *Light-Emitting Diodes*, 2nd edn. Cambridge: Cambridge University Press.

Sharma, B.L., ed. 1984. *Metal-Semiconductor Schottky Barrier Functions and Their Applications*. New York, NY: Plenum Press.

Sze, S.M. and K.K. Ng. 2007. *Physics of Semiconductor Devices*, 3rd edn. Hoboken, NJ: John Wiley & Sons.

Tu, K.N., R.D. Thompson, and B.Y. Tsaur. 1981. Low Schottky barrier of rare-earth silicide on n-Si. *Appl. Phys. Lett.* 38: 626–629.

Walukiewicz, W. 1989. Amphoteric native defects in semiconductors. *Appl. Phys. Lett.* 54: 2094–2096.

Crystal Growth and Doping

In the previous chapters, we discussed the properties of semiconductors and their defects. Doping, the controlled introduction of impurities into well-defined areas of a semiconductor, exploits these properties to produce useful devices. In this chapter, several methods of introducing dopants into semiconductors are summarized. These methods can be grouped into two broad categories: doping *during* and *after* crystal growth. After reviewing the basics of bulk crystal growth and doping, we describe several thin-film growth techniques. In addition to introducing precise concentrations of dopants, unwanted contaminants must be minimized. The degree of perfection of a single crystal depends on the purity of the starting materials and on the particular technique used to grow the crystal. Post-growth doping methods include diffusion, ion implantation, and neutron transmutation. Along with doping, we discuss annealing, a thermal processing step that is often required to repair crystal damage and electrically activate the dopants.

4.1 BULK CRYSTAL GROWTH

Crystal growth is an important field for many technologically important materials, including semiconductors (Pamplin, 1975). The most common way to form crystals of a material such as silicon or germanium is to cool a liquid, or *melt*, of the respective substance until it undergoes solidification. When the melt is cooled slowly from a temperature above the melting temperature T_m, we can distinguish between two cases:

- The melt contains at least one seed crystal. The seed crystal will begin to grow at $T = T_m$. The crystal growth depends strongly on the temperature gradients near the liquid/solid interface.
- The melt does not contain any seed and is very pure. In this case, no crystallization occurs upon decreasing the melt temperature below T_m. The melt gets into a metastable supercooled state, also called the Ostwald–Miers range (Bohm, 1985). Upon further cooling, supercooling becomes so large that spontaneous nucleation leads to many seeds simultaneously, resulting in a polycrystalline solid.

67

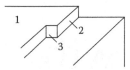

FIGURE 4.1 Three nucleation sites for single-crystal growth.

Figure 4.1 shows three different sites on a liquid/solid interface. Atoms condensing on top of a plane (1) release the least amount of energy; the ones lodging in the kink (3) release the most energy. The largest energy release leads to the highest growth velocity. The crystal planes with the smallest energy release per condensed atom grow more slowly than all the other planes and lag behind. This leads to the facets that give crystals their characteristic appearance.

The oldest and still most commonly used method for the growth of bulk semiconductor crystals is the *Czochralski* (CZ) technique (Czochralski, 1918; Moody and Frederick, 1983; Müller, 2007). A melt contained in a crucible (e.g., silica or graphite) is heated by radio frequency (RF) power or by resistors. A crystalline seed of well-defined crystal orientation is dipped into the melt. The balance of all heat flows determines the crystal growth rate and crystal diameter. Cylindrical symmetry is maintained by rotation of the seed. Stirring of the melt is achieved by natural convection and by forced convection caused by the seed rotation and the counter- or corotation of the crucible. There exists a large body of literature dealing with the modeling of heat flows in crystal pullers (Arizumi and Kobayashi, 1972; Langlois, 1981). Figure 4.2 shows a schematic of a typical CZ growth apparatus.

Most single-crystal silicon is grown using this technique. Typical crystal diameters range from 15 to 45 cm, and lengths reach over 2 m. The advantages of the CZ technique lie in the relative simplicity of the apparatus, the large size of the crystals that can be achieved, and ease of doping with impurities. Also, there are no special demands on the shape of the starting material. Polycrystalline pieces or powder can be melted down.

A disadvantage, in the case of silicon, is the reduction of the silica (SiO_2) crucible by the melt ($T \geq 1415°C$). The free oxygen produced by the reduction dopes the crystals with oxygen concentrations of 10^{17} to 10^{18} cm^{-3}. Oxygen further attacks the graphite susceptor, which has the function of supporting the soft silica crucible. CO and CO_2 dope the Si melt and lead to high carbon concentrations (4×10^{16}–5×10^{17} cm^{-3}). Boron, phosphorus, and other impurities in the silica are released by the reduction of the crucible, and they dope the melt and the crystal (Huff, 1983).

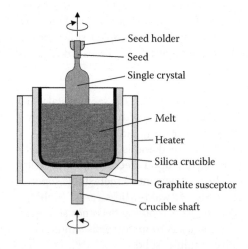

FIGURE 4.2 A Czochralski crystal puller.

It was recognized early on that a reduction of the melt flow velocities could lead to reduction in impurity concentrations, because the impurities released in the reduction of crucible material would no longer be transported to the growing crystal. Because a liquid semiconductor is a good electrical conductor, one can use Eddy currents to break the melt convection. A large magnet produces a strong magnetic field (a few kGauss) that causes Eddy currents in the flowing melt. The energy dissipated by the Eddy currents results in an increase of the effective viscosity, bringing the melt convection to a halt. However, the increase in equipment cost has limited "magnetic Czochralski" to the growth of special crystals (Thomas et al., 1990).

Germanium, with its melting point at 936°C, is much less affected by the silica reduction. This is the major reason why it has been possible to grow ultrapure germanium crystals with electrically active impurity concentrations of $\sim 10^{10}$ cm^{-3} and low compensation. The purity record lies around a few times 10^9 cm^{-3}. In these crystals, one finds one dopant impurity for every 10^{13} germanium atoms—in chemists' terms, "thirteen nines" pure material (Haller et al., 1981; Haller and Goulding, 1993). Doped germanium crystals can also be grown from a melt contained directly in the graphite susceptor because germanium and carbon do not react, and the small concentrations of boron and phosphorus introduced by the graphite susceptor are negligible compared to typical dopant concentrations.

For silicon, it is desirable to find a crystal growth method that does not rely on a crucible. The *floating zone* (FZ) technique was developed in response to this demand (Mühlbauer and Sirtl, 1974; Keller and Mühlbauer, 1981; Kramer, 1983). Figure 4.3 shows a typical FZ apparatus. Here, an RF coil inside the crystal grower melts a zone of the "feed" polycrystal. As the polycrystal moves vertically, the molten zone solidifies into single-crystal silicon. To ensure good mixing and uniform growth across the melt/solid interface, the feed polycrystal and the single crystal are counter-rotated. The density of liquid silicon is small (2.42 g cm^{-3}), and its surface tension is high (720 dyn cm^{-1}), allowing growth of large-diameter crystals. In addition, one can choose appropriate RF coil designs that help support the melt through Eddy-current forces. Near the end of the crystal pulling run, the entire crystal is balanced on the seed, with a few support clamps near the seed end (Rea, 1981).

Typical growth rates for the CZ and the FZ techniques are up to a few mm per minute. Inert gases or vacuum are used as the ambient environment. FZ combined with vacuum increases

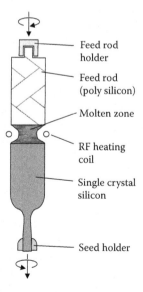

FIGURE 4.3 A float zone growth apparatus.

the effective purification for impurities that have high vapor pressures. This technique is used to remove phosphorus, a common impurity in polycrystalline silicon. Since the 1960s, FZ silicon single crystals have been produced with resistivities of ~100 kΩ cm ($n \approx 10^{10}$ cm^{-3}), which is close to intrinsic at room temperature. Oxygen and carbon concentrations ([O] < 10^{15} cm^{-3}, [C] < 10^{15} cm^{-3}) are also much lower than in CZ-grown crystals.

With the exception of InSb and GaSb, III-V compound semiconductors exhibit high dissociation pressures and component vapor pressures at the melting point. In the case of GaAs ($T_m = 1238°C$), the arsenic vapor pressure at the melting point is so high that a stoichiometric melt would rapidly become nonstoichiometric if one did not take countermeasures (Kamath, 1984). One method is to maintain an arsenic vapor pressure above the melt that corresponds precisely to the dissociation pressure (~0.9 atm). Adding a liquid seal in the form of a B_2O_3 layer further reduces arsenic evaporation, a method known as *liquid encapsulated Czochralski* (LEC).

As discussed in Chapter 2, excess arsenic leads to arsenic antisites, which are deep double donors. These antisites, along with gallium vacancies, move the Fermi level toward the center of the band gap, resulting in semi-insulating (SI) GaAs with resistivities as high as 10^9 Ω cm. For *n*-type GaAs, silica crucibles can be used because silicon contamination leads to donor doping. In the case of phosphides, very high pressures (60 atm for InP) are required, and the LEC technique is a necessity. Further discussions on semi-insulating III-V semiconductors can be found in edited volumes by Makram-Ebeid and Tuck (1982), Look and Blakemore (1984), and Grossmann and Ledebo (1988).

Other growth methods include the *Bridgman* technique, which uses a horizontal semicircular boat with a seed in contact with the polycrystalline material, or charge. Vertical Bridgman growth is similar but produces crystals fully enclosed by the crucible. Both techniques use a well-controlled temperature gradient to melt and slowly recrystallize the material. Because of this, another name for this growth process is the "gradient-freeze" technique. Bridgman growth is a preferred method for CdZnTe crystals used in radiation detectors (Szeles et al., 1996; Eisen and Shor, 1998).

4.2 DOPANT INCORPORATION DURING BULK CRYSTAL GROWTH

The CZ and FZ crystal growth techniques allow fairly uniform doping of bulk single crystals. Intentional doping of CZ crystals can be achieved by adding the dopant element in pure form to the melt. This doping method is often difficult to control due to the small quantities involved. To improve control, one can add a piece of heavily doped semiconductor, sometimes called the "master alloy," to the melt.

The upper limit of doping levels is normally given by the *solid solubility* of a particular impurity. Most impurities exhibit a retrograde solubility near the melting point (i.e., the solubility decreases sharply as the temperature approaches the melting point of the semiconductor) (Figure 4.4). At the phase boundary between solid and liquid, we observe an impurity concentration step described by the *segregation (or distribution) coefficient* k_0:

(4.1)
$$k_0 = \frac{c_s}{c_l}$$

where c_s and c_l are the impurity concentrations in the solid and liquid, respectively.

In certain cases, one needs to dope to the highest possible concentrations in order to obtain low electrical impedances. The solid solubility (Figure 4.4) is a key parameter for these situations. It corresponds to the maximum equilibrium dopant atom concentration that can be introduced at a certain temperature. Introduction of impurities at a high temperature and subsequent cooling leads to dopant supersaturation. If the impurities are immobile, the case for most impurities

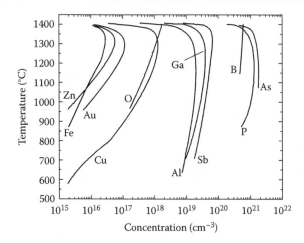

FIGURE 4.4 Equilibrium solubility of impurities in silicon. (After Trumbore, F.A. 1960. *Bell Syst. Tech. J.* 39: 205–234.)

at room temperature, the supersaturation can be "frozen" into the crystal. If the impurities are mobile, then precipitation will occur—usually an undesirable process.

Impurity *striations*, variations of the net-impurity concentration along the axis and the crystal radius, are an undesirable effect observed in almost all crystals grown from a melt (Carruthers and Witt, 1975). The impurity concentration fluctuations exhibit up to three periodicities. The first one is due to the crystal rotation in a group of isotherms that are not perfectly cylindrical. Once every revolution, each point at the solid/liquid interface passes through the coldest and the hottest spots. The crystal growth rate, and consequently the effective segregation coefficient, are modulated by this regular change, leading to regular impurity concentration fluctuations. A second effect is due to the fact that the incorporation of impurities changes the melting point. This can lead to oscillations of the growth rate, to a variation of the stagnation layer width at the solid/liquid interface, and to oscillations in the effective segregation coefficient. A third cause for impurity striations is the formation of convection cells in the melt, similar to magma convection in the Earth. The melt flow in these cells stirs the melt in distinct patterns that get "imaged" via the effective segregation coefficient directly into the local impurity concentration.

In heavily doped crystals, the dopant striations are made visible by cutting CZ boules along the growth axis all the way from the seed to the tail, lapping and polish etching the cut surface to remove saw damage, and then chemically attacking this surface with an etchant sensitive to the local doping concentration. The result of such a procedure is shown as black and white striations (Figure 4.5).

Along with the controlled introduction of dopants, it is important to keep out unwanted contaminants. *Purification* is essential for high-quality semiconductor materials. The first purification step is achieved by *chemical purification* of a substance containing the semiconductor element. Chemical purification can be performed at low temperatures such that contamination from containers is reduced to arbitrarily low levels. The processing can be done continuously and with very large quantities. The foremost purification method is distillation in multistage columns of $SiCl_4$, $GeCl_4$, $GaCl_3$, and so forth.

Impurity segregation between a liquid and a solid phase in contact is the principle used for *physical purification*. First, we consider the one-dimensional case of *normal freezing*. A long, constant cross-sectional area liquid begins to solidify at one end. Disregarding diffusion of impurities in the solid and assuming perfect mixing in the liquid, we derive an expression for the impurity concentration along the frozen material.

FIGURE 4.5 Impurity striations in Te-doped InSb, as revealed by chemical etching of a surface cut along the longitudinal growth axis. Picture taken with an optical microscope, magnification ~100. (Reprinted with permission from Witt, A.F. and H.C. Gatos. 1966. *J. Electrochem. Soc.* 113: 808–813. Copyright 1966, the Electrochemical Society.)

From Equation 4.1, the impurity concentrations at the solid–liquid interface are related by

$$c_s = k_0 c_l \tag{4.2}$$

Let I = the number of impurities in the melt, I_0 = the total number of impurities, V = the volume of the solid, V_0 = the total volume, and $c_0 = I_0/V_0$. The fraction that is frozen is denoted

$$x = V/V_0 \tag{4.3}$$

Consider what happens when we freeze an infinitesimal quantity dV. Impurities are removed from the melt and incorporated in the solid. The change in the number of impurities remaining in the melt is given by

$$dI = -k_0 c_l dV = -k_0 \frac{I}{V_0 - V} dV \tag{4.4}$$

Dividing through by I yields

$$\frac{dI}{I} = -k_0 \frac{dV}{V_0 - V} \tag{4.5}$$

We integrate both sides from 0 to x:

$$\int_{I_0}^{I(x)} \frac{dI}{I} = \int_0^{V(x)} -k_0 \frac{dV}{V_0 - V} \tag{4.6}$$

to obtain

(4.7) $$\ln\left(\frac{I(x)}{I_0}\right) = k_0 \ln\left(1 - \frac{V(x)}{V_0}\right) = \ln\left[(1-x)^{k_0}\right]$$

Therefore,

(4.8) $$I(x) = I_0(1-x)^{k_0}$$

To get an expression for $c_s(x)$, we use the fact that

(4.9) $$c_s(x) = k_0 c_l(x) = k_0 \left(\frac{I(x)}{V_0 - V(x)}\right)$$

Substituting Equation 4.8 for $I(x)$ leads to

(4.10) $$c_s(x) = k_0 c_0 (1-x)^{k_0 - 1}$$

Figure 4.6 shows the impurity concentration as a function of fraction solidified for several values of k_0 ranging from 10^{-2} to 5. Normal freezing is a nonrepetitive process and is not well-suited

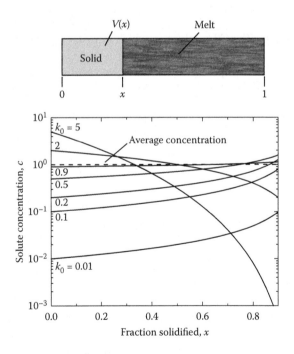

FIGURE 4.6 Curves for normal freezing, showing solute concentration in the solid versus fraction solidified, for $c_0 = 1$. (After Pfann, W.G. 1966. *Zone Melting*, 2nd ed., p. 12. New York, NY: John Wiley & Sons.)

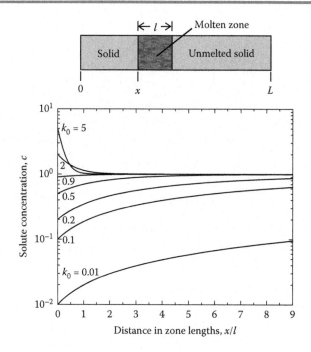

FIGURE 4.7 Curves for single-pass zone melting, showing solute concentration in the solid versus distance in zone lengths from beginning of charge, for $c_0 = 1$. (After Pfann, W.G. 1966. *Zone Melting*, 2nd ed., p. 12. New York, NY: John Wiley & Sons.)

for automatic purification. The CZ method, in which a crystal is grown from a depletable melt, is a kind of normal freezing.

The second type of physical purification is *zone purification* or *zone melting*. Instead of the whole quantity of material, only a short section is melted by a resistive or an RF heater coil (Figure 4.7). Let l = the length of the molten zone and L = the length of the bar. After a single zone pass, the impurity concentration is given by

$$c_s(x) = c_0 \left[1 - (1-k_0)\exp\left(\frac{-k_0 x}{l}\right) \right] \quad (4.11)$$

where x ranges from 0 to $L - l$ (here, unlike in Equation 4.10, x has units of length). For $x > L - l$, the normal freezing equation applies. Figure 4.7 shows a number of curves obtained from Equation 4.11 for various values of k_0. Zone purification lends itself to a repetitive, automatic purification process. The same substance can be "zoned" many times. For a given ratio of l/L, we obtain an ultimate impurity distribution after an infinite number of passes.

4.3 THIN FILM GROWTH

Epitaxy, the growth of a crystalline film on a crystalline substrate with a defined orientational relationship (Pashley, 1990), was first practiced and studied with alkali halides. A serious interest in silicon epitaxy arose when lightly doped layers on heavily doped substrates were required for the fabrication of junction field effect transistors. Growth of pure silicon on doped silicon is a form of *homoepitaxy*. Today there is great interest in multilayer growth of different semiconductors on top of each other. Thin-film growth with layers of different compositions is a form of *heteroepitaxy*.

Heteroepitaxy is used to form *quantum wells*, thin layers of one type of semiconductor sandwiched between another type of semiconductor with a higher band gap. A GaAs layer of a few tens of Angstroms thickness, located between two $Al_xGa_{1-x}As$ barriers, forms a quantum well. LEDs and laser diodes, discussed in Section 3.8, use multiple quantum wells to emit light. Quantum wells also give rise to unusual phenomena such as the integer (von Klitzing, 1980) and fractional (Tsui et al., 1982; Laughlin, 1983) quantum Hall effect. Resonant tunneling diodes use one quantum well for ultrafast switching purposes. Such diodes can turn on and off more than 6×10^{11} times per second (>600 GHz). *Superlattices* consist of a series of quantum wells grown on top of each other. Structures with several hundred layers have been grown successfully. Besides exhibiting interesting solid-state physics, such superlattices are used in electronic devices.

Every kind of crystal growth requires a way to make atoms mobile. Epitaxy makes use of essentially three techniques:

- *Liquid phase epitaxy* (LPE) uses a solvent to generate mobility of the growth substance (the solute).
- *Vapor phase epitaxy* (VPE), a form of chemical vapor deposition (CVD), uses a vapor of molecules containing the growth atoms. Metalorganic chemical vapor deposition (MOCVD), also called organometallic vapor phase epitaxy (OMVPE), belongs to the VPE family and is used for growth of III-V and II-VI compound semiconductor films.
- *Molecular beam epitaxy* (MBE) is a form of ultra-high vacuum evaporation in which the individual molecules fly along straight trajectories from a source to the substrate without interacting along the way. A subset of MBE is chemical beam epitaxy (CBE), also called metalorganic molecular beam epitaxy (MOMBE).

Each of the epitaxial growth techniques has advantages and disadvantages, discussed in the following sections. In all the techniques, dopant atoms can be introduced during growth, along with the atoms that make up the semiconductor.

Before moving on to discuss these techniques, we briefly mention two methods that are especially useful for high-throughput growth of oxide thin films. *Sputtering*, discussed briefly in Section 4.7, uses a plasma to dislodge atoms from a target. The atoms travel to the substrate, where they form a thin film. Sputtering is commonly used to deposit ZnO or indium tin oxide (ITO) for use as transparent contacts. A second method is *pulsed laser deposition* (PLD), in which an intense laser pulse impinges upon a target, producing a plume of atoms that are deposited onto the substrate. Sputtering and PLD are relatively inexpensive and do not require an ultra-high vacuum. The application of these techniques to the growth of transparent conducting oxides has been reviewed by Minami (2005). In general, sputtering has a higher throughput than PLD, while PLD preserves the stoichiometry of the source material. PLD can also be used to grow thin films, one atomic layer at a time.

4.4 LIQUID PHASE EPITAXY

LPE is a low-temperature process and can produce very pure epitaxial layers (Kuphal, 1991). The liquid solvent for semiconductor LPE is typically a metal. Gallium, bismuth, indium, tin, lead, and other metals have been used for silicon, germanium, and III-V compound semiconductor epitaxy. Since the 1980s, the importance of LPE has diminished because the interfaces between adjacent

4.4 Liquid Phase Epitaxy

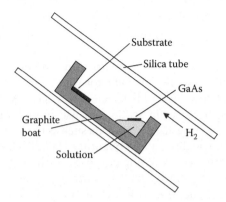

FIGURE 4.8 Tipping liquid phase epitaxy (LPE) system. (After Nelson, H. 1963. *RCA Review* 24: 603–615.)

layers are not atomically sharp, and the composition or the dopant concentrations cannot be conveniently varied in a continuous fashion. Nonetheless, LPE is a practical and inexpensive epitaxial growth technique for devices such as LEDs (Capper and Mauk, 2007).

The basic growth sequence starts with the saturation of the solvent with the materials to be deposited. A clean, flat substrate is brought in contact with the solvent. Upon slow cooling of the solvent, solute material will grow epitaxially onto the substrate. Figure 4.8 shows a simple growth apparatus (Nelson, 1963). A piece of GaAs floats on gallium, the solvent. The liquid gallium becomes saturated with GaAs at a given temperature. Tilting the apparatus so that the gallium flows over a GaAs single crystal substrate, followed by slowly lowering the temperature, starts the epitaxial growth. Layers of up to 100 μm thickness were grown using this apparatus. For layered structures, multisolvent epitaxy systems are used. A substrate wafer is held in a graphite slider. With a simple push rod, the wafer can be positioned under four different solvent containers, each of which holds a different combination of semiconductor alloy and dopant.

The crystal orientation has a strong influence on the growth mechanisms and epilayer morphology (Bauser and Strunk, 1984). LPE of silicon on a spherically ground and polished single crystal substrate results in at least four kinds of growth mechanisms (Figure 4.9). A central atomically flat facet of up to 1 cm in diameter develops perpendicular to the [111] orientation. This area is surrounded by a terraced surface. Once the surface normal deviates more than 2° from the [111] direction, the surface morphology again becomes very smooth.

Advantages of LPE are high purity and low equipment and operating costs. Disadvantages include the following: composition and doping of the films cannot be changed continuously;

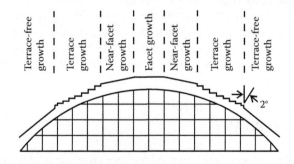

FIGURE 4.9 Cross section of a spherical substrate with a liquid phase epitaxy (LPE)–grown layer. The misorientation angle is exaggerated. (After Bauser, E. and H.P. Strunk. 1984. *J. Crystal Growth* 69: 561–580.)

selective growth through windows in masks is difficult, especially for small dimensions; and there is low throughput.

4.5 CHEMICAL VAPOR DEPOSITION

CVD is used extensively by the semiconductor industry to grow thin films of crystalline and polycrystalline silicon and silicides, compound semiconductors, metals, and insulators. We will focus on VPE, where gas molecules are used to deposit atoms onto a substrate.

A large number of studies have explored the thermodynamics and kinetics of VPE. These studies have led to a differentiation of the overall process into the following steps (Figure 4.10):

a. Reactant gas molecules AB are transported by a carrier gas (e.g., H_2) to the deposition region.
b. The reactants diffuse through the near-surface gas stagnant layer to the surface.
c. The reactants are adsorbed on the surface.
d. The molecules dissociate, and atom A goes to the proper site via surface diffusion.
e. The reaction products B are desorbed from the surface.
f. The reaction products out-diffuse to the gas stream.
g. The reaction products are removed.

The slowest step dictates the type of growth regime and the growth rate. Typically, we distinguish between mass transport limited processes (a, b, f, and g) and surface reaction limited processes (c through e).

Silicon VPE is performed with $SiCl_4$, SiH_4, or one of the chlorosilanes (SiH_xCl_{4-x}). A small concentration of these chemicals is transported in a H_2 gas stream to heated substrate wafers. For a liquid such as $SiCl_4$, a "bubbler" is used, in which the H_2 gas bubbles through the liquid and releases molecules of $SiCl_4$. The wafers are located inside a silica tube in a furnace. A surface reaction takes place, leaving silicon on the substrate while the other chemical constituents return to the gas stream.

Figure 4.11 shows a VPE apparatus. The overall chemical reaction for $SiCl_4$ proceeds as follows in the reactor chamber:

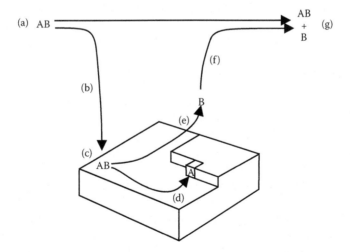

FIGURE 4.10 Sequence of events in vapor phase epitaxy (VPE) growth. (After Shaw D.W. 1975. *Crystal Growth and Characterization*, eds. R. Ueda and J.B. Mullin, p. 208. Amsterdam: North-Holland.)

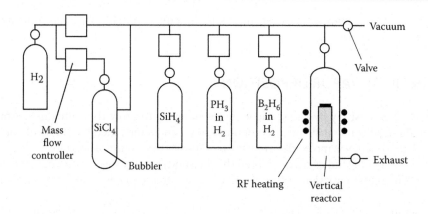

FIGURE 4.11 Vapor phase epitaxy (VPE) reactor. (After Springer Science+Business Media: *Halbleiter Technologie*, 1975, 121, Ruge, I, Figure 4.11.)

(4.12) $$SiCl_4 + 2H_2 \rightarrow Si + 4HCl$$

Substrates are heated to a temperature between 1150 and 1250°C. Deposition rates range from 0.5 to 1.5 µm/min. Doping is achieved by adding PH_3 or B_2H_6, stored in diluted form in H_2 gas cylinders. Because HCl is generated as a by-product of $SiCl_4$ epitaxy, there is a danger of local etching of silicon wafers. Etching of a heavily doped substrate leads to volatile dopant-containing molecules that inadvertently dope epilayers downstream in the reactor. This effect is called *autodoping*.

In the case of GaAs VPE, the following reactions produce arsenic, hydrogen, and gallium chloride:

$$2AsH_3 \rightarrow \frac{1}{2}As_4 + 3H_2$$

(4.13) $$2HCl + 2Ga \rightarrow 2GaCl + H_2$$

The epitaxial growth then takes place on the substrate, according to

(4.14) $$2GaCl + \frac{1}{2}As_4 + H_2 \rightarrow 2GaAs + 2HCl$$

One drawback to such a system is that it is rather difficult to control the generation rate of GaCl from Ga metal in a silica boat inside the reaction chamber.

A strong development effort resulted in an all-gas technique, the MOCVD growth method. Typical metalorganic molecules are trimethyl gallium (TMGa), $(CH_3)_3Ga$, trimethyl aluminum (TMAl), $(CH_3)_3Al$, and triethyl gallium (TEGa), $(C_2H_5)_3Ga$. An example of a dopant is dimethyl zinc (DMZn), $(CH_3)_2Zn$. These molecules are commonly called metalorganic *precursors*.

MOCVD has become the workhorse for multilayer III-V structures that do not need the ultimate atomic flatness provided by MBE. As an all-gas technique, MOCVD offers superb control of composition and doping. Nearly atomically sharp interfaces between layers of different composition and doping are achieved by using constant laminar gas flows in the reactor. When the flow of one gas component is abruptly increased or decreased, the ballast H_2 gas flow is changed accordingly, to maintain constant flow. The substrate temperature is held constant to within a few degrees. A typical reaction describing the formation of GaAs layers can be written as follows:

(4.15) $$(CH_3)_3 Ga + AsH_3 \rightarrow GaAs + 3CH_4$$

The TMGa and arsine (AsH_3) can be precisely metered with mass flow controllers. For GaN growth, typical precursors are TMGa and ammonia (NH_3) (Nakamura et al., 1994).

A variation on CVD that provides atomic flatness is *atomic layer deposition* (ALD) (Puurunen, 2005). ALD is especially useful for coatings and development of new dielectric materials. In ALD growth, the precursors are introduced one at a time. The first precursor deposits atom A until the surface is covered with one monolayer. Then, the chamber is purged and the second precursor flows in, covering the surface with one monolayer of atom B. By repeating the reaction cycle, an arbitrary number of monolayers may be grown. The chemical reactions must be self-terminating such that A will not grow on A and B will not grow on B. A similar approach is also used in MBE, discussed next.

4.6 MOLECULAR BEAM EPITAXY

MBE is conceptually the simplest but, paradoxically, the most expensive epitaxial growth technique. It is the premiere epitaxial growth method for atomic layer growth control. An MBE system consists of one or more large stainless steel, ultra-high-vacuum chambers, which represent a major part of the total cost of ~$1 million. Typical pressures are in the range of 10^{-10} to 10^{-12} Torr. A schematic of an MBE growth chamber is shown in Figure 4.12. A number of effusion ovens (Knudsen cells) provide beams of molecules or atoms by heating the various source substances to temperatures at which the required vapor pressures are reached. The molecular and atomic beams travel, collisionless, through the ultra-high-vacuum chamber to the substrate. Surface reactions and migration, which are necessary for epitaxial growth, are promoted through substrate heating. Each of the effusion ovens, or sources, is shrouded with liquid-nitrogen-cooled baffles in order to keep the inside surface of the MBE system as cold as possible to maintain the required vacuum. In front of each source sits a computer-controlled shutter that is used to turn the source "on" or "off."

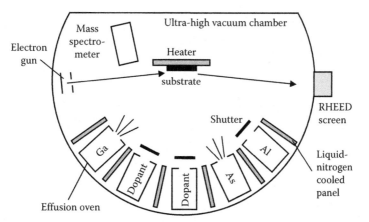

FIGURE 4.12 Typical molecular beam epitaxy (MBE) chamber for the growth of pure and doped GaAs and AlGaAs layers. (After Foxon, C.T. and B.A. Joyce. 1990. *Growth and Characterisation of Semiconductors*, eds. R.A. Stradling and P.C. Klipstein, pp. 35–64. New York, NY: Adam Hilger.)

Two or more analytical techniques are used in situ to monitor and control MBE growth. A monoenergetic electron beam is directed at the growing surface at a very shallow angle. The diffracted beam generates a visible interference light pattern on a fluorescent screen. This technique, *reflection high-energy electron diffraction* (RHEED), provides information that makes possible the growth of submonolayers (Neave et al., 1983). When the surface is atomically flat, the RHEED signal is maximized, while partial coverage leads to a minimum. This results in "RHEED oscillations" that are monitored to produce an extremely accurate thickness. As the film gets thicker, the surface tends to roughen and the amplitude of these oscillations diminishes (Figure 4.13). In addition to RHEED, a mass spectrometer continuously monitors the quality of the vacuum and the chemical composition of the various beams.

The MBE growth chamber is often separated from a second chamber by an ultra-high-vacuum valve. In research MBE machines, this second chamber is used for detailed studies of the epitaxial film. A number of analytical tools for thin film and surface studies are available in this chamber. A third chamber may be used as a substrate loading lock. The substrate can be moved between the various chambers with a system of magnetically coupled sliding rods.

A large variety of semiconductor epilayers have been grown by MBE. We choose to discuss growth of GaAs, AlAs, and their alloys because of the nearly perfect lattice match of the two compounds and because of their scientific and technical importance. Growth of these ternary alloys illustrates most of the critical aspects of MBE epitaxial growth, which are also encountered with other materials systems. Naturally, each combination of substrate and epilayer materials has its own set of complexities.

GaAs is grown with a Ga and an As_2 beam. As_2 can be formed by breaking up As_4 molecules by thermally (cracking) or with an energetic electron beam moving across the As_4 beam. When the As_2 molecules arrive at a Ga-covered GaAs surface, practically all of the As_2 molecules stick to this surface, dissociate, and form GaAs. The *sticking coefficient*, defined as the ratio of the number of As_2 molecules adsorbed on the surface to the number of As_2 molecules arriving at the surface, is nearly unity. As the surface gradually becomes As-terminated, the sticking coefficient goes to zero.

The difference in sticking coefficient for As arriving at a Ga- or As-terminated surface can be explained with basic vapor pressure arguments. A typical substrate temperature of 550 to 650°C is high enough for the additional As to re-evaporate but it is far too low to lead to GaAs dissociation. The Ga flux dominates the growth process, as long as the As flux is kept larger than the Ga flux. In practical GaAs epitaxial growth, the As_2 flux is adjusted to be at least twice the

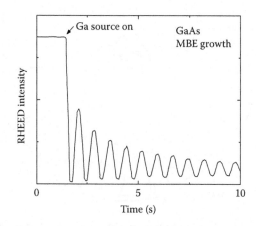

FIGURE 4.13 RHEED oscillations for MBE growth of GaAs. (After Neave, J.H. et al. 1983. *Appl. Phys. A* 31: 1–8.)

Chapter 4—Crystal Growth and Doping   **81**

FIGURE 4.14   Transmission electron micrograph of a GaAs (three monolayer)/AlAs (three monolayer) superlattice. (Grown by D. Hilton. Transmission electron micrograph, courtesy of J.P. Gowers.)

intensity of the Ga flux. This automatic stoichiometry stabilization has been of enormous help in the growth of GaAs epilayers.

Incorporation of dopants during MBE growth is generally straightforward. For each dopant, a separate source must be installed. The combination of source shutter control and vapor pressure for each dopant offers, in principle, great flexibility in dopant concentration profiles in the epilayers. Because the temperature of a large dopant source cannot be changed rapidly, one typically cannot use vapor pressure changes to control the dopant profile at the atomic-layer level. If such rapid changes are required, growth of the host material is stopped at every layer and the desired dopant atom quantity is added to each layer. An extreme case of this method is δ-doping, in which one heavily doped atomic layer is formed.

A structure that illustrates the capabilities of MBE growth is shown in the transmission electron micrograph of a superlattice consisting of a sequence of three monolayers each of GaAs and AlAs (Figure 4.14). Growth starts out at the bottom, where the layers are not yet perfectly flat. As growth proceeds over more than 100 individual layers, the interfaces become flatter and the quality of the superlattice improves. Using MBE, atomically abrupt interfaces can be grown routinely.

4.7 ALLOYING

The remaining sections in this chapter discuss ways to dope semiconductors after they have been grown. Alloying, a venerable method in materials science, has been used for the fabrication of "Ohmic" contacts on semiconductors and for fabricating early junction transistors. Soldering a wire to a piece of semiconductor single crystal can lead to an alloyed contact (though many experimentalists do not realize this).

The formation of an alloyed contact can be explained using the aluminum-silicon phase diagram as an example (Figure 4.15). First, a thin (few μm) layer of aluminum is evaporated onto a slice of single crystal silicon. Heating under an inert or reducing gas atmosphere leads to the formation of

82 4.7 Alloying

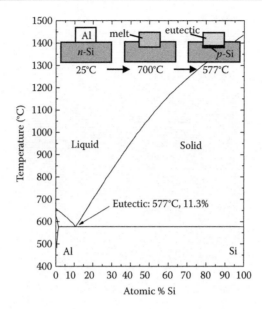

FIGURE 4.15 Aluminum-silicon phase diagram. (After Hansen, M. 1958. *Constitution of Binary Alloys.* New York, NY: McGraw-Hill.)

an aluminum-silicon eutectic at $T = 577°$ C. The eutectic composition corresponds to 89 atomic % aluminum and 11 atomic % silicon. Upon further heating, more silicon is dissolved because the amount of aluminum is limited. At 700°C, the silicon content has risen to 20%.

At 700°C, the solubility of aluminum in silicon is 9×10^{18} cm^{-3} (Figure 4.4). Cooling slowly from this temperature leads to epitaxial regrowth of silicon on the silicon substrate, with incorporation (doping) of aluminum acceptors in the epilayer. The heavily doped epilayer acts as a good Ohmic contact on *p*-type silicon but forms a *p-n* junction on *n*-type silicon. The described sequence is shown in Figure 4.15 for an *n*-type silicon wafer. The first widely used junction silicon and germanium diodes and transistors were fabricated using indium-alloyed emitter and collector regions.

If one wants to separate the metallurgical properties from the doping, one can choose an appropriate metal that does not become incorporated in the epitaxial regrowth layer. By adding effective dopants to this metal, it is possible to control both the metallurgical and doping aspects of the process. One example is gold with 1% gallium or 1% arsenic, which is used for alloyed contacts on germanium devices. The gold-germanium eutectic forms at 356°C, and additional germanium dissolves upon further heating. During cooling, germanium crystallizes epitaxially on the bulk germanium. The solubility of gold in germanium at low temperatures is negligible. However, the epilayer incorporates either gallium or arsenic atoms, leading to p^+ or n^+ doping, respectively. The gold-silicon eutectic temperature is also very low (370°C), allowing low-temperature contact formation on silicon.

N-type contacts on GaAs are often formed by evaporating a film of gold containing a few percent germanium on a surface with a very thin (~200 Å) nickel "wetting" layer. Upon heating for two minutes to ~450°C, the various layers and the GaAs interact. It is generally believed that gallium moves out of the GaAs into the gold layer, while germanium replaces the gallium atoms, forming shallow donors.

Various methods have been developed to deposit thin metal films onto semiconductors (Mahajan and Harsha, 1999). *Evaporation*, or physical vapor deposition, involves heating a metal in a bell jar at a high vacuum (~10^{-7} Torr). The evaporated metal atoms coat the surface of the semiconductor substrate, as well as the inside of the bell jar and other surfaces. A quartz gauge is used to monitor the thickness of the deposited layer. A second technique is sputtering, in which atoms from a metal "target" are ejected by high-energy ions (e.g., Ar⁺) in a partial vacuum (typically 1–100 mTorr). The ejected metal atoms travel to the substrate and form a film (Ohring, 1992). When depositing metals, a negative DC voltage is applied to the target, while the substrate is affixed to a grounded anode. A sufficiently high voltage will create a plasma of argon ions and electrons. The positive argon ions are attracted to the target, where they dislodge, or sputter, the metal atoms. If a magnetic field is applied, electrons are bent into helical paths, enabling them to ionize a large number of argon atoms. This technique, called *magnetron sputtering*, is often preferred because it yields high deposition rates.

4.8 DOPING BY DIFFUSION

A common post-growth doping method is the *diffusion* of dopant atoms into the semiconductor at an elevated temperature (Runyan, 1965). This section summarizes several practical aspects of diffusion. The details of diffusion processes and modeling are discussed in Chapter 8.

In order to obtain predictable, well-defined diffusion profiles in semiconductor wafers, careful consideration must be given to the following points:

- Doping homogeneity requires constant temperatures to within less than 0.5°C and perfect initial distribution of the dopant.
- Cleanliness is required to avoid contamination with fast diffusers.
- Extended defects such as grain boundaries and dislocations should be minimized.
- Ease of control of the diffusion parameters, and reproducibility, are important.

In general, diffusion is performed inside a tube furnace (Figure 4.16). Appropriate design of the diffusion furnace and tube, and method of dopant introduction, have been developed to the point where the above conditions can be fulfilled. Multizone furnaces, electrically heated and with electronically controllable temperature/time profiles, are typically used for diffusion. Silica or polysilicon tubes (for MOS work) contain the wafers, which may have SiO_2 masks that block dopants from specific areas. One end of the tube serves as the gas inlet, and the other end serves as the outlet and loading/unloading opening. The latter area is usually surrounded by a laminar flow box so that very low particulate-contamination conditions can be maintained. Gas manifolds are used for solid, liquid, and gaseous dopant sources.

The choice of dopants depends on the diffusion constant, the maximum solubility, and the effectiveness of SiO_2 masks. Dopants may be used in their elemental form, but more often they are in chemical compounds (e.g., oxides, halides, or hydrogen compounds). Most oxides have sufficiently high vapor pressures to be used as vapor-phase dopants. The dopant oxide is reduced

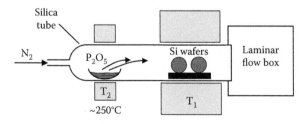

FIGURE 4.16 Diffusion furnace design. The phosphorus source P_2O_5 is kept at its own temperature T_2, and the source vapor is transported to the wafers with N_2 carrier gas.

directly on the semiconductor surface. For boron and phosphorus doping of silicon, typical over-all reactions are

$$2B_2O_3 + 3Si \rightarrow 4B + 3SiO_2$$

(4.16)

$$2P_2O_5 + 5Si \rightarrow 4P + 5SiO_2$$

Halides and hydrogen compounds are converted to oxides by the addition of oxygen gas or water vapor. Typical overall reactions are

$$4POCl_3 + 3O_2 \rightarrow 2P_2O_5 + 6Cl_2$$

(4.17)

$$2PH_3 + 4O_2 \rightarrow P_2O_5 + 3H_2O$$

$$4BBr_3 + 3O_2 \rightarrow 2B_2O_3 + 6Br_2$$

$$B_2H_6 + 3O_2 \rightarrow B_2O_3 + 3H_2O$$

The oxide is then reduced by the silicon as described in Equation 4.16. Table 4.1 lists some of the common dopant sources and the temperatures required to obtain sufficiently high vapor pressures.

In many practical cases, a *predeposition* step is followed by a *drive-in* step. The predeposition step introduces the dopant and oxide onto the semiconductor surface, according to Equation 4.16. The phosphosilicate or borosilicate glass that is formed during predeposition improves the uniformity of the distribution of dopant atoms on the surface. The temperature and time are chosen so that the dopant (e.g., boron or phosphorus) diffuses into a near-surface region. The surface concentration is given by the solubility limit at the predeposition temperature. During the drive-in step, the atoms are diffused into the semiconductor at elevated temperatures.

The following advantages can be realized with two-step diffusion:

- Choice of a particular near-surface impurity concentration that is typically lower than the solid solubility at the drive-in temperature can be made.
- The doping gradient at a *p-n* junction at a given depth can be controlled.
- The oxide acts as a mask, which prevents out-diffusion during the following diffusion step.

TABLE 4.1 Dopant Sources and the Temperatures Needed to Provide Sufficient Vapor Pressure for Doping

	T (°C)
SOLIDS	
B_2O_3	600–1200
P (elemental red phosphorus)	200–300
P_2O_5	200–300
LIQUIDS	
$P(CH_3O)_3$	10–30
BBr_3	0–30
$POCl_3$	2–40
PCl_3	170
GASES	
BCl_3	15
B_2H_6	Room temperature
PH_3	Room temperature

If a predeposition/deglaze/drive-in cycle is used, then the oxide is removed during the deglaze step.

The *spin-on* process is a simple method for the application of dopants. A drop of an organic silicon compound containing dopant elements dissolved in ethanol is placed on the center of a silicon wafer. The wafer is spun to ~1000 rpm, which leads to a uniform film on the surface. After a drying process, the wafers are heated to the desired diffusion temperature. Control of the various parameters is not as good as in the processes described above. However, because there is no need for high-purity gas manifolds, the spin-on method is fast and economical.

4.9 ION IMPLANTATION

Ion implantation is a direct spin-off from nuclear science. It has become a preferred semiconductor doping technology because it gives the operator great freedom in the choice of the dopant species and depth profile (Mayer et al., 1970; Ryssel and Ruge, 1986). In contrast to diffusion, ion implantation is a process far from equilibrium. Ions are accelerated to high energies and injected into the surface of a semiconductor. The ions are stopped by various interactions, some of which lead to the displacement of host material atoms. For most applications, this damage must be repaired by thermal annealing in order to restore the properties of the crystalline semiconductor.

An ion implantation machine is a compact nuclear accelerator taking up the space of a medium size laboratory. Figure 4.17 gives a schematic top view of a late 1970s/early 1980s Varian-Extrion implanter. The ions are generated in a gas-discharge ion source. The appropriate gases are stored in lecture bottles in the ventilated gas box. The preaccelerated ion beam (~25 keV) travels in a stainless steel high-vacuum pipe into the analyzer magnet. Depending on the ions to be implanted, the magnetic field is set to a value that bends these ions through a total angle of 90°. Heavier or lighter ions are bent at different angles and hit the vacuum chamber wall. The resolving aperture allows precise mass selection and the following variable slit controls the beam intensity. The ions that pass through the variable slit travel through the acceleration tube. The voltage across the acceleration tube can be varied between 0 and 175 kV.

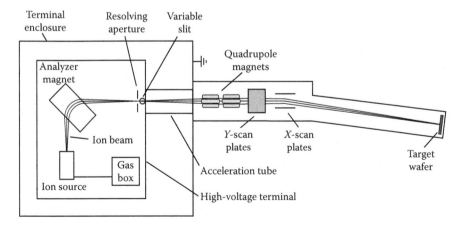

FIGURE 4.17 Extrion medium-current implanter; maximum energy 200 keV, maximum current 1.5 mA.

After the electrostatic acceleration, the ion beam is refocused by a magnetic quadrupole lens. In order to obtain an even implantation density across the full wafer, the ion beam is deflected electrostatically by a pair of X and Y electrodes (plates). Similar to a TV picture, the ion beam raster covers the semiconductor wafer. The ion implant dose is measured directly by integrating the current flowing into the wafer, as long as care is taken to consider the secondary electrons created by the energetic ion beam hitting the semiconductor wafer. These electrons are typically bent sideways by small permanent magnets and collected with a small positive potential on a ring electrode surrounding the wafer. Modern end stations contain silicon wafer carousels that rapidly move wafers into position for implantation while maintaining high vacuum.

Three aspects of ion implanters are crucial for semiconductor doping. First, the *ion source* should provide a high-intensity, pure ion beam and should require minimal maintenance. A typical ion source for gases is shown in Figure 4.18. An RF coil is wrapped around a silica tube containing a dopant gas (e.g., BF_3, B_2H_6, AsH_3, PH_3). The RF field ionizes and partially cracks the gas molecules, forming singly and multiply charged ionic species. The density of the plasma can be increased by a longitudinal magnetic field that keeps the free electrons on small-radius spirals, increasing the chance for ionizing collisions. The ions are extracted by a negative DC bias applied to the electrode.

Second, *ion beam apertures* are critical components of an implanter, especially those downstream where the ions have their full energy. Sputtering of the aperture material can lead to serious iron and chromium contamination of the semiconductor surface. Obviously, stainless steel is a poor choice for apertures. Carbon or silicon apertures, in contrast, suppress contamination.

The third point concerns the *implant homogeneity*. Consider Lissajous figures that are formed on an oscilloscope screen when the X and Y deflection are run by sine-wave voltages. The Lissajous figures become stationary when the ratio of the two sine-wave frequencies is simple (e.g., 1:2). When this happens with the X and Y plates in an implanter, we call it a "scan lock-up." The result is a standing-wave ion pattern, shown in Figure 4.19 as a grid of fluctuating ion dose. It is essential that the X- and Y-frequencies never form a simple ratio. When that requirement is satisfied, a well-tuned implanter can provide an ion dose homogeneity of $\pm 0.5\%$.

The detailed interactions between the implanted ions and the host material will be discussed in Chapter 11. The following basic concepts give an overview of important terms in the field of ion implantation science and technology (Carter and Grant, 1976):

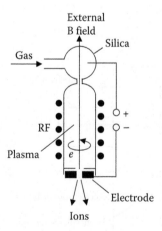

FIGURE 4.18 Example of an ion source. It is mostly useful for gases, because other materials can coat the silica envelope and degrade its performance. (After Stephens, K.G. 1984. *Ion Implantation Science and Technology*, ed. J.F. Ziegler, p. 414. New York, NY: Academic Press.)

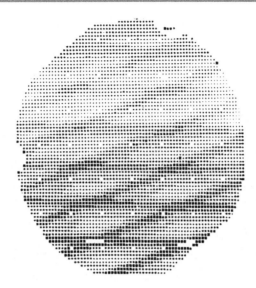

FIGURE 4.19 Doping variations from C-V measurements on a 3-inch (7.6 cm) wafer. The standing wave pattern resulted from a scan lock-up condition with an electrostatic-scan system. The doping variation average was 13% over the entire wafer. (Reprinted with permission from Current, M.I. and M.J. Markert. 1984. *Ion Implantation Science and Technology*, ed. J.F. Ziegler, p. 505. New York, NY: Academic Press. Copyright 1984, Elsevier.)

- *Binary collisions*: Although ion implantation involves a beam of many ions interacting with a solid containing many atoms, the collision between one ion and one target atom is of basic importance. In many circumstances, the problem of an ion entering a solid can be treated as a succession of binary collisions in which the ion interacts or collides with only one target atom at a time.
- *Interatomic potential*: The collision between atoms depends intrinsically on the interatomic potential $V(R)$. The forces acting on both particles, and hence their trajectories, are derived from $V(R)$.
- *Energy loss*: As an ion penetrates a solid, it loses its energy in a series of collisions with the target atoms until it eventually comes to rest. The amount of energy lost in each collision will determine the total path length, or range, of the ion. Scattering of the incoming particle occurs by either elastic or inelastic processes. In the former, kinetic energy is conserved; in the latter, it is converted into another form such as excitation of atomic electrons.
- *Ion range*: The range of an ion is determined by the rate at which it loses energy. Having described and evaluated energy loss processes, we can calculate ion ranges and predict the spatial distribution of implanted impurities within surface layers of targets. In noncrystalline or amorphous materials, energetic implanted ions have a Gaussian-shaped concentration profile. In single crystals, the picture becomes complicated because the crystal structure assumes a role in determining the range profile.
- *Channeling*: In a single crystal, the lattice atoms are arranged periodically in space. In certain directions, the structure has open channels bounded by densely packed walls. Ions entering these channels do not make a random sequence of collisions, as in an amorphous material, but rather are steered by a succession of correlated interactions with atoms in the channel walls.
- *Damage*: As ions slow down, their energy is transferred to the target atoms. An atom receiving sufficient kinetic energy will be displaced from its lattice site, and the target

will suffer "radiation damage." Because the energy required to displace an atom is typically ~25 eV, ions with energies of a few keV cause considerable damage. A recoiling target atom may be sufficiently energetic to act as a secondary projectile, which displaces further atoms and produces a cascade of displacement collisions.

- *Annealing*: Atoms displaced during implantation may return to their usual positions. This annealing may occur over a short timescale (e.g., during the period that the collision cascades are moving through the lattice) or over much longer periods following completion of the implantation. Annealing is used to thermally *activate* the dopants (i.e., move them onto proper lattice sites where they act as donors or acceptors).
- *Doping*: Doping by ion implantation should be contrasted to other methods, principally that of thermally activated diffusion. In the latter technique, the dopant concentration profile is dependent on thermal equilibrium criteria for a particular dopant/solid combination. Ion implantation is a nonequilibrium process, the dopant atoms being driven into the solid by kinetic energy. As implanted, the atoms are not necessarily on their substitutional sites.

In practice, shallow implants are used to produce a well-defined Gaussian doping profile (Chapter 11). Channeling, which increases the implantation range, is not desirable. Most channeling events can be avoided by a deliberate misorientation of the wafer and beam. This means that the ion beam is not parallel to any major crystal axis. There are, however, still a few ions that, by random collisions, get scattered into a channel. This leads to a fraction of the ions that penetrate deep into the crystal. Figure 4.20 displays the indium-impurity profile obtained in crystalline silicon (filled circles) compared to amorphous silicon (open circles). Due to channeling, the crystalline silicon shows a significant deviation from a Gaussian profile.

A sufficiently high ion implant dose will render the crystal amorphous, and primary as well as secondary channeling disappears. Preamorphization of silicon, by implanting with a large dose of silicon ions, has been used to obtain Gaussian dopant profiles. Implantation at elevated temperatures leads to a decrease in channeling due to the increase in the amplitude of the target atom

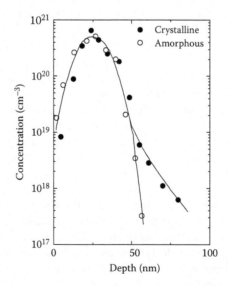

FIGURE 4.20 Comparison of room temperature implants of indium into amorphous and crystalline silicon. (After Lindhard, J. 1965. *Mat. Fys. Medd. Dan. Vid. Selsk.* 34(14): 1–64.)

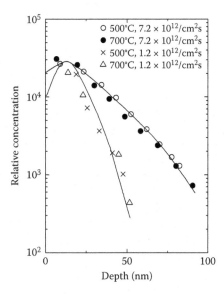

FIGURE 4.21 Effect of dose rate on distribution of ion-implanted antimony in silicon. The high dose rate results in radiation enhanced diffusion. (After Gamo, K. 1970. *Appl. Phys. Lett.* 17: 391–393.)

vibrations. On the other hand, much higher ion doses are needed to render a crystal amorphous at elevated temperatures, where annealing occurs during the implantation. Because channeling is strongly dependent upon the degree of crystallinity, it can be suppressed by a thin oxide layer on the wafer. Implantation through a thin oxide is also used to move the maximum dopant concentration closer to the surface in order to minimize metal contact resistance.

In addition to channeling, another important effect is *radiation-enhanced diffusion,* which occurs at temperatures far below typical diffusion temperatures. This effect can be explained by the large number of vacancies, interstitials, and electron–hole pairs that are formed during implantation. The recombination of an electron and a hole frees energy of the order of the semiconductor band gap (e.g., ~1 eV in silicon). This energy is released in a small volume, leading to local "heating" that promotes diffusion. The diffusion coefficient of many impurities has been found to increase with the implantation dose rate. An example is shown in Figure 4.21, where an antimony profile is plotted as a function of depth for two different dose rates and two substrate temperatures. The profiles are clearly dominated by dose rate and not by temperature.

4.10 ANNEALING AND DOPANT ACTIVATION

Dopants must occupy the correct substitutional lattice position to become electrically active donors or acceptors. Heating, or thermal activation, is the principal process used to move impurities into their substitutional sites. In all discussions of dopant activation, it should be remembered that the solid solubility (Figure 4.4) of any particular impurity cannot be exceeded for equilibrium conditions. If implantation leads to concentrations larger than the solubility limit, then precipitation and cluster formation will occur during thermal annealing, given sufficient time.

During annealing, complicated macroscopic defect structures appear and vanish. Dislocation loops and stacking faults are the major structural defects. Defects formed during hot implants are harder to remove than those created during room-temperature implants. For example, silicon implanted with phosphorus at 350°C requires an annealing temperature of 900°C for full activation, whereas the room-temperature counterpart requires only 600°C (Yoshihiro, 1980). Micrographs of a phosphorus-implanted silicon wafer after 800°C annealing and three implant temperatures are displayed in Figure 4.22. These structural defect pictures show the enormous complexity of damage formation associated with ion implantation.

As discussed in Section 4.9, amorphization is achieved through implantation at low temperatures where in situ annealing is effectively suppressed. Amorphization can be obtained through implantation of dopants (but this requires rather large doses), noble gases, or self-implantation (Si in Si, Ge in Ge, etc.). The amorphous layer can regrow, still in a solid state, starting from the amorphous–crystalline interface and proceeding toward the free surface. This process is called *solid phase epitaxy* (SPE), a kind of epitaxial crystal growth (Csepregi et al., 1996). The layer crystallized through SPE can be highly perfect, and if it contains dopants, one finds very high activation as long as the maximum solubility is not exceeded. High boron activation results are found for preamorphization using Ne, Si, BF_2, or B (at 77 K) followed by SPE/annealing (Ruge et al., 1972). In all four cases, full activation is achieved after annealing at ~600°C.

Nonequilibrium thermal annealing approaches have been developed that make use of powerful lasers or heating lamps. In these processes, rapid heating and cooling cycles are used to anneal the implanted layer. We consider two time regimes:

- Short, high-energy pulses (ns) that may involve melting and regrowth.
- Energy pulses (up to s) that lead to rapid thermal annealing but do not involve melting.

The idea behind *pulsed laser annealing* is simple. In order to restrict thermal annealing to the ion-implanted near-surface region, one needs a short pulse of photons, which is absorbed near the surface. The large power density raises the temperature of this region in a time much shorter than the lattice phonons need to move away from the surface into the interior. By the time significant thermal transport into the bulk begins to occur, the laser pulse has been turned off. With sufficient laser intensity, the near-surface region melts and undergoes SPE (Auston et al., 1978). Rapid surface layer heating has been achieved with pulsed lasers such as ruby, Nd:glass, and excimer lasers. Mode locking, spatial filtering, and a Q-switch are used to obtain very short pulses with a uniform distribution of the power over the semiconductor wafer area. Delivery of ~1 Joule cm^{-2} in ~100 ns leads to surface melting of silicon.

Figure 4.23 shows implant profiles of boron in silicon before and after laser annealing. The implant energies and doses were chosen such that the resulting concentration profiles exceed the

FIGURE 4.22 Transmission electron micrographs of defects after annealing at 800°C. These samples were implanted at 200 to 400°C. (Reprinted with permission from Tamura, M. Secondary defects in phosphorus-implanted silicon. *Appl. Phys. Lett.* 23: 651–653, Copyright 1973, American Institute of Physics.)

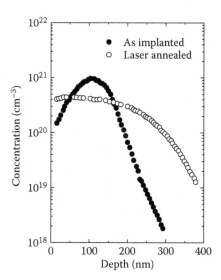

FIGURE 4.23 Depth profile of boron-implanted silicon, as implanted and after laser annealing. (After White, C.W. et al. 1978. *Appl. Phys. Lett.* 33: 662–664.)

maximum solubility limit. The postannealing profiles show that the substitutional fraction of the implanted dopant species remained above the maximum solubility limit, not too surprising considering that we are dealing with a process far from equilibrium. High dopant concentrations can also be achieved using *gas immersion laser deposition*, where BCl_3 gas is adsorbed on the silicon surface before each laser pulse (Kerrien et al., 2002). Boron diffuses into the near-surface region during each melting/regrowth cycle. This technique leads to boron concentrations of 6×10^{20} cm^{-3}, high enough to make silicon a superconductor with $T_c \sim 0.35$ K (Bustarret et al., 2006).

Despite impressive results, laser annealing in the ns time regime does not lead to improvements that justify the large investment in equipment. As it turns out, *rapid thermal annealing* on the timescale of seconds is short enough to lead to (1) significant reduction in the in-diffusion of undesirable contaminants such as iron and gold, and (2) effective activation of the implanted dopants. Instead of lasers, two banks of high-intensity quartz-halogen lamps are used as a photon source. Confocal mirrors guide the photons into a silica chamber that contains the semiconductor wafer and can be purged with a gas during annealing. The silicon wafer absorbs much of the radiation, while the silica chamber does not get hot. The total mass of all materials to be heated is kept to a minimum. This enables heating from room temperature to 1000°C in a few seconds. Rapid thermal processors of this basic design are commercially available and are widely used.

The preceding section dealt mostly with doping of elemental semiconductors by ion implantation. Compound semiconductors offer additional challenges. For example,

- How can one influence the implanted dopant atoms to assume a position on the appropriate sublattice?
- How is the local stoichiometry affected by implantation and the resulting damage, especially for compounds with very heavy and very light components (e.g., InP or InN)?
- Does thermal annealing restore stoichiometry?
- How is decomposition during thermal annealing suppressed?

Rather than addressing all of these points, we will review one technique, *co-implantation*, which is important for dopant activation in compound semiconductors (Pearton, 1993). Co-implantation

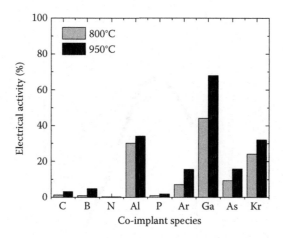

FIGURE 4.24 Electrical activity of implanted carbon in GaAs as a function of co-implant species for samples annealed at 800°C and 950°C. Electrical activity is the ratio of sheet carrier concentration to implant dose.

is the sequential implantation of two different types of ions into the same near-surface region. Specifically, we will discuss co-implantation of carbon with other elements in GaAs.

It has been observed that implanted carbon cannot be effectively activated by thermal annealing. Only around 1 percent of the implanted carbon atoms become shallow acceptors. One might assume that carbon, a group-IV element, is amphoteric and will occupy both sublattices, forming donors and acceptors that compensate each other. This does not turn out to be the case, as determined by local vibrational mode spectroscopy and mobility measurements.

Co-implantation with other elements strongly affects the level of carbon activation. Moll et al. (1992) conducted systematic studies of this co-implantation technique. Care was taken that the two ion species have identical depth distribution profiles in the host semiconductor. This requires that the two ions be implanted at different energies because they have different masses. The histogram in Figure 4.24 summarizes the activation efficiency for several co-implanted elements after rapid thermal annealing at 800°C and 950°C. Co-implantation with specific elements clearly improves the activation.

The large differences in co-implantation effectiveness can be interpreted based on two observations: (1) the heavier the co-implanted ion species, the higher the carbon activation, and (2) group-III co-implants are much more effective than group-V co-implants. The first trend suggests that heavier implant damage leads to higher activation, due to SPE. Carbon, with atomic mass of 12, does relatively little damage as compared to gallium with an average atomic mass of 70. The second trend can be explained with stoichiometry arguments. Implantation of group-III elements leads to a deviation from stoichiometry, with either group-III elements assuming group-V sites (antisites) or with excess group-V vacancies. Carbon in such an environment will preferably assume sites that restore stoichiometry. These are the group-V sublattice sites on which carbon forms shallow acceptors.

4.11 NEUTRON TRANSMUTATION

About 80% of all chemical elements consist of more than one isotope, atoms with equal numbers of protons but different numbers of neutrons. When an isotope absorbs a thermal neutron, it becomes a heavier isotope of the same element. The heaviest isotopes of an element are unstable and may decay by beta (nuclear electron) emission, increasing the atomic number by one. Alternatively, an isotope can decrease its atomic number by one via electron capture. By neutron irradiation, some of the host atoms are transformed into dopants upon radioactive decay. These processes are called neutron transmutation.

As an example, consider the stable isotopes of silicon. In the neutron flux of a nuclear reactor, these isotopes undergo reactions shown in Table 4.2. The thermal neutron capture cross section σ_c for the last reaction is 0.11 barn (1 barn $= 10^{-24}$ cm^2). This reaction is used to dope silicon with phosphorus. For a total neutron dose n (cm^{-2}), the concentration of phosphorus atoms [P] is given by

$$(4.18) \qquad\qquad [P] = n\sigma_c[^{30}Si]$$

where [^{30}Si] is the concentration of ^{30}Si in the crystal.

In 1973, a German high-power silicon device manufacturer showed for the first time that neutron-transmutation-doped silicon is superior to standard bulk doped crystals for the fabrication of 4 kV, high-current thyristors and diodes (Schnöller, 1974). The advantage of neutron transmutation doping (NTD) lies in the extreme doping homogeneity which can be achieved (Larrabee, 1984; Meese, 1979). The doping fluctuations observed in FZ crystals lead to high-field spots in reverse-biased, high-voltage devices. These spots cause a severe reduction of the average maximum attainable reverse breakdown fields.

Unfortunately, there are several radiation damage processes that occur during transmutation doping:

- Fast neutron knock-on displacements
- Fission gamma-ray-induced damage
- Gamma ray recoil damage
- Beta recoil damage
- Charged particle knock-on

Fast neutron knock-ons are the most detrimental reactions. Even where thermal neutrons ($E_{neutron} \leq 3k_BT/2$) are dominant, a case that is approached by modern reactors with special moderation, we are still faced with substantial gamma-ray recoil damage. This damage consists mainly of displaced silicon atoms. Thermal annealing to $\sim$900°C is necessary to remove most of the damage. The lifetime of minority carriers remains reduced after NTD and annealing, because of residual deep level centers.

A further problem with silicon NTD is the neutron capture by ^{31}P that was produced by neutron capture of ^{30}Si. The ^{32}P isotope obtained upon neutron capture decays into ^{32}S, a deep double donor in silicon. The half-life is about two weeks, much longer than the half-life of the ^{31}Si decay, leading to prolonged radioactivity of the Si samples. This secondary reaction limits the NTD method to

---

TABLE 4.2   Stable Isotopes of Silicon and Their Neutron Transmutation Products

Isotope	Natural Abundance	Thermal Neutron Capture Cross Section (10^{-24} cm^2)	Neutron Transmutation Product
^{28}Si	92.23%	0.17	^{29}Si (stable)
^{29}Si	4.67%	0.10	^{30}Si (stable)
^{30}Si	3.10%	0.11	^{31}P ($\tau_{1/2} = 2.62$ h)

resistivity levels of $\sim 5\ \Omega$ cm ($[P] \sim 10^{15}$ cm^{-3}). Using very high neutron flux reactors reduces the relative ^{32}P production because the ^{31}P has not yet fully formed during the irradiation time.

Whereas silicon is at present the only semiconductor that is neutron transmutation doped on a commercial scale, other semiconductors can be doped by NTD as well. Germanium becomes doped both with donors and acceptors by neutron transmutation. This leads to a fixed compensation ratio if one starts with high-purity crystals of natural isotopic composition. This fact has been used by Fritzsche and Cuevas (1960) to study conduction mechanisms at high homogenous doping levels and constant compensation. Germanium crystals engineered from pure isotopes can be doped n- or p-type, with an arbitrary compensation level (Haller, 1995).

PROBLEMS

4.1 A Ge crystal is grown by a Czochralski puller. The melt contains 10^{16} cm^{-3} boron ($k_0 = 17$) and 10^{16} cm^{-3} phosphorus ($k_0 = 0.08$). The grown crystal is 1.0 m long. Find the impurity concentrations 0.1 m and 0.8 m from the seed.

4.2 A 1.0-m long silicon crystal is grown by the Czochralski technique. The melt contained 10^{15} cm^{-3} phosphorus. The segregation coefficient of phosphorus in silicon is 0.35. How much of the crystal (m) has $[P] < 5 \times 10^{14}$ cm^{-3}?

4.3 A Ge crystal is 50 cm long and has a uniform Si ($k_0 = 5.5$) concentration of 10^{14} cm^{-3}. A molten zone of width 5 cm passes once through the crystal. Sketch the dopant profile versus x, from $x = 0$ to 45 cm.

4.4 Sketch a Czochralski apparatus for growing InP crystals.

4.5 What are the advantages of FZ growth? What unique characteristics does Si have that make FZ growth of large diameters feasible?

4.6 Match each growth technique with the description.
- Techniques: LPE, CVD, MOCVD, MBE
- Descriptions: Workhorse for III-V devices such as LEDs
  Economical way to grow pure Si on a doped Si substrate
  Gives the ultimate atomic-layer control
  A very cheap way to grow pure GaAs thin films

4.7 An MBE system is used to grow a GaAs thin film. Sketch the RHEED oscillations, and briefly explain how they come about.

4.8 State two thin-film growth techniques. For each, draw a simple sketch and give a 1–2 sentence description.

4.9 A research lab manager wants to study dopants in semiconductors and is trying to decide between ion implantation and diffusion. Give one advantage (good feature) for each technique.

4.10 Isotopically and chemically pure ^{70}Ge is exposed to 10^{15} neutrons/cm^2. What is the concentration of donors or acceptors (cm^{-3}) after the radioactivity has died down?
- ^{70}Ge neutron thermal capture cross section $= 3.25 \times 10^{-24}$ cm^2
- ^{71}Ge $\rightarrow$ ^{71}Ga by electron capture ($p + e \rightarrow n$) with a half-life of 11 days
- Lattice constant of Ge is $a = 5.7$ Å

REFERENCES

Arizumi, T. and N. Kobayashi. 1972. Theoretical studies of the temperature distribution in crystal being grown by the Czochralski method. *J. Crystal Growth* 13/14: 615–618.

Auston, D.H., C.M. Surko, T.N.C. Venkatesan, R.E. Slusher, and J.A. Golovchenko. 1978. Time-resolved reflectivity of ion-implanted silicon during laser annealing. *Appl. Phys. Lett.* 33: 437–439.

Bauser, E. and H.P. Strunk. 1984. Microscopic growth mechanisms of semiconductors—Experiments and models. *J. Crystal Growth* 69: 561–580.

Bohm, J. 1985. The history of crystal growth. *Acta Physica Hungarica* 57: 161–178.

Bustarret, E., C. Marcenat, P. Achatz et al. 2006. Superconductivity in doped cubic silicon. *Nature* 444: 465–468.

Capper, P. and M. Mauk, eds. 2007. *Liquid Phase Epitaxy of Electronic, Optical, and Optoelectronic Materials*. New York, NY: John Wiley & Sons.

Carruthers, J.R. and A.F. Witt. 1975. Transient segregation effects in Czochralski growth. In *Crystal Growth and Characterization*, eds. R. Ueda and J.B. Mullin, pp. 107–154. Amsterdam: North Holland.

Carter, G. and W.A. Grant. 1976. *Ion Implantation of Semiconductors*. New York, NY: John Wiley & Sons.

Csepregi, L., J. Gyulai, and S.S. Lau. 1996. The early history of solid phase epitaxial growth. *Mater. Chem. Phys.* 46: 178–180.

Current, M.I. and M.J. Markert. 1984. Mapping of ion implanted wafers. In *Ion Implantation Science and Technology*, ed. J.F. Ziegler, p. 505. New York, NY: Academic Press.

Czochralski, J. 1918. Ein neues Verfahren zur Messung der Kristallisationsgeschwindig-heit der Metalle. *Z. Phys. Chemie* 92: 219–221.

Eisen, Y. and A. Shor. 1998. CdTe and CdZnTe materials for room-temperature X-ray and gamma ray detectors. *J. Cryst. Growth* 184: 1302–1312.

Foxon, C.T. and B.A. Joyce. 1990. Growth of thin films and heterostructures of III-V compounds by molecular beam epitaxy. In *Growth and Characterisation of Semiconductors*, eds. R.A. Stradling and P.C. Klipstein, pp. 35–64. New York, NY: Adam Hilger.

Fritzsche, H. and M. Cuevas. 1960. Impurity conduction in transmutation-doped *p*-type germanium. *Phys. Rev.* 119: 1238–1245.

Gamo, K. 1970. Enhanced diffusion of high-temperature ion-implanted antimony into silicon. *Appl. Phys. Lett.* 17: 391–393.

Grossmann, G. and L. Ledebo, eds. 1988. *Semi-Insulating III-V Materials*. Boca Raton, FL: Taylor & Francis.

Haller, E.E. 1995. Isotopically engineered semiconductors. *J. Appl. Phys.* 77: 2857–2878.

Haller, E.E. and F.S. Goulding. 1993. Nuclear radiation detectors. In *Handbook on Semiconductors*, Vol. 4, 2nd ed., ed. C. Hilsum, pp. 937–963. New York, NY: Elsevier.

Haller, E.E., W.L. Hansen, and F.S. Goulding. 1981. Physics of ultra-pure germanium. *Adv. Phys.* 30: 93–138.

Hansen, M. 1958. *Constitution of Binary Alloys*. New York, NY: McGraw-Hill.

Huff, H.R. 1983. Chemical impurities and structural imperfections in semiconductor silicon, Part 1. *Solid State Techn.* 26: 89–95.

Kamath, G.S. 1984. Growth of GaAs single crystals for optoelectronic applications. *Solid State Techn.* 27: 173–175.

Keller, W. and A. Mühlbauer. 1981. *Floating-Zone Silicon*. New York, NY: Marcel Dekker.

Kerrien, G.J. Boulmer, D. Débarre et al. 2002. Ultra-shallow, super-doped and box-like junctions realized by laser-induced doping. *Appl. Surf. Sci.* 186: 45–51.

Kramer, H.G. 1983. Float-zoning of semiconductor silicon: A perspective. *Solid State Techn.* 26: 137–142.

Kuphal, E. 1991. Liquid-phase epitaxy. *Appl. Phys. A* 52: 380–409.

Langlois, W.E. 1981. Convection in Czochralski growth melts. *Physicochem. Hydrodynamics* 2: 245–261.

Larrabee, R.D., ed. 1984. *Neutron Transmutation Doping of Semiconductor Materials*. New York, NY: Plenum Press.

Laughlin, R.B. 1983. Anomalous quantum Hall effect: An incompressible quantum fluid with fractionally charged excitations. *Phys. Rev. Lett.* 50: 1395–1398.

Lindhard, J. 1965. Influence of crystal lattice on motion of energetic charged particles. *Mat. Fys. Medd. Dan. Vid. Selsk.* 34(14): 1–64.

Look, D.C. and J. S. Blakemore, eds. 1984. *Semi-Insulating III-V Materials: Kah-nee-ta*. Nantwich, UK: Shiva.

Mahajan, S. and K.S. Sree Harsha. 1999. *Principles of Growth and Processing of Semiconductors*. New York, NY: WCB/McGraw-Hill.

Makram-Ebeid, S. and B. Tuck, eds. 1982. *Semi-Insulating III-V Materials: Evian*. Nantwich, UK: Shiva.

Mayer, J.W., L. Eriksson, and J.A. Davis. 1970. *Ion Implantation in Semiconductors*. New York, NY: Academic Press.

Meese, J. ed. 1979. *Neutron Transmutation Doping in Semiconductors*. New York, NY: Plenum Press.

96   References

Minami, T. 2005. Transparent conducting oxide semiconductors for transparent electrodes. *Semicond. Sci. Technol.* 20: S35–S44.

Moll, A.J., K.M. Yu, W. Walukiewicz, W. L. Hansen, and E.E. Haller. 1992. Co-implantation and electrical activity of C in GaAs: Stoichiometry and damage effects. *Appl. Phys. Lett.* 60: 2383–2385.

Moody, J.W. and R.A. Frederick. 1983. Developments in Czochralski silicon crystal growth. *Solid State Techn.* 26: 221–225.

Mühlbauer, A. and E. Sirtl. 1974. Lamellar growth phenomena in <111>-oriented dislocation-free float-zoned silicon single crystals. *Phys. Stat. Sol. (a)* 23: 555–565.

Müller, G. 2007. The Czochralski method—Where we are 90 years after Jan Czochralski's invention. *Cryst. Res. Technol.* 42: 1150–1161.

Nakamura, S., T. Mukai, and M. Senoh. 1994. Candela-class high-brightness InGaN/AlGaN double-heterostructure blue-light-emitting diodes. *Appl. Phys. Lett.* 64: 1687–1689.

Neave, J.H., B.A. Joyce, P.J. Dobson, and N. Norton. 1983. Dynamics of film growth of GaAs by MBE from Rheed observations. *Appl. Phys. A* 31: 1–8.

Nelson, H. 1963. Epitaxial growth from the liquid state and its application to the fabrication of tunnel and laser diodes. *RCA Review* 24: 603–615.

Ohring, M. 1992. *The Materials Science of Thin Films.* New York, NY: Academic Press.

Pamplin, B.R. ed. 1975. *Crystal Growth.* Oxford, UK: Pergamon Press.

Pashley, D.W. 1990. The basics of epitaxy. In *Growth and Characterisation of Semiconductors*, eds. R.A. Stradling and P.C. Klipstein, pp. 1–16. New York, NY: Adam Hilger.

Pearton, S.J. 1993. Ion implantation in III-V semiconductor technology. *Int. J. Mod. Phys. B* 7: 4687–4761.

Pfann, W.G. 1966. *Zone Melting*, 2nd ed., p. 12. New York, NY: John Wiley & Sons.

Puurunen, R.L. 2005. Surface chemistry of atomic layer deposition: A case study for the trimethylaluminum/water process. *J. Appl. Phys.* 97: 121301 (52 pages).

Rea, S.N. 1981. Czochralski silicon pull rate limits. *J. Cryst. Growth* 54: 267–274.

Ruge, I. 1975. *Halbleiter Technologie*, p. 121. Berlin: Springer-Verlag.

Ruge, I., H. Müller, and H. Ryssel. 1972. Die Ionenimplantation als Dotiertechnologie. *Festkörperprobleme (Adv. Solid State Phys.)* 12: 23–106.

Runyan, W.R. 1965. *Silicon Semiconductor Technology.* New York, NY: McGraw-Hill.

Ryssel, H. and I. Ruge. 1986. *Ion Implantation.* New York, NY: John Wiley & Sons.

Schnöller M.S. 1974. Breakdown behavior of rectifiers and thyristors made from striation-free silicon. *IEEE Trans. Electr. Dev.* ED-21: 313–314.

Shaw D.W. 1975. Kinetics and mechanisms of epitaxial growth of semiconductors. In *Crystal Growth and Characterization*, eds. R. Ueda and J.B. Mullin, p. 208. Amsterdam: North-Holland.

Stephens, K.G. 1984. An introduction to ion sources. In *Ion Implantation Science and Technology*, ed. J.F. Ziegler, p. 414. New York, NY: Academic Press.

Szeles C., Y.Y. Shan, K.G. Lynn, and E.E. Eissler. 1996. Deep electronic levels in high-pressure Bridgman $Cd_{1-x}Zn_xTe$. *Nucl. Instrum. Meth. Phys. Res. A* 380: 148–152.

Tamura, M. 1973. Secondary defects in phosphorus-implanted silicon. *Appl. Phys. Lett.* 23: 651–653.

Thomas, R.N., H.M. Hobgood, P.S. Ravishankar, and T.T. Braggins. 1990. Melt growth of large diameter semiconductors. *Solid State Technol.* 33: April (pp. 163–167), May (pp. 121–127), and June (pp. 83–86).

Trumbore, F.A. 1960. Solid solubilities of impurity elements in germanium and silicon. *Bell Syst. Tech. J.* 39: 205–234.

Tsui, D.C., H.L. Stormer, and A.C. Gossard. 1982. Two-dimensional magnetotransport in the extreme quantum limit. *Phys. Rev. Lett.* 48: 1559–1562.

Von Klitzing, K., G. Dorda, and M. Pepper. 1980. New method for high-accuracy determination of the fine-structure constant based on quantized Hall resistance. *Phys. Rev. Lett.* 45: 494–497.

White, C.W., W.H. Christie, B.R. Appleton et al. 1978. Redistribution of dopants in ion-implanted silicon by pulsed-laser annealing. *Appl. Phys. Lett.* 33: 662–664.

Witt, A.F. and H.C. Gatos. 1966. Impurity distribution in single crystals. *J. Electrochem. Soc.* 113: 808–813.

Yoshihiro, N. 1980. Damage dependent electrical activation of phosphorus implanted in silicon. *Radiation Effects and Defects in Solids* 47: 85–90.

Electronic Properties

5

Through controlled doping, the electronic properties of semiconductors can be tuned over a wide range, making them an indispensible part of modern technology. This chapter deals with the electronic properties of dopants, defects, and free carriers in semiconductors. We begin with a discussion of hydrogenic impurities, deep levels, and methods of calculating the electronic properties of defects. Then, we investigate how the defect levels affect the population of electrons in the conduction band and holes in the valence band. Finally, we discuss electronic transport and the principal scattering mechanisms that affect carrier mobility.

5.1 HYDROGENIC MODEL

As discussed in Chapter 2, donor (acceptor) levels that lie close to the conduction band minimum (valence band maximum) are called "shallow." The binding energy is low such that the electron (hole) is readily given up to the conduction (valence) band at moderate temperatures. Shallow donors and acceptors are also called "hydrogenic" or hydrogen-like because they can be described with a simple model. The shallow impurity is treated analogously to a hydrogen atom embedded in a homogeneous medium with a relative dielectric constant ε and an electron/hole mass m^* that corresponds to the effective mass of the particular band extremum.

Figure 5.1 shows a classical picture of an electron (charge $-|e|$) orbiting a positively charged phosphorus donor (charge $+|e|$) in a silicon crystal. The Coulomb force on the electron is given by

$$(5.1) \qquad F = \frac{1}{4\pi\varepsilon\varepsilon_0}\frac{e^2}{r^2}$$

where the orbital radius is r. From Newton's second law,

$$(5.2) \qquad F = m^*a = m^*v^2/r$$

where v is the speed of the electron. Inserting Equation 5.1 for F yields

$$(5.3) \qquad v^2r = \frac{e^2}{4\pi\varepsilon\varepsilon_0 m^*}$$

97

5.1 Hydrogenic Model

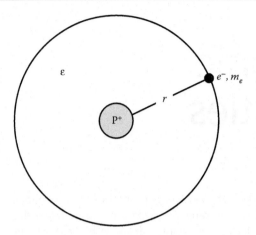

FIGURE 5.1 Bohr model for the phosphorus donor in silicon (or germanium), where P⁺ is the ion and e⁻ is the electron.

In the Bohr model, the angular momentum L is quantized in units of $\hbar$:

$$L = m^* v r = \hbar n \tag{5.4}$$

where n is a positive integer. The orbital radius is found by inserting this expression for v into Equation 5.3:

$$r = \frac{4\pi \hbar^2 \varepsilon \varepsilon_0}{e^2 m^*} n^2 \tag{5.5}$$

The radius for the ground state, $n = 1$, is the *Bohr radius*:

$$r_0 = \left(0.53\,\text{Å}\right) \frac{\varepsilon}{m^*/m} \tag{5.6}$$

where m is the fundamental electron mass. The Bohr radius for a donor is typically much larger than that of a hydrogen atom. GaAs, for example, has $\varepsilon = 13.2$ and an electron effective mass $m^*/m = 0.067$, yielding $r_0 \approx 100$ Å.

The total energy (potential plus kinetic) of the electron is

$$E = \frac{1}{2} m^* v^2 - \frac{1}{4\pi\varepsilon\varepsilon_0} \frac{e^2}{r} \tag{5.7}$$

Inserting Equation 5.3 for v^2 gives us

$$E = -\frac{1}{2} \frac{1}{4\pi\varepsilon\varepsilon_0} \frac{e^2}{r} \tag{5.8}$$

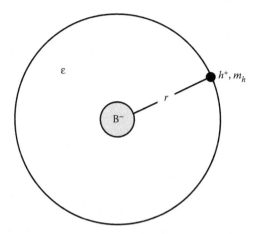

FIGURE 5.2 Bohr model for the boron acceptor in silicon (or germanium), where B⁻ is the ion and h^+ is the hole.

and inserting Equation 5.5 for r yields

(5.9) $$E = -\frac{1}{2}\frac{m^*e^4}{(4\pi\varepsilon\varepsilon_0\hbar)^2}\frac{1}{n^2}$$

The donor binding energy E_d is the energy required to bring the electron from the ground state ($n = 1$) to the conduction-band minimum, here set to be $E = 0$:

(5.10) $$E_d = \frac{1}{2}\frac{m^*e^4}{(4\pi\varepsilon\varepsilon_0\hbar)^2} = (13.6\,\text{eV})\left(\frac{m^*}{m}\right)\left(\frac{1}{\varepsilon^2}\right)$$

On an energy band diagram, the donor level (0/+) is indicated by a horizontal line an amount E_d below the conduction-band minimum (Figure 2.2).

The hydrogenic model also applies to acceptors. Substitutional boron in silicon, for example, takes an electron from the valence band, leaving behind a positively charged hole (Figure 5.2). The hole orbits around the negatively charged boron acceptor. The acceptor binding energy E_a is calculated with Equation 5.10, where m^* is the hole effective mass. The acceptor level (0/−) lies above the valence band maximum by an amount E_a (Figure 2.2).

For germanium ($\varepsilon = 16$, $m^* \approx 0.2\,m$), we find a binding energy for electrons and holes of ~10 meV, over 1000 times smaller than that for the hydrogen atom. The large differences in the electron and hole masses in silicon and GaAs lead to different values for the electron and hole binding energies:

$$\text{Si}\,(\varepsilon = 11.7): E_d \approx 33\,\text{meV}, E_a \approx 75\,\text{meV}$$

$$\text{GaAs}\,(\varepsilon = 13.2): E_d \approx 6\,\text{meV}, E_a \approx 28\,\text{meV}$$

The fact that the hydrogenic model, or *effective mass theory*, works so well is related to the large Bohr radii for electrons and holes bound to shallow-level impurities. As we noted, donor electrons in GaAs have a Bohr radius of $r_0 \approx 100$ Å. There are over 10^5 atoms within a sphere with radius r_0! The electrons or holes of shallow impurities are strongly *delocalized* (i.e., their wave functions are spread over a large volume compared to the space taken up by each host atom). This averaging over many host atoms allows us to approximate the crystal as a continuous medium with an average dielectric constant ε. The result is self-consistent: the large calculated value of r_0 justifies the effective mass approach.

As with the hydrogen atom, we can go beyond the Bohr model and examine the wave functions that describe the impurity electron or hole. In the nearly free electron approach (Section 1.6), electrons in an intrinsic semiconductor have an infinite spatial extent. The electron wave function is given by a *Bloch wave function*,

(5.11) $$\psi_\mathbf{k} = u_\mathbf{k}(\mathbf{r})e^{i\mathbf{k}\cdot\mathbf{r}}$$

where $u_\mathbf{k}(\mathbf{r})$ is a function that has the same periodicity and symmetry as the crystal lattice.

Let us consider the example of a nearly free electron in the conduction band of a one-dimensional (1D) crystal, at $k = 0$. From Equation 1.14, at $k = 0$,

(5.12) $$u_0(x) \sim \sin(2\pi x/a)$$

This is a periodic function that knows no bound—the wave function extends throughout the entire crystal. A donor electron, however, experiences the Coulomb attraction of the positively ionized donor. Assuming the attractive potential is a slowly varying function, the electron wave function is given by

(5.13) $$\psi(x) = F(x)u_0(x)$$

where $F(x)$ is called the *envelope function* (Luttinger and Kohn, 1955).

Figure 5.3 shows an illustration of a donor electron with Gaussian envelope function in a 1D crystal. The rapidly varying function $u_0(x)$ is multiplied by the slowly varying $F(x)$. Unlike the free electron, the donor electron wave function has a finite spatial extent, characterized by an uncertainty Δx. Using Fourier transformation, we can reproduce this wave function with a superposition of Bloch wave functions with k values near $k = 0$. The range of k values required to obtain

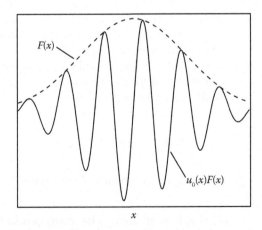

FIGURE 5.3 Donor electron wave function in a one-dimensional crystal (solid line, Equation 5.13). The dashed line is a Gaussian envelope function.

Chapter 5—Electronic Properties **101**

the wave function is denoted Δk. The inverse relationship between real space and k space is given by the Heisenberg uncertainty relation:

$$\Delta x \Delta k \approx 1 \tag{5.14}$$

A delocalized wave function has a large Δx and a small Δk. The hydrogenic model applies to donor or acceptor wave functions that are delocalized in real space (large Bohr radii) and hence localized in k space. Deep levels, on the other hand, are localized in real space but spread out in k space.

Equation 5.13 applies for a donor wave function that is composed of Bloch wave functions with $k \approx 0$. In general, for a conduction-band minimum at k, a hydrogenic wave function is given by a superposition of Bloch functions near k. We can write the donor electron wave function as

$$\psi(\mathbf{r}) = F(\mathbf{r}) u_{\mathbf{k}}(\mathbf{r}) e^{i\mathbf{k}\cdot\mathbf{r}} \tag{5.15}$$

The envelope function is a solution to Schrödinger's equation:

$$-\frac{\hbar^2}{2m^*} \nabla^2 F(\mathbf{r}) + V(\mathbf{r}) F(\mathbf{r}) = E\, F(\mathbf{r}) \tag{5.16}$$

where $V(r)$ is the potential energy perturbation introduced by the impurity atom. If V is a $1/r$ potential, then $F(\mathbf{r})$ is given by the familiar wave functions for the hydrogen atom. The $1s$ and $2p$ wave functions, for example, are written in spherical coordinates as follows:

$$
\begin{aligned}
F(1s) &\sim \exp(-r/r_0) \\
F(2p, m=0) &\sim r\cos\theta \exp(-r/2r_0) \\
F(2p, m=\pm1) &\sim r\sin\theta \exp(-r/2r_0)\exp(\pm i\varphi)
\end{aligned}
\tag{5.17}
$$

where m is an angular momentum quantum number. By taking linear combinations of the $m = \pm 1$ wave functions, the following p orbitals are obtained:

$$
\begin{aligned}
F(2p_x) &\sim x\exp(-r/2r_0) \\
F(2p_y) &\sim y\exp(-r/2r_0) \\
F(2p_z) &\sim z\exp(-r/2r_0)
\end{aligned}
\tag{5.18}
$$

As the particle approaches the donor or acceptor ion, the continuum model for ε breaks down and the potential deviates from the simple $1/r$ form. At short distances, the screening of the ion charge diminishes, resulting in a strongly negative potential energy. One can model this effect as a perturbation that is localized to the immediate vicinity of the ion ($r = 0$). The $1s$ ground-state wave function is nonzero at $r = 0$ and therefore experiences an energy shift, called a *central cell correction*. The p wave functions, on the other hand, are zero at $r = 0$ and therefore experience no first-order energy shift.

Natural extensions of hydrogenic impurities are "helium-like" centers. In silicon or germanium, substitutional sulfur or selenium impurities are double donors (Section 2.2). There are two electrons in excess of what is required to obtain a Lewis octet. In analogy with the helium atom, the two electrons orbit a donor ion of charge $+2|e|$. Consider the excitation of the first electron. The electron in the excited state experiences the Coulomb field of approximately $+|e|$, because the second electron left in the ground state screens the ion. The second electron, however, experiences the full +2|e| charge. Therefore, the donor binding energy for the second electron exceeds that of the first. Double acceptors are modeled similarly; examples include beryllium and zinc in silicon or germanium, and copper in GaAs (Figure 2.2).

5.2 WAVE FUNCTION SYMMETRY

It is often useful to classify impurity wave functions according to their symmetry. Section 2.6 discussed defect classifications according to the symmetry of the atoms, where the symmetry is defined by a set of point-group operations that do not change the defect. The specific set of rotation, reflection, and inversion operations defines the group. In diamond and zincblende semiconductors, substitutional impurities have tetrahedral (T_d) symmetry. In wurtzite semiconductors, these impurities have trigonal (C_{3v}) symmetry.

In addition to providing a way to classify the symmetry of the defect, the point group also tells us about the degeneracy of electronic states. This is important for the interpretation of absorption spectra. Consider the $2p$ orbital of a substitutional donor electron in a simple cubic crystal (Figure 5.4). The unperturbed wave functions are proportional to x, y, or z (Equation 5.18). Due to the cubic symmetry, all of these wave functions have the same energy. Now, consider what happens when we apply uniaxial stress so that the crystal becomes compressed along the z direction. The p_z orbital now has a different energy than the p_x and p_y orbitals. The lowering of symmetry lifts the degeneracy of the p orbitals.

Group theory provides a systematic way to classify the symmetry of wave functions and determines which energy levels are degenerate. These classifications are listed in Table 5.1 for several common point groups. In this table, the symmetry of the wave function is represented by Mulliken symbols (A_1, E, etc.) and by the Koster notation (Γ_1, Γ_2, etc.). In the Koster notation, Γ_1 is always the "fully symmetric" representation, which has the same symmetry as the defect. The s orbitals, which are spherically symmetric for a hydrogen atom, have Γ_1 symmetry. The other symbols represent symmetries that are lower than Γ_1. Examples of functions that have a given symmetry, such as x or $x^2 + y^2$, are given in the last column. Parentheses are used to indicate functions that are equivalent (i.e., degenerate) by symmetry.

To illustrate these principles, we consider three defects. First, let's revisit the p orbitals of a substitutional impurity atom. In a zincblende crystal, the substitutional impurity has T_d symmetry. The p_x, p_y, and p_z orbitals are proportional to x, y, and z, respectively. From Table 5.1, the orbitals have T_2 symmetry and are threefold degenerate. In a wurtzite crystal, however, the c axis is distinct from the a axes. The point-group symmetry lowers to C_{3v}, causing the orbitals to split into two degenerate E states (x,y) and one A_1 state (z). Suppose the symmetry is lowered to C_s by the application of stress. This would cause the levels to split into two nondegenerate A' states and one A'' state.

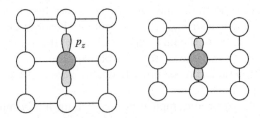

FIGURE 5.4 A p_z orbital in a simple cubic crystal. The p_x and p_y orbitals, not shown in the diagram, are orthogonal to the p_z orbital. As the crystal is compressed along the z direction, the p_z energy level shifts relative to that of the p_x and p_y energies.

TABLE 5.1 Common Point Groups and Wave Function Symmetries

Point Group	Mulliken	Koster	Degeneracy	Functions
T_d	A_1	Γ_1	1	$x^2 + y^2 + z^2$
	A_2	Γ_2	1	
	E	Γ_3	2	$(2z^2 - x^2 - y^2, x^2 - y^2)$
	T_1	Γ_4	3	
	T_2	Γ_5	3	$(x, y, z); (xy, xz, yz)$
C_{3v}	A_1	Γ_1	1	$z; x^2 + y^2; z^2$
	A_2	Γ_2	1	
	E	Γ_3	2	$(x, y); (x^2 - y^2, xy); (xz, yz)$
C_s	A'	Γ_1	1	$y; z; yz; x^2; y^2; z^2$
	A''	Γ_2	1	$x; xy; xz$
C_1	A	Γ_1	1	(All functions)
D_{2d}	A_1	Γ_1	1	$x^2 + y^2; z^2$
	A_2	Γ_2	1	
	B_1	Γ_3	1	$x^2 - y^2$
	B_2	Γ_4	1	$z; xy$
	E	Γ_5	2	$(x, y); (xz, yz)$
D_{3d}	A_{1g}	Γ_1^+	1	$x^2 + y^2; z^2$
	A_{2g}	Γ_2^+	1	
	E_g	Γ_3^+	2	$(x^2 - y^2, xy); (xz, yz)$
	A_{1u}	Γ_1^-	1	
	A_{2u}	Γ_2^-	1	z
	E_u	Γ_3^-	2	(x, y)

Source: Cotton, F.A., *Chemical Applications of Group Theory*, 3rd edn., John Wiley & Sons, New York, 1990.

Note: The symmetries are labeled with the Mulliken and Koster notations. Examples of functions with the given symmetry are listed. A constant (e.g., 1) always transforms as the fully symmetric representation (Γ_1 or Γ_1^+).

The ways that levels split are summarized in *compatibility tables* (Cotton, 1990). Using these tables, we can reproduce the results of the previous example. From Table 5.2, one can see that the T_2 states split into A_1 and E states when the symmetry is lowered from T_d to C_{3v}. From Table 5.3, we find that when the symmetry is lowered to C_s, the A_1 state becomes A' and the E state splits into A' and A''. Note that group theory tells us when levels may split, but it does not tell us the energies of the levels.

The second case, important for transition metals, is that of the d orbitals for a substitutional impurity (Clerjaud, 1985). The five d orbitals may be written as functions proportional to xy, yz, xz, $x^2 - y^2$, and $2z^2 - x^2 - y^2$. In T_d symmetry, these states split into two energy levels, E and T_2 (Table 5.1). An example of this splitting is seen in the case of substitutional copper, which for cubic ZnS has T_d symmetry. The neutral copper acceptor (Cu^{2+}) has one unfilled d orbital. A photon with the right energy can promote an electron from the E level to the T_2 level. One can also model this absorption process as a transition of a single *hole* from T_2 to E. When one accounts for spin-orbit coupling and spin-spin interactions, the T_2 and E levels split into multiplets. In ZnO, hole transitions from T_2 to E result in IR absorption peaks at 716.9 and 721.6 meV (Figure 5.5). Similar transitions in InP:Fe lead to four peaks near 350 meV (Bishop, 1986).

TABLE 5.2 Compatibility Table for the T_d Point Group

T_d	C_{3v}	D_{2d}
A_1	A_1	A_1
A_2	A_2	B_1
E	E	$A_1 \oplus B_1$
T_1	$A_2 \oplus E$	$A_2 \oplus E$
T_2	$A_1 \oplus E$	$B_2 \oplus E$

Note: This table shows how levels split as the symmetry is reduced to C_{3v} or D_{2d}. The $\oplus$ symbol indicates multiple levels.

TABLE 5.3 Compatibility Table for the C_{3v} Point Group

C_{3v}	C_s	C_1
A_1	A'	A
A_2	A''	A
E	$A' \oplus A''$	$A \oplus A$

Note: This table shows how levels split as the symmetry is reduced to C_s or C_1. In both cases, the twofold degeneracy of the E level is lifted.

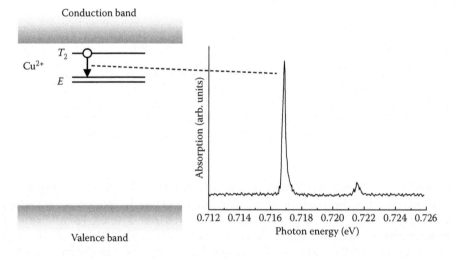

FIGURE 5.5 Cu^{2+} energy levels in ZnO. Light can cause a transition of a hole from the T_2 level to one of the E levels, resulting in two infrared (IR) absorption peaks. (After McCluskey, M.D. and S.J. Jokela. 2009. *J. Appl. Phys.* 106: 071101 (13 pages). IR data courtesy of L. Halliburton.)

The third example is the silicon vacancy (Watkins, 1986). As a first approximation, we simply remove a silicon atom and keep the surrounding atoms fixed. Each of the four nearest neighbors now has an unpaired electron, or dangling bond (Figure 5.6). Let the wave function of a dangling bond be

$$\phi(\mathbf{r} - \mathbf{r}_i) \equiv \phi_i \tag{5.19}$$

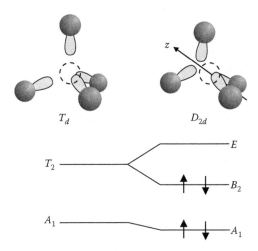

FIGURE 5.6 Ball-and-stick model and energy-level diagram for the neutral silicon vacancy. A Jahn–Teller distortion lowers the symmetry from T_d to D_{2d}.

where $\mathbf{r}_i$ is the position of the ith silicon nucleus. The possible wave functions for an electron in the vacancy are approximated by a linear combination of the individual dangling-bond orbitals. For T_d symmetry, there is one state with A_1 symmetry and three degenerate states with T_2 symmetry. The A_1 wave function is fully symmetric (i.e., it has the same symmetry as the defect):

$$\psi(A_1) \sim \phi_1 + \phi_2 + \phi_3 + \phi_4 \tag{5.20}$$

The threefold degenerate T_2 wave functions are given by

$$\psi(T_2) \sim \begin{Bmatrix} \phi_1 + \phi_2 - \phi_3 - \phi_4 \\ \phi_1 - \phi_2 + \phi_3 - \phi_4 \\ \phi_1 - \phi_2 - \phi_3 + \phi_4 \end{Bmatrix} \tag{5.21}$$

The A_1 wave function has the lowest energy (Figure 5.6).

As discussed in Section 2.6, the dangling bonds in a silicon vacancy reconstruct and the silicon atoms relax. This Jahn–Teller distortion lowers the symmetry from T_d to D_{2d}. To figure out what happens to the energy levels, we consult Table 5.2. As one goes from T_d to D_{2d}, the label for the A_1 state does not change—the A_1 level may shift in energy, but it does not split. The T_2 states, on the other hand, split into a nondegenerate B_2 level and a doubly degenerate E level. A neutral vacancy has four electrons. Two electrons (spin up and spin down) occupy the A_1 state, and two electrons occupy the B_2 state (Figure 5.6).

Electron spin is important in systems, such as acceptors or transition metals, which have spin-orbit coupling. Spin has the strange property that it gets multiplied by -1 when it is rotated through 360°. One must rotate the spin through 720° before it resumes its original state. Because of this, rotations greater than 360° must be added to the point group operations. These larger groups are called double groups (Falicov, 1989; Snoke, 2009). They can be used to determine the splitting of levels where spin-orbit coupling or spin-spin interactions are present.

5.3 DONOR AND ACCEPTOR WAVE FUNCTIONS

The total impurity wave function is the product of the Bloch wave function and the envelope function (Equation 5.15). For *donor wave functions* in a direct-gap semiconductor such as GaAs, the situation is relatively simple. The envelope functions are given by the hydrogen wave functions (Equation 5.17) and the energies are given by the Rydberg series (Equation 5.9), with a central cell correction for the $1s$ ground state. Infrared (IR) light can excite transitions from the $1s$ ground state to the p states or into the conduction-band continuum.

For indirect-gap semiconductors such as silicon or germanium, one must consider conduction-band minima with nonzero $\mathbf{k}$ values. We will examine the case of silicon, which has conduction-band minima along the <100> directions, near the X point (Figure 1.16). These minima have electron effective masses $m_{e//}$ and $m_{e\perp}$ for directions parallel and perpendicular to the <100> direction, respectively (Table 1.2). The effect of this anisotropy is to lift the degeneracy of states with different m quantum numbers (Kohn and Luttinger, 1955; Faulkner, 1969). For p orbitals, the $m = \pm 1$ states have a different energy than the $m = 0$ state. These states are labeled $p_{\pm}$ and p_0, respectively.

A second complication involves the central cell correction of the ground state (Yu and Cardona, 1996). From Equation 5.15, the ground-state wave functions are proportional to $\exp(i\mathbf{k} \cdot \mathbf{r})$; here, $\mathbf{k}$ is defined to point in a <100> direction. Because there are six <100> directions ([100], [$\bar{1}$00], [010], etc.), there are six basis functions:

$$(5.22) \qquad e^{ikx}, e^{-ikx}, e^{iky}, e^{-iky}, e^{ikz}, e^{-ikz}$$

Using the Euler identities (Appendix A), we can construct linear combinations of the basis functions that have the appropriate symmetries for the T_d group. There is one fully symmetric A_1 wave function,

$$(5.23) \qquad \psi(A_1) \sim \cos(kx) + \cos(ky) + \cos(kz)$$

two degenerate E wave functions,

$$(5.24) \qquad \psi(E) \sim \begin{cases} \cos(kx) - \cos(ky) \\ 2\cos(kz) - \cos(kx) - \cos(ky) \end{cases}$$

and three degenerate T_2 wave functions,

$$(5.25) \qquad \psi(T_2) \sim \begin{cases} \sin(kx) \\ \sin(ky) \\ \sin(kz) \end{cases}$$

The A_1, E, and T_2 wave functions overlap the central cell differently. Due to this *valley-orbit splitting*, the $1s$ ground state splits into three levels. The A_1 wave function has the maximum overlap and is usually lowest in energy. Figure 5.7 shows an IR spectrum for phosphorus donors in silicon. The sharp peaks correspond to electron transitions from the $1s$ (A_1) ground state to the p_0 and $p_{\pm}$ levels.

Acceptor wave functions are composed of valence-band states near $k = 0$. For simplicity, we assume that the split-off band is low enough in energy so that it does not contribute to the wave

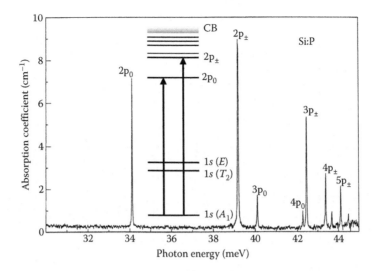

FIGURE 5.7 Infrared absorption spectrum of phosphorus donors in silicon. The peaks correspond to transitions from the ground state to the p_0 and $p_\pm$ excited states. Two such transitions are indicated by the arrows in the energy-level diagram. (Reprinted with permission from Jagannath, C., Z.W. Grabowski, and A.K. Ramdas. 1981. *Phys. Rev. B* 23: 2082–2098. Copyright 1981 by the American Physical Society.)

function. As we discussed in Section 1.6, the valence band states have an angular momentum of $j = 3/2$. Therefore, we can model a free hole as having $j = 3/2$. When this hole is bound by an acceptor, it has an envelope function, similar to the donor electrons. Let $l =$ the angular momentum of the envelope function (s is $l = 0$, p is $l = 1$, etc.). The total angular momentum of the hole is given by f. Following the rules of quantum mechanics,

(5.26) $$f = |j - l| \ldots j + l$$

For $l = 0$, there is only one value for the total angular momentum, $f = 3/2$. For $l = 1$, there are three possible values: $f = 1/2, 3/2,$ or $5/2$. These states are described with spectroscopic notation, where a capital letter represents the l value and the subscript is the f value. The $n = 1$ ground state is denoted

(5.27) $$1S_{3/2}$$

The $n = 2$ excited states are

(5.28) $$2S_{3/2}, 2P_{1/2}, 2P_{3/2}, 2P_{5/2}$$

What happens to these states in T_d symmetry? It turns out that the only level that splits is the $P_{5/2}$ level (Balderseschi and Lipari, 1973). Light can excite a hole from the $1S_{3/2}$ ground state to the P states, or into the valence-band continuum. These transitions are shown for Cu and Ag acceptors in ZnTe (Figure 5.8). The two acceptors give rise to a similar set of spectral lines. The only difference is the central cell correction of the $1S_{3/2}$ ground state.

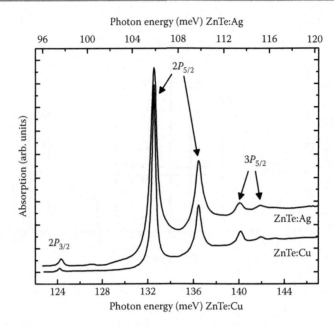

FIGURE 5.8 Infrared absorption spectrum of Cu and Ag acceptors in ZnTe. The bottom energy scale is for Cu and the top energy scale is for Ag. Peaks correspond to transitions from the ground state to the P levels. The energy shifts of the two spectra are due to different central-cell corrections for Cu and Ag. (After Chen, G., Miotkowski, I. and Ramdas, A.K. 2012. *Phys. Rev. B* 85: 165210 (7 pages). Data courtesy of A.K. Ramdas.)

In summary,

- The donor or acceptor ground state is shifted by a central cell correction that depends on the specific impurity.
- For indirect gap semiconductors, the donor ground state is split due to valley-orbit splitting and p states are split due to anisotropic effective masses.
- Acceptors have a $1S_{3/2}$ ground state, and $S_{3/2}$, $P_{1/2}$, $P_{3/2}$, and $P_{5/2}$ bound excited states. For substitutional acceptors in diamond or zincblende semiconductors, the $P_{5/2}$ level splits into two levels.

Hydrogenic theory is extremely useful to characterize the electronic and optical properties of shallow-level impurities. Additional details and examples may be found in Pajot (2010).

5.4 DEEP LEVELS

The hydrogenic model applies when the Bohr radius is large enough to allow us to treat the crystal as a continuum with a uniform dielectric constant and effective mass. Deep levels, in contrast, have localized wave functions. The energy levels depend on the details of the defect. In this section, we first consider theoretical approaches that rely on a model for the defect potential. We then summarize "first principles" techniques that, ideally, calculate electronic energies without the need for adjustable constants.

The following discussion illustrates how the hydrogenic model breaks down when the central cell correction becomes too large. Let the potential energy of a donor electron be

Chapter 5—Electronic Properties **109**

$$(5.29) \qquad V(r) = -\frac{1}{4\pi\varepsilon\varepsilon_0}\frac{e^2}{r} + \delta V(r)$$

where $\delta V(r)$ is the central cell potential, modeled as a "square well" of radius b:

$$(5.30) \qquad \delta V(r) = \begin{cases} -V_1 & r < b \\ 0 & r \geq b \end{cases}$$

Here, $E = 0$ is defined to be the conduction-band minimum.

Due to the addition of δV, the hydrogen wave functions in Equation 5.17 are no longer valid. Instead, we use the variational principle to arrive at an approximate wave function. Let the ground-state wave function be approximated by the hydrogen $1s$ orbital:

$$(5.31) \qquad F(r) = \frac{1}{\sqrt{\pi r_1^3}}\exp(-r/r_1)$$

where r_1 is a free parameter. From Equation 5.16, the energy E is

$$(5.32) \qquad E = -\frac{\hbar^2}{2m^*}\frac{1}{F(r)}\nabla^2 F(r) + V(r)$$

Because $F(r)$ is not an exact solution, E is not a constant but rather a function of r. We calculate the "average energy" by taking the integral of E times the probability density:

$$(5.33) \qquad \langle E \rangle = \int_0^\infty E(r)F(r)^2 4\pi r^2 dr$$

It is straightforward (but tedious) to obtain an analytical solution to Equation 5.33. According to the variational principle, the correct wave function is the one that minimizes the average energy. Following that procedure, we vary r_1 and choose the value that minimizes $<E>$.

The results are shown in Figure 5.9. As V_1 is increased from zero, the radius r_1 gradually decreases, but the wave function remains delocalized. The donor electron energy $<E>$ experiences a small central cell correction. The wave function remains delocalized until V_1 reaches a critical value ($\sim$3 eV). At that point, the radius r_1 suddenly shrinks, and the wave function becomes localized. As the wave function becomes localized around the central cell, the energy plummets and becomes very sensitive to V_1. The donor has undergone a *shallow-to-deep transition*. For shallow (hydrogenic) states, the properties are primarily determined by the effective mass and dielectric constant and are weakly perturbed by the central cell. For deep (localized) states, the properties depend on the central cell and are insensitive to the conduction-band minima or valence-band maxima.

Even though this model illustrates several aspects of the shallow-to-deep transition, it does not attempt to describe the properties of the deep state accurately. As pointed out in Section 5.1, the envelope function is really only valid for a slowly varying potential, which is not true for a localized potential well. To address that problem, one can use the *Green's function* approach (Appendix F). This approach assumes that the energies and wave functions of the intrinsic semiconductor are known. We label the **k**-points of the conduction band as $\mathbf{k}1$, $\mathbf{k}2$, ... $\mathbf{k}N$, where N is

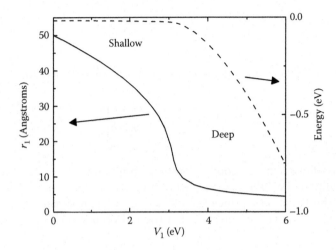

FIGURE 5.9 Bohr radius r_1 (solid line) and donor electron energy $\langle E \rangle$ (dashed line) as a function of the central-cell perturbation V_1. Zero energy corresponds to the conduction-band minimum. Parameters are $b = 5$ Å, $\varepsilon = 9.4$, and $m^* = 0.1\, m$. A shallow-to-deep transition occurs at $V_1 \approx 3$ eV.

the number of **k**-points in the first Brillouin zone. We consider only the lowest conduction band, although the analysis can be extended to include higher bands.

For an unperturbed semiconductor, the electron wave functions are given by Bloch wave functions $\psi_\mathbf{k}$, and the energies $E_{\mathbf{k}n}$ are defined by the conduction band. The energies are the eigenvalues of a diagonal matrix:

$$(5.34) \quad \begin{vmatrix} E_{\mathbf{k}1} - E & 0 & \cdots & 0 \\ 0 & E_{\mathbf{k}2} - E & \cdots & 0 \\ \vdots & \vdots & \ddots & \vdots \\ 0 & 0 & \cdots & E_{\mathbf{k}N} - E \end{vmatrix} = 0$$

This equation is solved by

$$(5.35) \quad E = E_{\mathbf{k}n}$$

where n is any integer between 1 and N.

When we introduce a defect potential $V(\mathbf{r})$, the matrix is no longer diagonal. Instead, we must solve

$$(5.36) \quad \begin{vmatrix} E_{\mathbf{k}1} - E + V_{11} & V_{12} & \cdots & V_{1N} \\ V_{21} & E_{\mathbf{k}2} - E + V_{22} & \cdots & V_{2N} \\ \vdots & \vdots & \ddots & \vdots \\ V_{N1} & V_{N2} & \cdots & E_{\mathbf{k}N} - E + V_{NN} \end{vmatrix} = 0$$

where

$$(5.37) \quad V_{ij} = \int \psi_{\mathbf{k}i}^* V(\mathbf{r}) \psi_{\mathbf{k}j} d^3\mathbf{r}$$

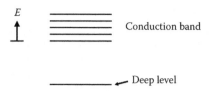

FIGURE 5.10 Energy levels calculated by the Green's function approach.

and the asterisk indicates the complex conjugate (i is replaced by $-i$). This integral depends on the details of the potential and the Bloch wave functions. To simplify things, we assume that the potential is highly localized such that it is zero everywhere except $\mathbf{r} = 0$. Furthermore, we assume that the Bloch wave functions are given by complex exponentials. With these approximations, V_{ij} becomes a *constant* that does not depend on i or j. For such a potential, all the energies in the conduction band play an important role, not just points near the conduction-band minimum. In other words, a potential $V(\mathbf{r})$ that is localized in real space is delocalized in k space (V_{ij}).

The energies E obtained by solving Equation 5.36 are shown schematically in Figure 5.10, for the case of an attractive potential ($V_{ij} < 0$). The effect of the defect potential is that one energy level is removed from the conduction band and placed in the band gap. This is the deep defect level. One desirable feature of the Green's function approach is that it clearly shows changes in the electronic structure that are caused by the introduction of a defect (Pantelides, 1986). The disadvantage is that it is not a first-principles technique. One assumes a *known* potential, which may not be accurate.

Several theoretical methods have been developed to calculate the electronic properties of deep-level defects accurately (Drabold and Estreicher, 2007). The crystal is commonly modeled with *supercells* that contain hundreds of host atoms along with the defect atoms. The supercell repeats, just like the unit cell of a crystal, allowing one to use superpositions of plane waves to describe the electrons. In practice, electrons are separated into valence electrons and ion core electrons. The ion potentials are modeled with *pseudopotentials*, which are designed to produce accurate wave functions for the valence electrons outside the core (Austin et al., 1962). Note that in semiconductors such as GaN or ZnO, the Ga or Zn $3d$ electrons are actually treated as valence electrons (Janotti and Van de Walle, 2007).

Today, the majority of defect calculations involve *density functional theory* (DFT). This method uses the fact that the ground-state electron density uniquely determines the external potential that acts on these electrons (Hohenberg and Kohn, 1964). The total energy of a system of electrons is given by

(5.38) $$E = T[n] + V_{ee}[n] + V_{ext}[n] + E_{xc}[n]$$

where $n(\mathbf{r})$ is the electron density, T is the kinetic energy, V_{ee} is the electron-electron potential energy, V_{ext} is the electron-ion potential energy, and E_{xc} is the exchange-correlation term. The square brackets indicate that the terms are "functionals" whose values depend on the function $n(\mathbf{r})$. The Born–Oppenheimer approximation is used (i.e., the ion cores are assumed to be frozen with respect to the fast-moving valence electrons).

The first three terms on the right-hand side of Equation 5.38 are relatively straightforward to calculate numerically. The kinetic energy T is for a single particle in a "mean field" potential. The potential energy terms V_{ee} and V_{ext} are given by the classical Coulomb forces. The hardest term to calculate is the exchange-correlation functional, $E_{xc}[n]$. The *exchange* part of this functional comes from the symmetry of the wave function. For electrons, the total wave function must be antisymmetric with respect to particle exchange, a mathematical feature that satisfies the Pauli exclusion principle. Electrons with the same spin will tend to avoid each other, thereby lowering their Coulomb repulsion. The *correlation* part is due to the fact that, in a many-body wave function, the position of one electron affects the positions of all the other electrons.

Obtaining an exact solution for the exchange-correlation functional is impractical for all but the simplest systems. One such case is the uniform electron gas, where $n(\mathbf{r})$ is a constant. The *local density approximation* (LDA) estimates the exchange-correlation functional by partitioning the system into boxes, each of which is treated as a uniform electron gas (Kohn, 1999). $E_{xc}[n]$ is calculated by summing the exchange-correlation energies from all the boxes.

Although parameters such as lattice constants are calculated with good accuracy, LDA underestimates band gaps, which in turn affects the defect energies. Various corrective remedies have been proposed to get around this problem. One method is to use "hybrid functionals" to estimate the exchange-correlation term. The hybrid functionals are linear combinations of the LDA value with values obtained by different techniques. The coefficients in the linear combination are determined empirically or by comparison with other calculations. Another method is the "LDA + U" method, which artificially shifts the energies of certain states (e.g., Zn d states) by an amount U.

The DFT-LDA method calculates the total electron energy (Equation 5.38) in the supercell. Atoms are allowed to relax in order to minimize the energy. To determine the acceptor or donor levels, the energy is calculated for different defect charge states. As an example, consider a donor impurity. In its neutral charge state, it has an energy $\varepsilon(0)$. In the positive charge state, the donor energy is $\varepsilon(+)$ and the donor electron energy is equal to the Fermi energy E_F. The (0/+) level is given by the value of E_F for which these energies are equal; hence,

(5.39)
$$E(0/+) = \varepsilon(0) - \varepsilon(+)$$

For values of E_F above $E(0/+)$, the neutral charge state is energetically preferred. If E_F lies below $E(0/+)$, then the defect gives up its electron and becomes positively charged.

What happens when $E(0/+)$ is calculated to lie in the conduction band? In that case, it is energetically favorable for the donor to give up its electron. However, the electron will be weakly attracted to the positively charged donor, and we are back to the hydrogenic model. A similar argument applies to an acceptor level $E(0/-)$ that lies in the valence band. Because LDA calculations underestimate the band gap, though, care must be taken when using this method to determine whether a level is shallow or deep. This problem is the focus of ongoing theoretical efforts.

An important example of donor and acceptor level calculations is that of interstitial hydrogen. As discussed in Section 2.5, interstitial hydrogen is a negative-U center (i.e., the total energy of neutral hydrogen is always higher than that of the positive or negative charge state). Therefore, hydrogen has a (+/−) level. When the Fermi level is above the (+/−) level, hydrogen exists as H^-. When the Fermi level is below the (+/−) level, hydrogen is a proton (H^+). The H(+/−) level was calculated for various semiconductors (Van de Walle and Neugebauer, 2003). The relative positions of the H(+/−) levels and valence band maxima were calculated, using the DFT-LDA approach, while the conduction band minima were determined from experimental band gaps. The results of the calculations are striking: the H(+/−) level is constant, relative to the vacuum level, regardless of the semiconductor (Figure 5.11). This "universal alignment" of hydrogen has interesting implications. For materials with high electron affinities (ZnO, InN), the H(+/−) level lies in the conduction band, making hydrogen a shallow donor. For a semiconductor like GaSb, the H(+/−) level is predicted to lie in the valence

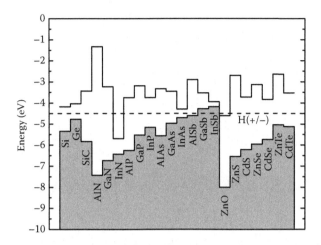

FIGURE 5.11 Valence band maxima and conduction band minima (solid lines), and the H(+/−) level (dashed line), as calculated by density functional theory (DFT) for various semiconductors. Here, zero energy corresponds to the vacuum level. We shifted the Si and Ge levels down by 0.5 eV to give better agreement with experimental data. (After Van de Walle, C.G. and J. Neugebauer. 2003. *Nature* 423: 626–628.)

band, which suggests that it is a shallow acceptor. For most semiconductors, the H(+/−) level is somewhere in the gap.

5.5 CARRIER CONCENTRATIONS AS A FUNCTION OF TEMPERATURE

The previous sections discussed methods to calculate defect energy levels. In this section, we show how the ground-state energy levels of dopants and defects affect the concentration of free carriers (electrons and holes) in a semiconductor. We first consider an intrinsic semiconductor and then consider a semiconductor with donors and acceptors.

The Fermi distribution function (Equation 1.21) tells us the probability that a state with energy E will be populated with an electron. Let us assume that the Fermi level is below the conduction band minimum such that, in the conduction band, we can approximate the Fermi distribution function by the Boltzmann tail (Equation 1.22):

$$f(E) = e^{(E_F - E)/k_B T} \tag{5.40}$$

where $E = 0$ is set arbitrarily at the valence band maximum. The conduction band is modeled as a parabolic band (Equation 1.16) with effective mass m_e. Such a band has a *density of states* given by

$$D(E) = \frac{V}{2\pi^2} \left(\frac{2m_e}{\hbar^2} \right)^{3/2} (E - E_g)^{1/2} \tag{5.41}$$

where V is the volume of the crystal and E_g is the band gap (Appendix E). Here, the density of states is the number of electronic states per unit of energy.

The number of electrons in the conduction band, N_{el}, is obtained by integrating the number of states $D(E)dE$ times the probability $f(E)$ that a state is filled:

5.5 Carrier Concentrations as a Function of Temperature

(5.42)
$$N_{el} = \int_{E_g}^{\infty} D(E)f(E)dE$$

$$= \frac{V}{2\pi^2}\left(\frac{2m_e}{\hbar^2}\right)^{3/2} e^{E_F/k_BT} \int_{E_g}^{\infty} (E - E_g)^{1/2} e^{-E/k_BT} dE$$

Making the substitution $x = (E - E_g)/k_BT$ yields

(5.43)
$$n = \frac{1}{2\pi^2}\left(\frac{2m_e}{\hbar^2}\right)^{3/2} e^{E_F/k_BT} (k_BT)^{3/2} \left(\int_0^{\infty} x^{1/2} e^{-x} dx\right) e^{-E_g/k_BT}$$

where $n = N_{el}/V$ is the density (number per unit volume) of free electrons. Using the fact that the integral equals $\sqrt{\pi}/2$, the free electron density is written

(5.44)
$$n = N_C e^{(E_F - E_g)/k_BT}$$

where

(5.45)
$$N_C = 2\left(\frac{m_e k_B T}{2\pi\hbar^2}\right)^{3/2}$$

is the "effective density of states" of the conduction band. Compared to the exponential term, it has a weak temperature dependence ($\sim T^{3/2}$). For $T = 300$ K and $m_e = m$, $N_C = 2.5 \times 10^{19}$ cm^{-3}, a value that is significantly higher than typical dopant concentrations.

One can obtain the density of holes by considering that the probability of finding a hole in the valence band is equal to the probability of *not* finding an electron, $1 - f(E)$. The calculation yields

(5.46)
$$p = N_V e^{-E_F/k_BT}$$

and

(5.47)
$$N_V = 2\left(\frac{m_h k_B T}{2\pi\hbar^2}\right)^{3/2}$$

where p is the density of free holes, and N_V is the effective density of states of the valence band. For simplicity, the valence bands are approximated by a single parabolic band with effective mass m_h (Appendix E).

From Equations 5.44 and 5.46, one can see that raising the Fermi energy increases the number of electrons, whereas lowering the Fermi energy increases the number of holes. The product of n and p is

(5.48)
$$np = N_C N_V e^{-E_g/k_BT}$$

which is independent of the Fermi energy. Equation 5.48 is called a *mass-action law*.

Chapter 5—Electronic Properties **115**

For an *intrinsic* semiconductor, electrons in the conduction band can only come from the valence band, so $n = p$. Applying this condition to Equation 5.48, the intrinsic carrier concentration is given by

$$n_i = \sqrt{N_C N_V}\, e^{-E_g/2k_B T} \tag{5.49}$$

Inserting this expression into Equation 5.44 yields the intrinsic Fermi level:

$$E_{F,i} = \frac{1}{2} E_g + \frac{3}{4} k_B T \ln(m_h/m_e) \tag{5.50}$$

Because the second term on the right-hand side is typically much smaller than the first, an intrinsic semiconductor has its Fermi level near the middle of the gap.

In a real semiconductor, several donor and acceptor species will exist. Suppose we have a silicon crystal with phosphorus as the majority donor dopant and some residual boron as a minority acceptor dopant. Let N_d and N_a be the concentrations of donors and acceptors, respectively. To minimize energy, the donor electrons will fill all available energy states from the lowest energy upwards. Because the valence band is full, no donor electrons can be accommodated by valence band states. Boron acceptors, however, can bind donor electrons. After all the acceptors have an extra electron (and have become negatively charged), there are only $N_d - N_a$ donors left that can give their electrons to the conduction band. $N_d - N_a$ is called the "net donor concentration." One defines the *compensation ratio k* as

$$k = \frac{N_{\text{minority}}}{N_{\text{majority}}} \tag{5.51}$$

With this definition, k can lie between 0 (uncompensated) and 1 (fully compensated).

Consider a semiconductor that contains one kind of donor with binding energy E_d and concentration N_d, one kind of acceptor with concentration N_a, and $N_d > N_a$. Each donor is able to accommodate, at most, one electron. For simplicity, we ignore the spin degeneracy of the donor ground state. From the Fermi distribution function (Equation 1.21), the concentration of occupied donor states (neutral donors) is given by

$$N_d^{(0)} = \frac{N_d}{1 + e^{(E-E_F)/k_B T}} \tag{5.52}$$

where E is the position of the donor level relative to the valence band maximum. Using Equation 5.44 to solve for E_F, we obtain

$$N_d^{(0)} = \frac{N_d}{1 + e^{-E_d/k_B T}(N_C/n)} \tag{5.53}$$

where $E_d = E_g - E$.

Donors can become ionized by giving their electrons to compensating acceptors or to the conduction band. Let us assume that the semiconductor is not in the intrinsic regime (i.e., a negligible number of electrons are excited from the valence band into the conduction band). In that case, the concentration of neutral donors is

$$N_d^{(0)} = N_d - N_a - n \tag{5.54}$$

Equating the right sides of Equations 5.53 and 5.54 yields

$$(5.55) \qquad N_d - N_a - n = \frac{N_d}{1 + e^{-E_d/k_BT}(N_C/n)}$$

From this equation, one can derive an expression for the free-electron concentration:

$$(5.56) \qquad \frac{(N_a + n)n}{(N_d - N_a - n)} = N_C e^{-E_d/k_BT}$$

Defining

$$(5.57) \qquad \phi \equiv N_C e^{-E_d/k_BT}$$

the free-electron concentration is given by

$$(5.58) \qquad n = \frac{1}{2}(N_a + \phi)\left[\sqrt{1 + \frac{4(N_d - N_a)\phi}{(N_a + \phi)^2}} - 1\right]$$

5.6 FREEZE-OUT CURVES

It is useful to examine the carrier concentration over certain ranges in temperature. First, we consider temperatures that are high ($k_BT \gg E_d$) but not high enough to excite electrons across the band gap. In that regime, the exponential in Equation 5.56 approaches one, so that

$$(5.59) \qquad (N_a + n)n \approx N_C(N_d - N_a - n)$$

For moderate doping levels, (N_d, $N_a \ll N_C$), the solution to Equation 5.59 can be approximated as

$$(5.60) \qquad n = N_d - N_a$$

This is called the *saturation regime*. In this regime, all the donors are ionized. The free-electron concentration is equal to the net donor concentration. This regime is useful for many semiconductor devices because the carrier concentration is stable over a range of temperatures and is controlled by the doping.

As the temperature decreases, the electrons in the conduction band start to relax back into the donor levels. First, we consider the range of concentrations $N_d \gg n \gg N_a$. In this range, Equation 5.56 simplifies to

$$(5.61) \qquad n = \sqrt{N_C N_d}\, e^{-E_d/2k_BT}$$

In an Arrhenius plot, where $\log(n)$ is plotted as a function of inverse temperature, Equation 5.61 is a nearly straight line with a slope proportional to $-E_d/2k_B$. This is the "half-slope" *freeze-out regime*. Heavily compensated semiconductors, which have comparable concentrations of donors and acceptors, do not have a distinct half-slope freeze-out regime.

As the temperature decreases more, n becomes small such that $n \ll N_d, N_a$. In that case, Equation 5.56 is approximated as

$$(5.62) \qquad n = \left(\frac{N_d - N_a}{N_a}\right) N_C e^{-E_d/k_B T}$$

This is the "full-slope" freeze-out regime. In an Arrhenius plot, Equation 5.62 is a nearly straight line with a slope proportional to $-E_d/k_B$.

In summary, for n-type semiconductors, an Arrhenius plot yields the following information:

- The electron saturation concentration corresponds to $N_d - N_a$.
- The concentration at which the half-slope regime turns into the full-slope regime corresponds to N_a.
- The slopes of the freeze-out curves yield E_d.

We cannot determine E_a from the Arrhenius plot. We simply assumed that the acceptor level lies at least several $k_B T$ below the donor level. In p-type semiconductors, the roles of donors and acceptors are exchanged.

What happens to the freeze-out curve for higher temperatures? Eventually, the intrinsic carrier concentration will first become comparable to, then larger than, $N_d - N_a$. The semiconductor assumes intrinsic properties from this temperature on upwards. In this *intrinsic regime*, the free electron and hole concentrations are equal and given by Equation 5.49. In an Arrhenius plot, the slope of the freeze-out curve in the intrinsic regime is proportional to $-E_g/2k_B$. This fact provides researchers with a way to determine the band gap from electrical measurements. The intrinsic, saturation, and freeze-out regimes are shown in Figure 5.12.

Let us discuss a slightly more complicated case in which a deep level is present in addition to the shallow donors and acceptors. Such a situation is found for dislocation-free, ultra-pure, p-type

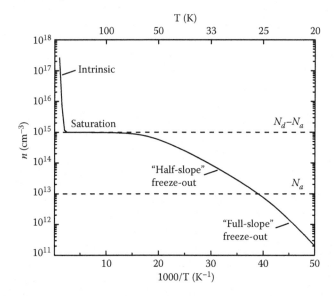

FIGURE 5.12 Arrhenius plot of free-electron concentration versus inverse temperature, for an n-type semiconductor. For this plot, $N_d = 10^{15}$ cm^{-3}, $N_a = 10^{13}$ cm^{-3}, $E_g = 1.1$ eV, $E_d = 33$ meV, and $m_e = m_h = m$.

5.6 Freeze-Out Curves

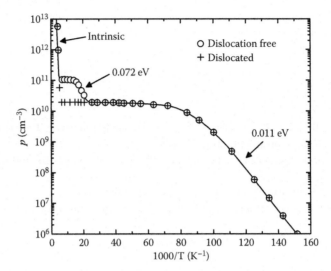

FIGURE 5.13 Arrhenius plots of free-hole concentration versus inverse temperature, for p-type germanium. Both samples contain acceptors with a binding energy of 0.011 eV. The dislocation-free sample also contains deep acceptors (0.072 eV), attributed to divacancy-hydrogen complexes. (After Haller, E.E., W.L. Hansen, and F.S. Goulding. 1981. *Adv. Phys.* 30: 93–138.)

germanium (Haller et al., 1981). Figure 5.13 shows the Arrhenius plots for a dislocation-free and a dislocated piece of germanium single crystal cut from a slice of a partially dislocated crystal. We find $N_a - N_d = 2 \times 10^{10}$ cm^{-3}. The low-temperature freeze-out slope is 11 meV, which is the full slope for shallow acceptors. No half-slope regime can be seen, which tells us that N_d is slightly smaller than N_a (i.e., we have a highly compensated semiconductor). Above 50 K, the hole concentration rises again for the dislocation-free piece. This indicates that holes from an acceptor level, which is deeper than the shallow acceptor level, are promoted to the valence band. Careful analysis yields $E_{a,deep} = 72$ meV. This acceptor level is likely due to divacancy–hydrogen (V_2H) complexes that form during the slow cooling of the crystal after growth in a H$_2$ atmosphere (Haller, 1991). The excess vacancies cannot find sinks in the dislocation-free crystal and form V_2H centers. In the dislocated crystal, the excess vacancies condense at dislocations, causing dislocation climb.

Free carrier statistics offer a powerful semiconductor characterization tool (Blakemore, 1987). When precisely doped and compensated crystals are developed for a specific application, free carrier freeze-out curves provide quantitative analysis. When an unknown impurity is introduced into a crystal, one can determine the electron or hole binding energy. Many of the energy levels shown in Figure 2.2 were determined in this way.

Our discussion of freeze-out curves has focused on carrier concentrations. It is also useful to examine the position of the Fermi level as a function of temperature. One can calculate the Fermi level from the number of free electrons (Equation 5.44) or holes (Equation 5.46) at a given temperature. At temperatures sufficiently high for the semiconductor to be intrinsic, the Fermi level lies close to the middle of the band gap (Equation 5.50).

At lower temperatures, the semiconductor becomes extrinsic and the dopants dominate the free-carrier concentration. As the temperature drops, the Fermi level moves either toward the conduction band edge for n-type semiconductors or toward the valence band top for p-type semiconductors. When 50% of the majority dopants are frozen out (neutral), the Fermi level coincides with the majority dopant energy level. In an uncompensated crystal at low temperatures, the Fermi level would approach a position close to halfway between the dopant energy level and the band extremum. In compensated crystals, though, a fraction of the majority dopants remains

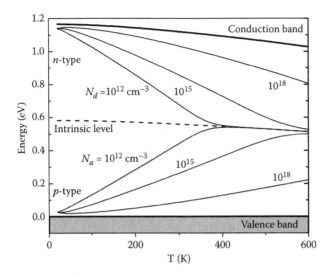

FIGURE 5.14 Fermi energy versus temperature for different doping concentrations in silicon. Zero energy corresponds to the valence band maximum. The intrinsic Fermi level and conduction band edge are also shown. (After Grove, A.S. 1967. *Physics and Technology of Semiconductor Devices.* New York, NY: John Wiley & Sons.)

ionized at the lowest temperatures. With the Fermi distribution becoming abrupt as we go to lower temperatures, the Fermi level becomes pinned to the dopant energy level. Figure 5.14 displays the position of E_F for a number of p- and n-type dopant concentrations as a function of temperature for silicon. The reduction of the band gap with temperature is also shown.

5.7 SCATTERING PROCESSES

Sections 5.5 and 5.6 described how to calculate the free-carrier concentration in a semiconductor, given the donor/acceptor energy levels and dopant concentrations. In this section, we discuss how carriers are scattered, in three ways: phonons, ionized impurities, and neutral impurities. These scattering processes limit the carrier mobility.

The drift velocity **v**, defined to point along the direction of the applied electric field, is an average velocity superimposed on the random thermal motion of the free carriers (Section 1.10). For low electric fields, the average carrier speed v is dominated by the thermal energy:

$$\frac{1}{2}m^*v^2 \approx \frac{3}{2}k_B T \tag{5.63}$$

In this discussion, we will focus on the temperature dependence of scattering processes. It is sufficient to note that

$$v \sim T^{1/2} \tag{5.64}$$

and, because the momentum $\hbar k$ is proportional to velocity, the mean value for k is

$$k \sim T^{1/2} \tag{5.65}$$

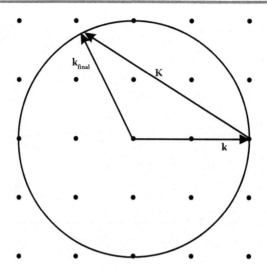

FIGURE 5.15 *K*-space diagram of phonon scattering. A carrier with momentum $\hbar\mathbf{k}$ is scattered by a phonon of momentum $\hbar\mathbf{K}$. For an elastic process, $\mathbf{k}_{final}$ must lie on the surface of constant energy, shown here as a circle. The **k** points are depicted as black dots; in reality, they are much denser than in the figure.

As discussed in Section 1.4, phonons can be treated as particles with momentum and energy. Consider a collision between a carrier and a phonon where the phonon is destroyed (absorbed). In that process, the phonon momentum is transferred to the electron. Here, we assume that the energy of the phonon is low, compared to that of the carrier, such that the collision is nearly *elastic*. To a good approximation, the *direction* of the carrier momentum $\hbar\mathbf{k}$ is altered but the *magnitude* remains unchanged.

Figure 5.15 shows an example of *phonon scattering* in *k* space, where the **k**-points are indicated by a grid of dots. Because the carrier energy depends only on the magnitude *k*, the surface of constant energy is a sphere. For an elastic scattering process, therefore, the final electron **k** value must lie on the surface of the sphere. The number of possible final states is proportional to the surface area of the sphere:

$$\text{\# of states} \sim 4\pi k^2 \tag{5.66}$$

The population of phonons *n* is given by the Planck distribution:

$$n = \frac{1}{e^{\hbar v_s K / k_B T} - 1} \tag{5.67}$$

where v_s is the sound speed for an acoustical phonon (Equation 1.10). In the limit of low phonon energies $\hbar v_s K$, Equation 5.67 can be approximated:

$$n \approx k_B T / \hbar v_s K \tag{5.68}$$

As shown in Figure 5.15, *K* ranges from 0 to 2*k*. As a rough approximation, we replace *K* with an "average" value *k*, so that

$$n \sim T / k \tag{5.69}$$

The scattering rate is proportional to the number of states (Equation 5.66) multiplied by the number of phonons (Equation 5.69):

$$(5.70) \qquad 1/\tau \sim (4\pi k^2)(T/k)$$

From Equation 5.65, we obtain the temperature dependence:

$$(5.71) \qquad 1/\tau \sim T^{3/2}$$

Finally, because the mobility is proportional to τ (Equation 1.34), phonon scattering leads to a mobility

$$(5.72) \qquad \mu_{\text{phonon}} = A T^{-3/2}$$

where the constant A contains fundamental and materials parameters (Seeger, 1982).

Qualitatively, phonon scattering limits the mobility at high temperatures, due to the larger population of phonons. In addition to phonon absorption, electrons can create (emit) phonons, a process that also has a $T^{-3/2}$ dependence. In general, the mobilities of carriers in direct-gap semiconductors follow Equation 5.72 rather accurately at high temperatures. Deviations in the temperature exponent are found for indirect semiconductors such as germanium and silicon. For pure germanium, one finds a $T^{-5/3}$ dependence for electrons above 70 K. Intervalley scattering is the major cause for these deviations.

At low temperatures, Coulomb scattering by *ionized impurities* becomes dominant. Consider N ionized impurities per cm^3, each with a cross section σ. We define σ to be an effective area, centered on the impurity, such that carriers passing through the area experience significant deflection. The carrier scattering rate is given by

$$(5.73) \qquad 1/\tau = Nv\sigma$$

Early in the last century, Rutherford solved this fundamental problem when he analyzed the scattering of α particles by charged nuclei (Eisberg and Resnick, 1985). The cross section is proportional to Z^2 and inversely proportional to the carrier kinetic energy squared:

$$(5.74) \qquad \sigma \sim \left(\frac{1}{4\pi\varepsilon\varepsilon_0} \frac{Ze^2}{\frac{1}{2}mv^2} \right)^2$$

Inserting this expression into Equation 5.73 yields

$$(5.75) \qquad 1/\tau \sim NZ^2/v^3 \sim NZ^2 T^{-3/2}$$

Because the mobility is proportional to τ, ionized impurity scattering leads to a mobility

$$(5.76) \qquad \mu_{\text{ion}} = B\frac{T^{3/2}}{NZ^2}$$

where the constant B contains fundamental and materials parameters (Seeger, 1982). Qualitatively, ionized impurity scattering is reduced as the thermal energy of carriers increases.

The third scattering mechanism that we consider is *neutral impurity scattering*. This is a rather weak interaction and can only be observed in doped semiconductors with very low compensation. At low temperatures, the dopant impurities are neutral (the free carriers are frozen out), and few ionized impurities remain because of the low compensation. In the low-temperature regime, neutral impurity scattering is temperature independent:

$$\mu_{\text{neutral}} = \frac{C}{N_0} \tag{5.77}$$

where N_0 is the neutral impurity concentration, and C is a materials constant.

To obtain the total mobility in a semiconductor, we note that the scattering rates $1/\tau$ are additive. Therefore, we add the *inverse* of the various mobility terms:

$$\frac{1}{\mu} = \frac{1}{\mu_{\text{phonon}}} + \frac{1}{\mu_{\text{ion}}} + \frac{1}{\mu_{\text{neutral}}} \tag{5.78}$$

Equation 5.78, which is analogous to adding parallel resistors, is called *Matthiessen's rule*. In this discussion, we assumed an average thermal energy given by Equation 5.63. In reality, there is a distribution of carrier kinetic energies. A more accurate expression is given by

$$\mu = \frac{4}{3\sqrt{\pi}} \int_0^\infty \frac{x^{3/2} e^{-x}}{1/\mu_{\text{phonon}} + 1/\mu_{\text{ion}} + 1/\mu_{\text{neutral}}} dx \tag{5.79}$$

where x is the ratio of the carrier kinetic energy to $k_B T$.

A plot of hole mobility in NTD-doped ^{70}Ge:Ga is shown in Figure 5.16 (Itoh et al., 1994). The solid line is the theoretical mobility, without any adjustable parameters. The dashed lines show the contributions from phonon, ionized impurity, and neutral impurity scattering that went into Equation 5.79. At high temperatures, phonon scattering limits mobility. In addition, high temperatures result in more ionized gallium acceptors, causing significant ionized impurity scattering. At low temperatures, neutral impurity scattering dominates. Ionized impurity scattering also plays a role at low temperatures, because the cross section increases as the carrier kinetic energy decreases (Equation 5.74).

In this section, our discussion of carrier mobility has focused on *majority* carriers (i.e., electrons in n-type semiconductors and holes in p-type semiconductors). In devices such as p-n junctions, the transport of *minority* carriers is important. Processes involving minority carriers will be treated in Chapter 9.

5.8 HOPPING AND IMPURITY BAND CONDUCTION

So far, we have been discussing lightly doped semiconductors, in which the average distance between dopants is large such that interactions may be neglected. In those cases, conduction is achieved through electrons in the conduction band or holes in the valence band. As temperature is lowered, the carriers freeze out. The decrease in carrier concentration results in a decrease in conductivity.

When we consider interactions between dopants, however, new conduction pathways become possible. The electrical conductivity is given by

$$\sigma = ne\mu + \sigma_{\text{hopping}} + \sigma_{\text{impurity band}} \tag{5.80}$$

where n is the free-carrier concentration; μ is the mobility; σ_{hopping} and $\sigma_{\text{impurity band}}$ are due to hopping and impurity band conduction, discussed in this section. We consider donors in an n-type

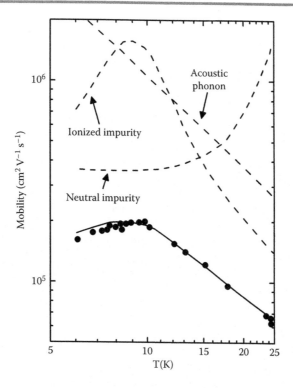

FIGURE 5.16 Mobility of holes in ^{70}Ge:Ga. The solid line is the theoretical calculation. The dashed lines show contributions due to phonon, ionized impurity, and neutral impurity scattering. (After Itoh, K.M. et al. 1994. *Phys. Rev. B* 50: 16995–17000.)

semiconductor, but the same arguments apply to acceptors in *p*-type semiconductors. Similar to the LCAO approach to band structure, interactions between donors lead to a broadening of the donor level into a range of energies. A particular donor state has a random energy as well as position.

In *hopping conduction*, carriers tunnel from occupied to empty impurity levels, without needing to go into the conduction band. Because empty levels are required, hopping conduction depends strongly on the number of compensating impurities. In the variable range hopping model (Mott, 1972), the temperature dependence of the hopping conductivity is given by

$$\sigma_{\text{hopping}} = \sigma_0 \exp\left[-(T_0/T)^\beta\right] \qquad (5.81)$$

where $\beta = \frac{1}{4}$; σ_0 and T_0 are empirical constants. "Variable range" refers to the idea that electrons can hop between states of random energies and positions. In an energetic picture, an electron is promoted from a filled donor state below the Fermi level to an empty state above the Fermi level, leaving behind a hole (Figure 5.17). The Coulomb interaction between the electron and hole reduces the density of states near the Fermi level, which is important for low-temperature excitations. This effect results in an exponent of $\beta = \frac{1}{2}$ at low temperatures (Shklovskii and Efros, 1984).

To investigate hopping conduction, ultra-pure germanium was doped via NTD to a wide range of doping levels (Itoh et al., 1996). Dopant concentrations were chosen to be high enough for hopping conduction to occur at low temperatures. Figure 5.18 shows a plot of resistivity ($\rho = 1/\sigma$)

5.8 Hopping and Impurity Band Conduction

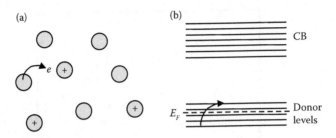

FIGURE 5.17 Hopping conduction in (a) real space and (b) energy space. An electron (e) travels from a neutral donor to an unfilled, positive donor.

versus $T^{-1/2}$ for several samples. The experimental points follow Equation 5.81, with $\beta = \frac{1}{2}$, over up to ten orders of magnitude.

Astronomers and particle astrophysicists have taken advantage of the well-controlled resistivity versus temperature dependence of NTD germanium. By applying Ohmic contacts to millimeter-sized cubes of NTD germanium, they fabricated variable temperature resistors (thermistors) as photon detectors in the focal planes of far-IR telescopes, on the ground and in space (Rieke et al., 2004). The search for neutrino-less double beta decay and dark matter particles is performed in an experiment called Cryogenic Underground Observatory for Rare Events, with an array of over 2000 NTD germanium thermistors (Fiorini, 1998). Each thermistor is glued to a $5 \times 5 \times 5$ cm^3 TeO$_2$ single crystal cooled down to $\sim 10^{-2}$ K. The NTD germanium is doped precisely to impurity levels just short of the metal-to-insulator transition. These are just two examples that illustrate the broad applicability of semiconductors in many fields of science.

Impurity band conduction is the process by which a carrier moves from one dopant level to a filled dopant level. Because it does not require empty states, impurity band conduction is insensitive to the concentration of compensating impurities. Instead, it relies on the fact that donors can weakly bind a second electron, becoming negatively charged. As the donor wave functions

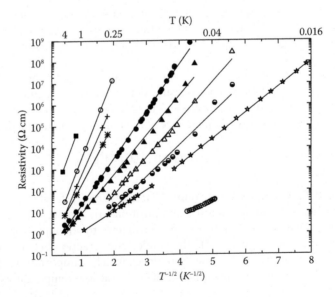

FIGURE 5.18 Hopping conduction in several germanium samples doped by neutron transmutation doping (NTD). The straight lines indicate that the conductivity is given by Equation 5.81, with $\beta = 1/2$. (After Itoh, K. et al. 1996. *Phys. Rev. Lett.* 77: 4058–4061.)

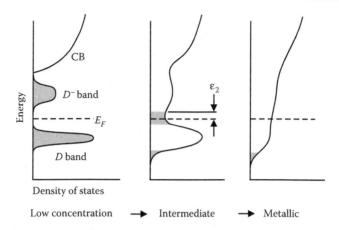

FIGURE 5.19 Impurity band conduction. At low donor concentrations, the D and D^- bands are distinct. At intermediate concentrations, the bands overlap. Shaded regions indicate localized (immobile) states. The minimum energy required to promote an electron into an extended (mobile) state is ε_2. At high concentrations, $\varepsilon_2 = 0$ and the semiconductor is metallic. (After Kamimura, H. and H. Aoki. 1989. *The Physics of Interacting Electrons in Disordered Systems*. Oxford, UK: Clarendon Press.)

overlap, the neutral and negatively charged donor levels broaden into bands, denoted D and D^- (Hubbard, 1964; Dubon et al., 1997). At a low donor concentration, these bands are narrow, and the states are localized. No impurity band conduction can occur. At higher concentrations, the bands broaden (Figure 5.19). From Anderson localization theory (Anderson, 1958), the edges of each band correspond to localized states, whereas the center consists of extended states. A minimum energy ε_2 is required to promote an electron from the filled D band to an extended state in the empty D^- band, where it is mobile and able to conduct. The thermal activation is given by

$$\sigma_{\text{impurity band}} = \sigma_0 \exp(-\varepsilon_2/k_B T) \tag{5.82}$$

As the donor concentration increases, the bands broaden and ε_2 decreases. At a critically high donor concentration n_c, ε_2 goes to zero and the system becomes metallic. In the metallic phase, we have a *degenerate semiconductor*. The free-electron concentration is essentially independent of temperature. The concentration n_c defines the point at which such a *metal-insulator transition* occurs (Mott, 1990). On average, the distance between neighboring impurities is $\sim n_c^{-1/3}$. When this distance becomes comparable to the Bohr radius r_0, then the metal-insulator transition occurs. Empirically,

$$n_c^{1/3} r_0 \approx 0.26 \tag{5.83}$$

for a wide range of condensed-matter systems (Kamimura and Aoki, 1989).

5.9 SPINTRONICS

The ability to transport charge, via the motion of electrons or holes, makes semiconductors an indispensable part of modern electronics. More recently, researchers have proposed "spintronic" devices, in which the carrier *spins* are manipulated. In quantum computing devices, the spin states would be used to construct "qubits," theoretically enabling the manipulation of huge amounts of

126 5.9 Spintronics

data (Steane, 1998). For nonvolatile memory storage applications, ferromagnetic semiconductors could store data for extended periods of time.

Semiconductors doped with transition metals, called *dilute magnetic semiconductors*, have been shown to exhibit ferromagnetism at low temperatures. GaAs doped with Mn was an early, promising example. The substitutional Mn dopants serve two purposes. First, they act as acceptors (Mn^{2+}), increasing the hole concentration. Second, the Mn^{2+} atoms have unfilled d orbitals that give rise to an electronic spin. The localized Mn^{2+} spins interact with free holes through the exchange interaction (Dietl et al., 2000). The holes "mediate" the exchange interaction, causing the Mn^{2+} spins to align. Below the *Curie temperature*, the semiconductor becomes ferromagnetic and 100% of the Mn^{2+} spins align along a particular direction.

The Curie temperature increases with the concentration of substitutional Mn^{2+} dopants and hole concentration. $Ga_{1-x}Mn_xAs$ samples with $x = 0.07$ and hole concentrations exceeding 10^{20} cm^{-3}, grown by low-temperature MBE, exhibit a Curie temperature of 110 K (Ohno, 1998). Attempting to push the Mn concentration higher results in Mn interstitials and the precipitation of separate phases such as MnAs (Yu and Walukiewicz, 2002). The quest for room-temperature ferromagnetism led many researchers to pursue transition-metal doping of wide-band-gap semiconductors (Pearton et al., 2003). Unfortunately, many of the early reports of Curie temperatures above 300 K proved to be the result of separate phases (Chambers and Farrow, 2003). Future advances in semiconductor doping may enable researchers to surmount the "room-temperature barrier."

PROBLEMS

5.1 GaN has an electron effective mass of $0.2m$ and static dielectric constant $\varepsilon = 9$. From the hydrogenic model, find the donor binding energy and Bohr radius.

5.2 Explain why s states experience a central-cell correction while p states do not.

5.3 For a particular semiconductor, the hydrogenic model gives a donor binding energy of 28 meV. Find the energy of the $1s \rightarrow 2p$ transition for the following cases:
   a. A donor with $E_d = 28$ meV (perfectly hydrogenic).
   b. A donor with $E_d = 35$ meV (shift due to central-cell correction).

5.4 Consider a semiconductor with the diamond crystal structure and lattice constant a. A vacancy has four dangling bond orbitals (Equations 5.20 and 5.21). Let each orbital be given by a Gaussian function centered on an atom, $\phi_i = A\exp[-(\mathbf{r} - \mathbf{r}_i)^2/b^2]$. Find the probability density $|\psi|^2$ at the center of the vacancy, for the A_1 and T_2 wave functions. Assume no relaxation.

5.5 Derive the silicon donor wavefunctions (Equations 5.23 through 5.25) from appropriate linear combinations of the basis functions (Equation 5.22).

5.6 A silicon wafer is doped with $[P] = 10^{16}$ cm^{-3} and $[B] = 10^{14}$ cm^{-3}. Sketch the carrier concentration (log scale) versus $1000/T$. Label the regions (intrinsic, etc.) and indicate the values on the vertical axis.

5.7 A GaAs sample has 2×10^{16} cm^{-3} Zn acceptors ($E_a = 31$ meV) and 1×10^{13} cm^{-3} Si donors. Sketch the log of the hole concentration [$\log(p)$] as a function of inverse temperature ($1/T$ or $1000/T$). Label the slopes of the plot with correct numerical values. Indicate important points on the log(p) axis.

5.8 A semiconductor of band gap 2.0 eV has 5×10^{15} cm^{-3} donors and 1×10^{13} cm^{-3} acceptors. The donor binding energy is 50 meV.
   a. Where is the Fermi level at 0 K?
   b. The carrier mobility is observed to decrease when the sample is warmed above room temperature. Why?

Chapter 5—Electronic Properties   **127**

5.9   Electron mobility of a semiconductor is measured at various temperatures. At 324 K, when scattering due to ionized impurities is negligible, the mobility is 100 cm^2/Vs. At 9 K, when phonon scattering is negligible, the mobility is 270 cm^2/Vs. What is the mobility at 100 K? Assume Equations 5.72 and 5.76 are valid and neutral impurity scattering is negligible.

5.10   Which impurity will cause a metal-insulator transition at a lower concentration in germanium, B or Cu, and why?

REFERENCES

Anderson, P.W. 1958. Absence of diffusion in certain random lattices. *Phys. Rev.* 109: 1492–1505.

Austin, B.J., V. Heine, and L.J. Sham. 1962. General theory of pseudopotentials. *Phys. Rev.* 127: 276–282.

Baldereschi, A. and L.O. Lipari. 1973. Spherical model of shallow acceptor states in semiconductors. *Phys. Rev. B* 8: 2697–2709.

Bishop, S. 1986. Iron impurity centers in III-V semiconductors. In *Deep Centers in Semiconductors*, ed. S.T. Pantelides, pp. 541–626. New York, NY: Gordon and Breach.

Blakemore, J.S. 1987. *Semiconductor Statistics*, 2nd ed. New York, NY: Dover.

Chambers, S.A. and R.F. Farrow. 2003. New possibilities for ferromagnetic semiconductors. *MRS Bull.* 28: 729–733.

Chen, G., Miotkowski, I. and Ramdas, A.K. 2012. Lyman spectra of holes bound to Cu, Ag, and Au acceptors in ZnTe and CdTe. *Phys. Rev. B* 85: 165210 (7 pages).

Clerjaud, B. 1985. Transition-metal impurities in III-V compounds. *J. Phys. C* 18: 3615–3661.

Cotton, F.A. 1990. *Chemical Applications of Group Theory*, 3rd edn. New York, NY: John Wiley & Sons.

Dietl, T., H. Ohno, F. Matsukura, J. Cibert, and D. Ferrand. 2000. Zener model description of ferromagnetism in zinc-blende magnetic semiconductors. *Science* 287: 1019–1022.

Drabold, D.A. and S. Estreicher, eds. 2007. *Theory of Defects in Semiconductors*. Berlin: Springer.

Dubon, O.D., W. Walukiewicz, J.W. Beeman, and E.E. Haller. 1997. Direct observation of the Hubbard gap in a semiconductor. *Phys. Rev. Lett.* 78: 3519–3522.

Eisberg, R. and R. Resnick. 1985. *Quantum Physics of Atoms, Molecules, Solids, Nuclei, and Particles*, 90–95. New York, NY: John Wiley & Sons.

Falicov, L.M. 1989. *Group Theory and Its Physical Applications*. Midway Reprint ed. Chicago, IL: University of Chicago Press.

Faulkner, R.A. 1969. Higher donor excited states for prolate-spheroid conduction bands: A re-evaluation of silicon and germanium. *Phys. Rev.* 184: 713–721.

Fiorini, E. 1998. CUORE: A cryogenic underground observatory for rare events. *Phys. Rep.—Rev. Sec. Phys. Lett.* 307: 309–317.

Grove, A.S. 1967. *Physics and Technology of Semiconductor Devices*. New York, NY: John Wiley & Sons.

Haller, E.E. 1991. Hydrogen-related phenomena in crystalline germanium. In *Semiconductors and Semimetals*, Vol. 34, eds. J.I. Pankove and N.M. Johnson, pp. 113–137. San Diego, CA: Academic Press.

Haller, E.E., W.L. Hansen, and F.S. Goulding. 1981. Physics of ultra-pure germanium. *Adv. Phys.* 30: 93–138.

Hohenberg, P. and W. Kohn. 1964. Inhomogeneous electron gas. *Phys. Rev.* 136: B864–B870.

Hubbard, J. 1964. Electron correlations in narrow energy bands III. An improved solution. *Proc. Roy. Soc. A* 281: 401–419.

Itoh, K.M., W. Walukiewicz, H.D. Fuchs, J.W. Beeman, E.E. Haller, J.W. Farmer, and V.I. Ozhogin. 1994. Neutral impurity scattering in isotopically engineered Ge. *Phys. Rev. B* 50: 16995–17000.

Itoh, K., E.E. Haller, J.W. Beeman et al. 1996. Hopping conduction and metal-insulator transition in isotopically enriched neutron-transmutation-doped ^{70}Ge:Ga. *Phys. Rev. Lett.* 77: 4058–4061.

Jagannath, C., Z.W. Grabowski, and A.K. Ramdas. 1981. Linewidths of the electronic excitation spectra of donors in silicon. *Phys. Rev. B* 23: 2082–2098.

Janotti, A. and C.G. Van de Walle. 2007. Native point defects in ZnO. *Phys. Rev. B* 76: 165202 (22 pages).

Kamimura, H. and H. Aoki. 1989. *The Physics of Interacting Electrons in Disordered Systems*. Oxford, UK: Clarendon Press.

Kohn, W. 1999. Nobel lecture: Electronic structure of matter-wave functions and density functionals. *Rev. Mod. Phys.* 71: 1253–1266.

References

Kohn, W. and J.M. Luttinger. 1955. Theory of donor states in silicon. *Phys. Rev.* 98: 915–922.

Luttinger, J.M. and W. Kohn. 1955. Motion of electrons and holes in perturbed periodic fields. *Phys. Rev.* 97: 869–883.

McCluskey, M.D. and S.J. Jokela. 2009. Defects in ZnO. *J. Appl. Phys.* 106: 071101 (13 pages).

Mott, N.F. 1972. Conduction in non-crystalline systems IX. The minimum metallic conductivity. *Philos. Mag.* 26:1015–1026.

Mott, N.F. 1990. *Metal-Insulator Transition.* London, UK: Taylor & Francis.

Ohno, H. 1998. Making nonmagnetic semiconductors ferromagnetic. *Science* 281: 951–956.

Pajot, B. 2010. *Optical Absorption of Impurities and Defects in Semiconducting Crystals I: Hydrogen-Like Centres.* Berlin: Springer-Verlag.

Pantelides, S.T. 1986. Perspectives in the past, present, and future of deep centers in semiconductors. In *Deep Centers in Semiconductors*, ed. S.T. Pantelides, pp. 1–86. New York, NY: Gordon and Breach.

Pearton, S.J., C.R. Abernathy, M.E. Overberg et al. 2003. Wide band gap ferromagnetic semiconductors and oxides. *J. Appl. Phys.* 93: 1–13.

Rieke, G. H., E.T. Young, C.W. Engelbracht et al. 2004. The multiband imaging photometer for spitzer (MIPS). *Astrophys. J. Suppl. Ser.* 154: 25–29.

Seeger, K. 1982. *Semiconductor Physics: An Introduction*, 2nd ed. Berlin: Springer-Verlag.

Shklovskii, B.I. and A.L. Efros. 1984. *Electronic Properties of Doped Semiconductors.* Solid-State Series Vol. 45. Berlin: Springer-Verlag.

Snoke, D.W. 2009. *Solid State Physics: Essential Concepts.* San Francisco, CA: Addison-Wesley.

Steane, A. 1998. Quantum computing. *Rep. Prog. Phys.* 61: 117–173.

Van de Walle, C.G. and J. Neugebauer. 2003. Universal alignment of hydrogen levels in semiconductors, insulators, and solutions. *Nature* 423: 626–628.

Watkins, G.D. 1986. The lattice vacancy in silicon. In *Deep Centers in Semiconductors*, ed. S.T. Pantelides, pp. 147–183. New York, NY: Gordon and Breach.

Yu, P.Y. and M. Cardona. 1996. *Fundamentals of Semiconductors.* Berlin: Springer-Verlag.

Yu, K.M. and W. Walukiewicz. 2002. Effect of the location of Mn sites in ferromagnetic $Ga_{1-x}Mn_xAs$ on its Curie temperature. *Phys. Rev. B* 65: 201303 (4 pages).

Vibrational Properties

6

In addition to altering electronic properties, impurities also affect the vibrational properties of semiconductors. This chapter begins with a discussion of phonons in intrinsic semiconductors. We then discuss the new vibrational modes that arise when defects are introduced. These modes interact with light and the surrounding lattice, giving rise to well-defined peaks in the infrared (IR) or Raman spectrum. Vibrational spectra give important information about the atomic structure of defects and serve as useful "fingerprints" for identification. The final sections of this chapter discuss specific examples of defect modes that are of fundamental and technological interest.

6.1 PHONONS

Covalent or ionic bonds pull atoms together. As the atoms approach each other, the core electrons overlap, and the Pauli exclusion principle forces the overlapping electrons into separate high-energy states. This increase in energy results in repulsion at short interatomic distances. The balance between the attraction of bonding and Pauli repulsion results in an equilibrium bond length. The total electronic energy (kinetic plus potential) is calculated with the Born–Oppenheimer approximation, where the nuclei are assumed to be fixed relative to the speedy electrons. This "total energy" is then considered to be the potential energy U for the nuclei.

For a bound pair, U versus the distance between nuclei R always looks qualitatively the same (Figure 6.1):

- As R gets large, U approaches zero (no interaction).
- As R approaches zero, U gets large (Pauli exclusion).
- U has a minimum at R_0, the equilibrium bond length.

Generally, R_0 is a few Angstroms (1 Å $= 10^{-10}$ m).

The potential U determines the equilibrium separation at absolute zero temperature. As a crystal warms up, atoms begin to oscillate around their equilibrium positions. For small displacements around equilibrium, a Taylor expansion yields

$$(6.1) \quad U(R) \approx U(R_0) + \left.\frac{dU}{dR}\right|_{R_0}(R-R_0) + \frac{1}{2}\left.\frac{d^2U}{dR^2}\right|_{R_0}(R-R_0)^2$$

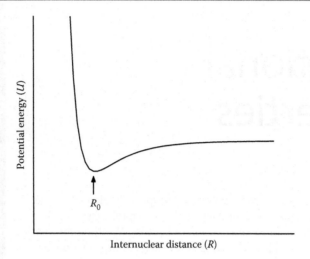

FIGURE 6.1 Potential energy (U) versus internuclear separation (R) for a pair of atoms.

Because $U(R_0)$ is an arbitrary constant, it can be set to zero. Also, the derivative of the potential is zero at its minimum:

$$\left.\frac{dU}{dR}\right|_{R_0} = 0 \tag{6.2}$$

The lowest-order term that survives is the quadratic term:

$$U(R) = \frac{1}{2}\left.\frac{d^2U}{dR^2}\right|_{R_0}(R-R_0)^2 \tag{6.3}$$

Letting

$$C \equiv \left.\frac{d^2U}{dR^2}\right|_{R_0} \tag{6.4}$$

we see that the potential is just that of a spring:

$$U(R) = \frac{1}{2}C(R-R_0)^2 \tag{6.5}$$

where C is the *force constant*. Therefore, for small oscillations, we can model a crystal as a system of springs and masses.

All solids have vibrational modes. The simplest model for crystal vibrations is a monatomic linear chain (Kittel, 2005). However, because many of the important semiconductors have two different atoms (e.g., GaAs and GaN), the *diatomic* linear chain is a better model. The diatomic linear chain has alternating masses ($M_1 \geq M_2$) connected by springs with a force constant C.

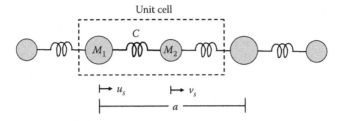

FIGURE 6.2 Diatomic linear chain model.

The distance between like atoms is a. In this one-dimensional (1D) chain, a is the length of the unit cell and the integer s denotes the unit cell number (Figure 6.2). For elemental semiconductors such as silicon, we simply set $M_1 = M_2$.

The motion of atoms around their equilibrium positions can be described as a linear superposition of lattice waves. These waves are given by complex exponentials:

(6.6)
$$u_s = u e^{i(Ksa - \omega t)}$$
$$v_s = v e^{i(Ksa - \omega t)}$$

where u_s and v_s are the displacements of M_1 and M_2, respectively, in cell number s; u and v are the amplitudes; $K = 2\pi/\lambda$ is the phonon "momentum"; and ω is the frequency. If $K > 0$, the wave travels to the right; if $K < 0$, it travels to the left. The frequency is given by (Appendix C)

(6.7)
$$\omega^2 = C(1/M_1 + 1/M_2) \pm C\sqrt{(1/M_1 + 1/M_2)^2 - \frac{2}{M_1 M_2}(1 - \cos Ka)}$$

The frequency versus K graph, or *dispersion relation*, is plotted in Figure 6.3 for the first Brillouin zone.

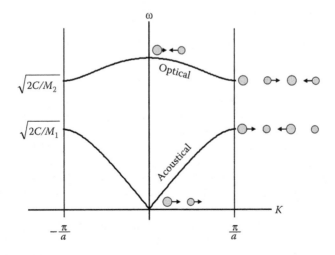

FIGURE 6.3 Dispersion relation for the diatomic linear chain ($M_1 > M_2$). The normal modes are shown for points at the zone center and edge.

132   6.1 Phonons

It is useful to discuss the results at the Brillouin zone center ($K = 0$) and edge ($K = \pi/a$). At $K = 0$, the vibrational motion of atoms in one cell is identical to that in another cell. Equation 6.7 yields two solutions as $Ka \to 0$. To lowest order,

$$(6.8) \qquad \omega = \left(\sqrt{\frac{C}{2(M_1 + M_2)}} \right) aK , \sqrt{2C(1/M_1 + 1/M_2)}$$

The low frequency corresponds to in-phase motion of M_1 and M_2 and is referred to as the *acoustical branch*. The slope of the acoustical branch near $K = 0$ is equal to the speed of sound. The high-frequency mode corresponds to neighboring atoms oscillating out of phase. If the two atoms have different charges, then this mode produces an oscillating dipole and can interact with electromagnetic radiation. The high-frequency branch is therefore called the *optical branch*.

At $K = \pi/a$, the wavelength of the lattice wave is $\lambda = 2a$. In other words, the pattern repeats every two unit cells. There are two ways that this happens: M_1 oscillates while M_2 rests, or vice versa. The respective frequencies are given by

$$(6.9) \qquad \omega = \sqrt{2C/M_1} , \sqrt{2C/M_2}$$

The high frequency corresponds to the oscillation of the lighter mass. The mass difference results in a *forbidden gap* in the phonon dispersion relation. In this range of frequencies, lattice waves cannot propagate. Examples of semiconductors with a forbidden gap are GaP, AlSb, and GaN. If the masses are identical, as in elemental semiconductors like silicon or germanium, then there is no forbidden gap. Some compound semiconductors with atoms of similar mass, like GaAs, also lack a forbidden gap.

The preceding example considered motion along one direction only. In three dimensions, the wave is characterized by a wavevector $\mathbf{K}$, where the magnitude is $K = 2\pi/\lambda$ and the vector points along the direction of propagation. For a *longitudinal* wave, the atomic motion is parallel to the propagation direction. Atoms can also oscillate perpendicular to the propagation direction, resulting in *transverse* waves. In three dimensions, there are two transverse modes for every longitudinal mode. The dispersion relation therefore consists of

- One longitudinal acoustical (LA) branch
- Two transverse acoustical (TA) branches
- One longitudinal optical (LO) branch
- Two transverse optical (TO) branches

Examples of TO and LO frequencies, at $K = 0$, are listed in Table 6.1. Ionic solids show strong reflection for photons with frequencies between the TO and LO phonon frequencies, a range known as the *reststrahlen* band. Nonpolar solids such as silicon have phonon modes, but they cannot absorb photons because the vibrational excitations have no induced dipole moment—they are "IR inactive." Combination modes can be excited if the sum of the phonon momenta equals zero.

A note about units: Commonly, vibrational frequencies are expressed in units of wavenumbers (cm^{-1}) defined as

$$(6.10) \qquad \frac{1}{\lambda} = \frac{\omega}{2\pi c}$$

TABLE 6.1 Zone-Center Optical Phonon Frequencies in Semiconductors

Structure	Material	TO (cm^{-1})	LO (cm^{-1})
Diamond	C	1332	1332
	Si	520	520
	Ge	301	301
Zincblende	AlSb	318	345
	CdTe	140	167
	GaAs	269	292
	GaP	366	402
	GaSb	225	236
	InAs	219	243
	InP	307	351
	InSb	174	183
	SiC	794	962
	ZnS	271	352
	ZnSe	209	250
	ZnTe	190	210
Wurtzite	CdS	228 (**E** // **c**), 235 (**E** $\perp$ **c**)	305 (**E** // **c**), 305 (**E** $\perp$ **c**)
	CdSe	166, 172	211, 210
	GaN	533, 560	744, 746
	SiC	790, 794	962, 962
	ZnO	380, 407	574, 583

Source: After Barker, A.S. Jr., and A.J. Sievers. 1975. *Rev. Mod. Phys.* 47: S1–179.

Note: For the diamond structure, the transverse optical (TO) and longitudinal optical (LO) frequencies are degenerate at $K = 0$. For wurtzite semiconductors, the frequencies are different for light polarized perpendicular or parallel to the c axis.

where ω is the frequency in rad/s and $c = 3 \times 10^{10}$ cm/s. In this book, impurity frequencies are the values at low temperatures (2–15 K), unless stated otherwise. By convention, absorbance is defined by the base-10 logarithm:

(6.11)
$$\text{Absorbance} = \log_{10}(I_0/I)$$

where I_0 and I are the incident and transmitted light intensities, respectively. This is in contrast to the absorption coefficient α, which is defined in terms of a natural logarithm (Equation 1.26).

6.2 DEFECT VIBRATIONAL MODES

When a defect is introduced into an otherwise perfect crystal, the translational symmetry is broken. In the case of electrically active defects, energy levels are introduced into the band gap (Chapter 5). For a *mass defect*, new vibrational modes will arise (Barker and Sievers, 1975). If a mass defect replaces a heavier atom, for example, its vibrational frequency will lie above the phonon bands. Unlike a phonon, the vibrational mode of such a defect is localized in real space and frequency space, and is referred to as a *local vibrational mode* (LVM). This mode is the vibrational analogue of a deep-level electronic state. Examples of LVMs are listed in Table 6.2.

6.2 Defect Vibrational Modes

TABLE 6.2 Examples of Local Vibrational Modes (LVMs) from Substitutional Impurities

Host	Impurity	T (K)	Frequency (cm^{-1})
Silicon	^{10}B	80	646
	^{11}B	80	622
	^{12}C	80	608
	^{13}C	80	589
	^{14}C	80	573
Germanium	^{10}B	80	571
	^{11}B	80	547
	^{28}Si	300	389
	^{31}P	80	343
GaAs	^{10}B$_{Ga}$	77	540
	^{11}B$_{Ga}$	77	517
	^{12}C$_{As}$	77	582
	^{13}C$_{As}$	77	561
	^{27}Al$_{Ga}$	80	362
	^{28}Si$_{Ga}$	80	384
	^{28}Si$_{As}$	80	399
GaP	^{10}B$_{Ga}$	77	594
	^{11}B$_{Ga}$	77	570
	^{12}C$_P$	20	606
	^{14}N$_P$	4.2	493

Source: After Barker, A.S. Jr., and A.J. Sievers. 1975. *Rev. Mod. Phys.* 47: S1–179.

Note: T is the temperature at which the measurement was taken.

To model these vibrational modes, numerical analysis is required. Let us revisit the diatomic linear chain. The equations of motion are given by (Appendix C)

$$(6.12) \qquad \begin{aligned} M_1\ddot{u}_s &= C(v_{s-1} + v_s - 2u_s) \\ M_2\ddot{v}_s &= C(u_s + u_{s+1} - 2v_s) \end{aligned}$$

Instead of assuming plane waves (Equation 6.6), we let the normal modes be given by

$$(6.13) \qquad \begin{aligned} u_s &= u_{s0}e^{-i\omega t} \\ v_s &= v_{s0}e^{-i\omega t} \end{aligned}$$

Substituting these expressions into Equation 6.12 yields

$$(6.14) \qquad \begin{aligned} (2C/M_1 - \omega^2)u_s - C/M_1(v_{s-1} + v_s) &= 0 \\ (2C/M_2 - \omega^2)v_s - C/M_2(u_s + u_{s+1}) &= 0 \end{aligned}$$

In matrix form, this set of equations is written

(6.15)
$$\begin{bmatrix} 2C/M_1 - \omega^2 & -C/M_1 & 0 & 0 & \cdots & 0 & -C/M_1 \\ -C/M_2 & 2C/M_2 - \omega^2 & -C/M_2 & 0 & \cdots & 0 & 0 \\ \vdots & \vdots & \vdots & \vdots & \ddots & \vdots & \vdots \\ -C/M_2 & 0 & 0 & 0 & \cdots & -C/M_2 & 2C/M_2 - \omega^2 \end{bmatrix} \begin{bmatrix} u_1 \\ v_1 \\ \vdots \\ v_N \end{bmatrix} = 0$$

where N is the number of unit cells, and periodic boundary conditions have been assumed. The eigenvalues ω and normal-mode displacements (u, v) are determined numerically.

A mass defect is introduced by replacing the M_1 or M_2 values, on the first or second line of the matrix, with the defect mass m. The phonon density of states is then represented by a histogram of the eigenvalues. As an example, we model GaP as a linear chain with $M_1 = 70$ and $M_2 = 31$. By numerically diagonalizing the matrix in Equation 6.15, the phonon density of states was calculated for a chain with $N = 128$ (Figure 6.4). The force constant C was chosen so that the maximum phonon frequency matched the experimental TO phonon mode at $K = 0$ (366 cm^{-1}). When a carbon atom ($m = 12$) replaces a phosphorus atom, a new mode arises at 513 cm^{-1}. This is an LVM; unlike a phonon, it has a decaying vibrational amplitude. Experimentally, the ^{12}C$_P$ LVM has a frequency of 606 cm^{-1} (Hayes et al., 1970). The higher experimental frequency implies that the Ga-C bond is stronger than the Ga-P bond.

If a phosphorus atom is replaced by a heavier atom, then a *gap mode* may appear in the forbidden gap between the acoustical and optical phonons. As an example, consider the isoelectronic As$_P$ impurity ($m = 75$). When arsenic replaces a phosphorus atom, a new mode appears at 240 cm^{-1} (Figure 6.5). Although the vibrations are localized around the impurity, gap modes are not as localized as LVMs. Experimentally, the As$_P$ gap mode has a frequency of 269 cm^{-1} (Grosche et al., 1995). IR measurements resolved fine structure arising from different combinations of the neighboring gallium isotopes.

The third case is one where the impurity atom is heavy or the force constants are weak, or both. In that instance, the defect vibrational mode lies in the phonon band and is called a *resonant mode*. More accurately, the resonant mode is a *band* of vibrational modes that have large amplitudes for the impurity atom. Resonant modes are not localized in real space. Instead, they involve significant motion of the surrounding host atoms. Because the frequency range is broad, resonant modes are often difficult to detect with IR or Raman spectroscopy. This is the case for native defects such as vacancies, which typically do not have LVMs.

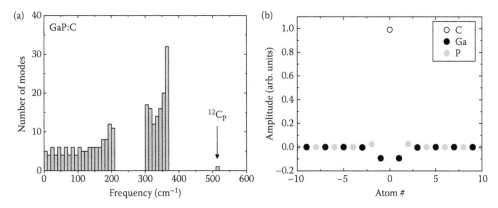

FIGURE 6.4 (a) Density of vibrational states for GaP:C, calculated with the linear-chain model. (b) Vibrational amplitudes for the ^{12}C local vibrational mode (LVM).

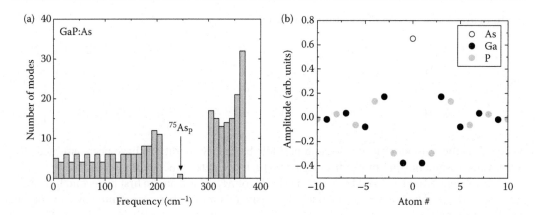

FIGURE 6.5 (a) Density of vibrational states for GaP:As, calculated with the linear-chain model. (b) Vibrational amplitudes for the ^{75}As gap mode.

The linear chain model can be extended to three dimensions. In the Keating model, bending force constants are added to the usual spring constants (Keating, 1966). The Keating model has been applied, for example, to study the spatial extent of hydrogen vibrational modes (McCluskey, 2009). Overall, the more sophisticated models replicate the qualitative results of the linear chain model.

In the case of LVMs, one can use a simple *diatomic model* to describe frequencies and their isotopic shifts (Newman, 1993). In this model, an impurity of mass m is attached by a spring C to a host atom of mass M (Figure 6.6). To account for the vibration of the other host atoms, the mass M is multiplied by an empirical constant χ. The bond-stretching frequency of the diatomic quasi-molecule is given by

$$\omega = \sqrt{C\left(\frac{1}{\chi M} + \frac{1}{m}\right)} \equiv \sqrt{\frac{C}{\mu}} \qquad (6.16)$$

where μ is the *reduced mass*. The relative displacements of the atoms are

$$x_m / x_M = -\chi M / m \qquad (6.17)$$

where x_m and x_M are the displacements of m and M, respectively. The displacements of the masses can be described in terms of a *normal-mode coordinate Q*:

$$x_m = \frac{\chi M/m}{1 + \chi M/m} Q, \quad x_M = -\frac{1}{1 + \chi M/m} Q \qquad (6.18)$$

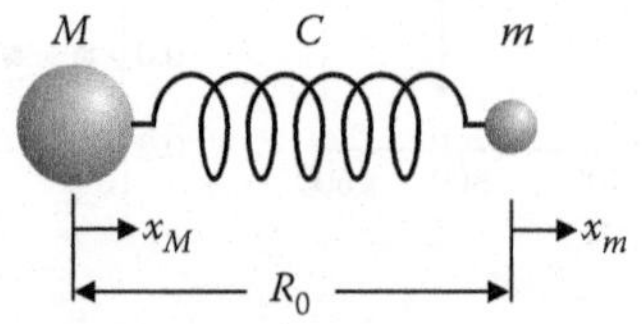

FIGURE 6.6 Diatomic model for impurity local vibrational modes (LVMs).

where the denominators were chosen to make $Q = R - R_0$. Here, Q is simply the change in the bond length. Similar to a phonon, the energy of the vibrational mode is

(6.19) $$E_n \equiv \hbar\omega_n = \left(n + \tfrac{1}{2}\right)\hbar\omega$$

The transition from $n = 0$ to $n = 1$ is called the "fundamental" mode or "first harmonic." The transition from $n = 0$ to $n = 2$ is the "overtone" or "second harmonic." (In the following discussion, we refer to ω_n as an energy level even though it has units of frequency.)

To verify the identity of an LVM experimentally, the impurity may be replaced with a different isotope, causing a well-defined shift in the vibrational frequency. The most dramatic isotope shift occurs when hydrogen ($m = 1$) is replaced by deuterium ($m = 2$). In that case, the isotopic frequency ratio is given by

(6.20) $$r = \omega_H/\omega_D = \sqrt{2\frac{\chi M + 1}{\chi M + 2}}$$

where ω_H and ω_D are the hydrogen and deuterium frequencies, respectively. Because M is finite, the ratio r is slightly less than $\sqrt{2}$.

Another effect that reduces r is the *anharmonicity* of the potential. The parabolic potential given by Equation 6.5 is a good approximation for small displacements. For larger displacements, however, the potential deviates from this ideal-spring law. The *Morse potential* is often used to model anharmonic effects (Morse, 1929):

(6.21) $$U(R) = D\{\exp[-\beta(R - R_0)] - 1\}^2$$

where D is the binding energy. For small displacements ($R \approx R_0$), the Morse potential approximates a harmonic potential (Figure 6.7), with a spring constant $C = 2D\beta^2$. A light isotope has a larger vibrational amplitude than a heavy isotope. Therefore, its frequency is affected more by anharmonicity.

The energy levels obtained from the Morse potential are

(6.22) $$\omega_n = \omega_e\left(n + \tfrac{1}{2}\right)\left[1 - x_e\left(n + \tfrac{1}{2}\right)\right]$$

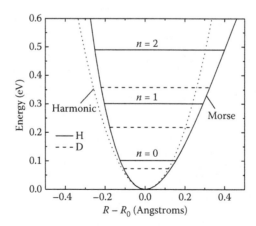

FIGURE 6.7 Harmonic (dotted line) and Morse (solid line) potentials. Energy levels for hydrogen (H) and deuterium (D) are shown by horizontal lines. The n values refer to the H levels.

138 6.3 Infrared Absorption

where

(6.23)
$$\omega_e = \beta\sqrt{2D/\mu}$$

is the harmonic component and

(6.24)
$$\omega_e x_e = \hbar\beta^2/2\mu$$

is an anharmonic term. The $n = 0 \to n = 1$ transition (fundamental) is given by

(6.25)
$$\omega_1 - \omega_0 = \omega_e - 2\omega_e x_e$$

The $n = 0 \to n = 2$ transition (overtone) is

(6.26)
$$\omega_2 - \omega_0 = 2\omega_e - 6\omega_e x_e$$

Higher harmonics can also be calculated from Equation 6.22. As shown in Figure 6.7, anharmonicity causes the spacing between levels to decrease slightly as n increases. Note that the anharmonic shift, $2\omega_e x_e$, is inversely proportional to the reduced mass. This effect reduces the isotopic frequency ratio r.

6.3 INFRARED ABSORPTION

A vibrational mode is IR active if the atomic oscillations cause a change in the dipole moment. This process leads to a peak in the IR absorption spectrum that corresponds to the vibrational frequency. Classically, an oscillating dipole can be modeled as two masses M and m, with electric charges $\pm q$, attached to each other by a spring C. For simplicity, we let M be large so that we can ignore its motion (Equation 6.17).

The equation of motion for m is given by (Appendix A)

(6.27)
$$\ddot{x} + \gamma\dot{x} + \omega_0{}^2 x = qE/m$$

where x is the displacement from equilibrium, γ is a damping constant, $\omega_0 = \sqrt{C/m}$, and E is the electric field. At the dipole, the electric field of a plane wave is given by the real part of

(6.28)
$$E = E_0 e^{-i\omega t}$$

Inserting the trial solution

(6.29)
$$x = x_0 e^{-i\omega t}$$

into Equation 6.27 yields the amplitude

(6.30)
$$x_0 = \frac{qE_0/m}{\omega_0{}^2 - \omega^2 - i\omega\gamma}$$

The velocity is obtained by differentiating Equation 6.29 with respect to time:

$$(6.31) \qquad v = -x_0 i\omega e^{-i\omega t}$$

The power absorbed by the dipole is given by force times velocity:

$$(6.32) \qquad P = q\,\mathrm{Re}(E)\mathrm{Re}(v)$$

Assuming that E_0 is real, Equation 6.32 becomes

$$(6.33) \qquad P = \frac{q^2 E_0^2 \omega}{m}\cos\omega t \frac{\gamma\omega\cos\omega t - (\omega_0^2 - \omega^2)\sin\omega t}{(\omega_0^2 - \omega^2)^2 + \gamma^2\omega^2}$$

Using the fact that

$$(6.34) \qquad \left\langle \cos^2 \omega t \right\rangle = 1/2 \quad \text{and} \quad \left\langle \sin\omega t \cos\omega t \right\rangle = 0$$

the time-averaged power absorption is given by

$$(6.35) \qquad \left\langle P \right\rangle = \frac{q^2 E_0^2}{2m\gamma}\frac{\gamma^2\omega^2}{(\omega_0^2 - \omega^2)^2 + \gamma^2\omega^2}$$

The peak in the absorption spectrum occurs near $\omega = \omega_0$, where the vibrational amplitude is greatest. The damping factor γ is approximately the full width at half maximum (FWHM) of the peak. Examples of two absorption peaks are shown in Figure 6.8. For a sharp peak ($\gamma \ll \omega_0$), the absorption spectrum is nonzero only in a narrow range around ω_0. In that case, Equation 6.35 can be approximated as a Lorentzian function:

$$(6.36) \qquad \left\langle P \right\rangle \approx \frac{q^2 E_0^2}{2m\gamma}\frac{\gamma^2}{4(\delta\omega)^2 + \gamma^2}$$

where $\delta\omega \equiv \omega - \omega_0$.

The cross section of absorption is given by the ratio of the power absorption to the intensity of the electromagnetic wave:

$$(6.37) \qquad \sigma = \frac{\left\langle P \right\rangle}{\frac{1}{2}\varepsilon_0 n c E_0^2} = \frac{q^2}{\varepsilon_0 n m c \gamma}\frac{\gamma^2}{4(\delta\omega)^2 + \gamma^2}$$

where n is the refractive index. Here, the cross section defines an effective area in which incident light is totally absorbed. The absorption coefficient is $\alpha = N\sigma$, where N is the density of dipoles (number per unit volume). The *integrated absorption* is defined as the area under the absorption peak. Using units of wavenumbers, the integrated absorption is expressed as

$$(6.38) \qquad A_I = \int_0^\infty \alpha(1/\lambda)\,d(1/\lambda) = \frac{N}{2\pi c}\int_0^\infty \sigma(\omega)\,d\omega$$

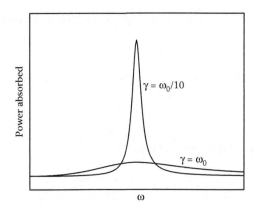

FIGURE 6.8 Absorption as a function of electric field frequency ω, for harmonic oscillators with different damping coefficients.

Substituting Equation 6.37 into this expression yields

$$A_I = \frac{Nq^2}{2\pi\varepsilon_0 nmc^2} \int_{-\infty}^{\infty} \frac{\gamma}{4(\delta\omega)^2 + \gamma^2} d(\delta\omega) \tag{6.39}$$

The integral is π/2. Inserting that value, and replacing m by the reduced mass, we have

$$A_I = \frac{Nq^2}{4\varepsilon_0 n\mu c^2} \tag{6.40}$$

where the expression is in SI units (Wilson et al., 1980). If q is the charge of an electron (e), μ is the mass of a proton, and $n = 3$, then

$$A_I(\text{cm}^{-2}) = 1.6 \times 10^{-16} N(\text{cm}^{-3}) \tag{6.41}$$

This simple result provides a reasonable order-of-magnitude estimate. The C-H stretch mode in GaAs, for example, has a calibration factor of 1.2×10^{-16} (Davidson et al., 1996). In general, calibration factors must be obtained experimentally for each particular defect.

6.4 INTERACTIONS AND LIFETIMES

Defect vibrational modes are affected by the surrounding host atoms. As the temperature of a semiconductor increases, the lattice expands and phonon modes are thermally populated. Although both of these factors affect LVMs, lattice expansion is typically not dominant (Vetterhöffer and Weber, 1996). Instead, *oscillations* of the neighboring host atoms perturb the LVM frequency and affect the vibrational lifetime.

The LVM absorption peak broadens as temperature is increased. This broadening is due to a reduction in the vibrational lifetime. In the absence of inhomogeneous broadening, the peak is described by a Lorentzian function (Equation 6.36) with an FWHM of γ. Consider a short laser

Chapter 6—Vibrational Properties   **141**

pulse that puts the oscillator into an excited state. After the pulse is turned off, the equation of motion for the mass is given by Equation 6.27:

$$(6.42) \qquad \ddot{x} + \gamma \dot{x} + \omega_0^2 x = 0$$

The solution to Equation 6.42 is an exponentially decaying oscillation (Appendix A):

$$(6.43) \qquad x = x_0 e^{-i\omega t} e^{-\gamma t/2}$$

For a "high-Q" ($\gamma \ll \omega_0$) oscillator, $\omega \approx \omega_0$. The lifetime $T_1 = 1/\gamma$ is the time that it takes for the oscillation to decay to $1/e$ of its original energy.

Quantum mechanically, T_1 is the time constant for the excited state ($n = 1$) to decay to the ground state ($n = 0$). T_1 is therefore called the *ground-state recovery* time. To conserve energy, an LVM decays into two or more lower-frequency vibrational modes. These modes may be localized (LVMs) or delocalized (resonant modes or gap modes). The decay rate depends on the multiphonon density of states (i.e., the number of possible decay channels). It also depends on how strongly the impurity couples to the lattice. Values of T_1 for hydrogen in silicon, for example, range from 4 to 295 ps (Lüpke et al., 2002, 2003).

A second important time constant is the *dephasing* time T_2^*. This is essentially the mean time between collisions with neighboring atoms. When a host atom bumps into the defect, the phase of the oscillation is altered abruptly. If this dephasing process occurs at a high rate, then the width of the absorption peak will increase significantly. As temperature increases, the thermal population of phonons causes a decrease in T_2^*. Combining the effects of ground-state recovery and dephasing, the homogeneous FWHM of the absorption peak (cm^{-1}) is

$$(6.44) \qquad \Gamma = \frac{1}{2\pi c T_1} + \frac{1}{\pi c T_2^*}$$

In addition to broadening, the peak center frequency is perturbed by the lattice vibrations. The frequency shift can be described classically. As before, the impurity is modeled as a mass m that is attached to a host atom M by a spring. Another host atom is some distance away. Due to thermal fluctuation, this neighboring atom is displaced a random amount δx from equilibrium at any given time. This displacement will change the LVM frequency by $\delta\omega$, which can be expanded in a Taylor series:

$$(6.45) \qquad \delta\omega = a(\delta x) + b(\delta x)^2 + \cdots$$

Averaged over time, the linear term does not contribute to the shift in frequency. To lowest order, the shift is given by

$$(6.46) \qquad \delta\omega = b \langle (\delta x)^2 \rangle$$

where the brackets indicate an average over time. In a classical system of springs and masses, $\langle(\delta x)^2\rangle$ is proportional to the lattice thermal energy $E(T)$. Therefore, the shift in LVM frequency is simply

$$(6.47) \qquad \delta\omega \sim E(T)$$

142   6.5 Raman Scattering

This model assumes that the LVM interacts with *all* the lattice phonons. It is a simplification because the oscillations of the surrounding atoms are really defect modes, not pure phonon modes. The quantum-mechanical derivation of Equation 6.47 was given by Elliot et al. (1965). The temperature-dependent broadening, due to dephasing (T_2^*), was calculated to be proportional to the square of the temperature.

A second model assumes that the LVM interacts with *one* phonon frequency (ω_0) preferentially (Persson and Ryberg, 1985). In the weak coupling limit, the LVM–phonon interaction leads to a temperature-dependent frequency shift given by

$$(6.48) \qquad \delta\omega = \frac{\delta\omega_0}{e^{\hbar\omega_0/k_BT} - 1}$$

and temperature-dependent linewidth broadening given by

$$(6.49) \qquad \delta\Gamma = \frac{2(\delta\omega_0)^2}{\eta} \frac{e^{\hbar\omega_0/k_BT}}{(e^{\hbar\omega_0/k_BT} - 1)^2}$$

where $\delta\omega_0$, ω_0, and η are adjustable parameters. Despite its simplicity, this model does a good job describing the temperature dependence of hydrogen LVMs in various semiconductors (Vetterhöffer and Weber, 1996; McCluskey and Haller, 1999). An example is shown in Figure 6.9, where the shift of a hydrogen LVM in ZnO was fit to Equation 6.48. The phonon frequency obtained from the fit was $\omega_0 \approx 100$ cm^{-1}. At that frequency, there is a peak in the ZnO phonon density of states. While interactions with other vibrational modes certainly occur, the large number of phonons near 100 cm^{-1} have the greatest effect on the LVM's behavior.

6.5 RAMAN SCATTERING

Complementary to IR absorption, Raman scattering involves inelastic scattering of monochromatic light (Raman, 1928). In *Stokes* Raman scattering, the incoming photon loses some of its energy by exciting a vibrational mode. In *anti-Stokes* Raman scattering, the photon gains energy from a thermally populated mode.

This process can be described by classical electromagnetic theory (Cardona, 1982). The electric field of the incoming light is given by

$$(6.50) \qquad \mathbf{E} = \hat{\mathbf{e}}_\mathbf{L} E_L \cos(\omega_L t)$$

where $\hat{\mathbf{e}}_\mathbf{L}$ is the polarization, E_L is the amplitude, and ω_L is the frequency. This field induces a polarization $\mathbf{p}$ in the defect through the polarizability tensor $\boldsymbol{\alpha}$:

$$(6.51) \qquad \mathbf{p} = \boldsymbol{\alpha} \cdot \mathbf{E} = \boldsymbol{\alpha} \cdot \hat{\mathbf{e}}_\mathbf{L} E_L \cos(\omega_L t)$$

Physically, the polarization arises from the electric field pushing electrons in one direction and nuclei in the other.

The value of $\boldsymbol{\alpha}$ changes when the atoms are displaced from equilibrium. Consider a mass defect that is displaced by a normal-mode coordinate Q. For small displacements,

$$(6.52) \qquad \alpha \approx \alpha_0 + \alpha_1 Q$$

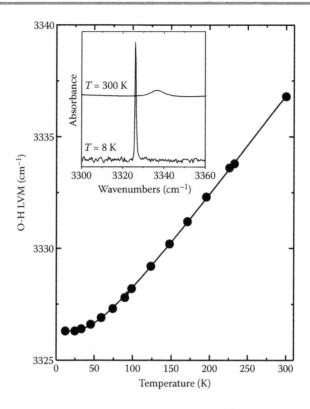

FIGURE 6.9 Temperature dependence of a hydrogen local vibrational mode (LVM) in ZnO. As temperature increases, the O-H LVM broadens and shifts to higher frequency. The solid line is a fit to Equation 6.48. Note that most LVMs shift to *lower* frequency with increasing temperature. (After McCluskey, M.D., S.J. Jokela, and W.M. Hlaing Oo. 2006. *Physica B*, 376–377: 690–693.)

where α_0 and α_1 are tensors that depend on details of the particular defect and normal mode. If the defect is undergoing an oscillation, then we can write

(6.53) $$Q = Q_0 \cos(\omega_0 t)$$

Putting together Equations 6.51 through 6.53 yields

(6.54) $$\mathbf{p} = \alpha_0 \cdot \hat{\mathbf{e}}_L E_L \cos(\omega_L t) + \alpha_1 \cdot \hat{\mathbf{e}}_L E_L \cos(\omega_L t) Q_0 \cos(\omega_0 t)$$

The first term on the right-hand side is elastic (Rayleigh) scattering. By a trigonometric identity, the second term can be written

(6.55) $$\mathbf{p}_{\text{Raman}} = \alpha_1 \cdot \hat{\mathbf{e}}_L E_L Q_0 \tfrac{1}{2} \{\cos[(\omega_L - \omega_0)t] + \cos[(\omega_L + \omega_0)t]\}$$

This dipole emits radiation with frequencies $\omega_L - \omega_0$ (Stokes) and $\omega_L + \omega_0$ (anti-Stokes).

Quantum mechanically, the Stokes line is due to the *creation* of a vibrational quantum, and the anti-Stokes line is due to the *destruction* of a vibrational quantum. At zero temperature, no vibrational excitations are thermally populated, so there is no anti-Stokes line—a phenomenon that is not captured by the classical derivation. In Chapter 10, we discuss how systematic variations in

6.6 WAVE FUNCTIONS AND SYMMETRY

As discussed in the previous sections, the vibrational motion of an impurity atom can be modeled as a harmonic oscillator with a frequency ω and reduced mass μ. For a 1D harmonic oscillator, the wave function is given by

$$\psi_n(x) \sim H_n(x)e^{-x^2/2} \tag{6.56}$$

where x is in units of $\sqrt{\hbar/\mu\omega}$ and $H_n(x)$ is a Hermite polynomial:

$$\begin{aligned} H_0(x) &= 1 \\ H_1(x) &= 2x \\ H_2(x) &= 4x^2 - 2 \end{aligned} \tag{6.57}$$

For a three-dimensional (3D) harmonic oscillator in free space, the solution to Schrödinger's equation is a product of three 1D wave functions:

$$\psi(x,y,z) = H_{n_x}(x)H_{n_y}(y)H_{n_z}(z)e^{-(x^2+y^2+z^2)/2} \tag{6.58}$$

where n_x, n_y, and n_z are integers. The energies of the 3D system are obtained by adding the energies of the 1D modes:

$$\omega = \left(n_x + \tfrac{1}{2}\right)\omega_x + \left(n_y + \tfrac{1}{2}\right)\omega_y + \left(n_z + \tfrac{1}{2}\right)\omega_z \tag{6.59}$$

In a crystal, the symmetry of a vibrational mode can be described similarly to that of an electronic state. The exponential factor in Equation 6.58 does not change upon rotation or reflection (i.e., it is fully symmetric), so we only need to examine the Hermite polynomials to determine symmetry. As can be seen in Equation 6.57, the ground state is fully symmetric. Like the s orbital, it will always have the Γ_1 representation. The first excited states are proportional to x, y, or z, similar to the p orbitals. The second excited states are proportional to quadratic functions such as xy or $x^2 + y^2$. These quadratic functions, listed in Table 5.1, are obtained from linear combinations of the wave functions in Equation 6.58.

As an example, consider a substitutional impurity in a diamond or zincblende crystal. The point-group symmetry is T_d (see Table 5.1). The ground state has A_1 symmetry. The first excited states have T_2 symmetry (x, y, z) and are degenerate. Physically, the force constants are the same in all three directions. The second excited states are A_1, E, and T_2. We can use the compatibility table (Table 5.2) to see what happens when one goes to a wurtzite crystal. When the point group is lowered from T_d to C_{3v}, the T_2 modes split into A_1 and E (Figure 6.10). This splitting occurs because the force constant along the z direction is different than that along the x and y directions.

A second example is the phosphorus–hydrogen complex in silicon (Figure 6.11), which has C_{3v} symmetry (see Table 5.1). The ground state has the A_1 representation. The first excited vibrational

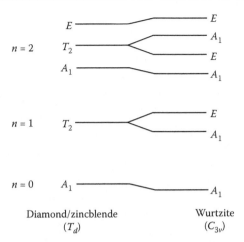

FIGURE 6.10 Symmetries of vibrational modes for substitutional impurities, for the ground state, first excited state, and second excited state.

state has E (x, y) or A_1 (z) symmetry. The A_1 mode is called the *stretch mode* because the hydrogen oscillation stretches (and compresses) the Si-H bond. The E modes are called *wag modes* (or "bending modes") because the hydrogen moves transverse to the Si-H bond. Typically, wag-mode frequencies are lower than stretch modes by at least a factor of two. The wag-mode second excited states are A_1 and E.

Given an anharmonic potential, modes with the same symmetry will influence each others' frequencies. For example, the stretch mode (A_1) and the wag-mode second excited state (A_1) can interact. In that case, the stretch mode acquires some transverse motion and the wag mode has some longitudinal motion. To model such interactions, we use perturbation theory. Consider two vibrational modes that have frequencies ω_1 and ω_2 for a harmonic (quadratic) potential. When we "turn on" the anharmonicity, the vibrational modes are calculated via

(6.60)
$$\begin{vmatrix} \omega_1 - \omega & A \\ A & \omega_2 - \omega \end{vmatrix} = 0$$

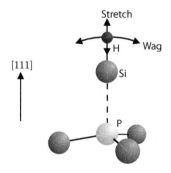

FIGURE 6.11 Hydrogen stretch and wag modes for the Si:P,H complex. The dashed line indicates a broken bond.

where

(6.61)
$$A = \int \psi_1^* V(\mathbf{r})\psi_2 d^3\mathbf{r}$$

The parameter A is nonzero only if the wave functions ψ_1 and ψ_2 have the same symmetry *and* $V(\mathbf{r})$ contains anharmonic terms. The eigenvalues ω are obtained by solving Equation 6.60:

(6.62)
$$\omega_\pm = \tfrac{1}{2}\left[\omega_1 + \omega_2 \pm \sqrt{(\omega_1 - \omega_2)^2 + 4A^2}\right]$$

The interaction causes the levels to repel each other. From Equation 6.62, when ω_1 and ω_2 are degenerate, the anharmonic potential splits the levels by an amount $2A$. The corresponding wave functions are linear combinations of ψ_1 and ψ_2 (i.e., the modes hybridize).

The interaction between the stretch mode (ω_1) and a wag-mode second excited state (ω_2) is an example of a *Fermi resonance*. The ω_2 mode involves the excitation of two vibrational quanta and therefore absorbs IR light very weakly. When ω_1 and ω_2 are nearly degenerate, however, the two modes hybridize. Then, both vibrational modes contain significant stretch-mode components and can absorb IR light, resulting in two IR peaks.

An example of this phenomenon was reported for donor-hydrogen complexes in silicon (Zheng and Stavola, 1996). The second-harmonic wag modes have frequencies of approximately $2 \times 810 = 1620$ cm^{-1}, a value that is close to that of the stretch mode. A Fermi resonance is also observed for boron–deuterium (B-D) complexes in silicon (Watkins et al., 1990). In that case, the second harmonic of the boron vibration interacts with the B-D stretch mode. As shown in Figure 6.12, this interaction causes the two modes to repel each other. The magnitude of the interaction depends on the boron isotope.

In AlSb:Se, the Al-H stretch mode interacts resonantly with a combination of *three* lower-frequency modes (Figure 6.13). This combination mode involves a transverse gap mode, in which

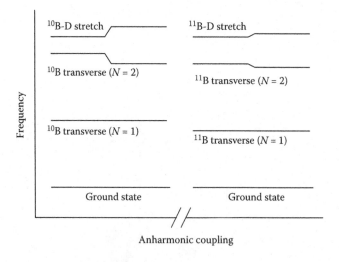

FIGURE 6.12 Energy levels for boron transverse modes and boron-deuterium stretch modes in Si:B,D. The transverse modes are labeled by $N = n_x + n_y$. Note that the resonant interaction between the modes is greater for ^{10}B than for ^{11}B. (After Watkins, G.D. et al. 1990. *Phys. Rev. Lett.* 64: 467–470.)

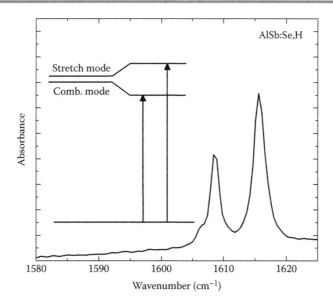

FIGURE 6.13 Resonant interaction between a stretch mode and a combination of three lower-frequency modes. The interaction results in two infrared (IR) absorption peaks. (After McCluskey, M.D. 2009. *Phys. Rev. Lett.* 102: 135502.)

the hydrogen and aluminum move in phase, plus a second-harmonic wag mode (McCluskey, 2009). The Al-D stretch mode, in contrast, is not close to any combination modes and therefore shows only one peak. Unlike the stretch mode, which is highly localized, the combination mode is delocalized, involving the motion of many atoms. The resonance between the stretch and combination mode results in a linear superposition of a local and extended mode. Because this vibrational excitation lies somewhere between a local mode and an extended lattice wave (phonon), it is referred to as a *localon*.

6.7 OXYGEN IN SILICON AND GERMANIUM

In the remainder of this chapter, we will discuss examples of defects that give rise to well-defined vibrational modes: interstitial oxygen in silicon and germanium, impurities in GaAs, and hydrogen in semiconductors. In addition to providing illustrations of vibrational properties, these systems have technological importance. The vibrational "signatures" are useful for optimizing the growth and processing of semiconductor devices.

As discussed in Section 2.4, oxygen is an omnipresent impurity in Czochralski-grown silicon. In as-grown crystals, oxygen exists primarily as interstitial oxygen, denoted O_i. In this form, oxygen binds to two silicon atoms along a <111> direction (Figure 2.8). Modeling the defect as a triatomic molecule, there are three vibrational modes: symmetric stretch mode (v_1), transverse mode (v_2), and asymmetric stretch mode (v_3). These modes are shown in Figure 6.14 for the case of a linear molecule.

The *asymmetric stretch mode* of O_i is the most extensively studied semiconductor defect vibration (Hrostowski and Kaiser, 1957). At liquid-helium temperatures, the $^{16}O_i$ mode gives rise to an IR absorption peak at 1136 cm^{-1} (Table 6.3). The rarer isotopes $^{17}O_i$ and $^{18}O_i$ have peaks at 1107 and 1084 cm^{-1}, respectively. These values are in agreement with a simple harmonic oscillator in which the frequency is inversely proportional to the square root of the reduced mass. Different combinations of silicon isotopes (28, 29, and 30) give rise to splittings of a few cm^{-1} (Figure 6.15).

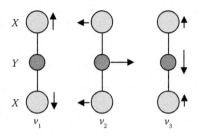

FIGURE 6.14 Vibrational modes of a linear triatomic molecule. (After Herzberg, G. 1945. *Molecular Spectra and Molecular Structure Vol. II – IR and Raman Spectra of Polyatomic Molecules*. New York, NY: Van Nostrand.)

The isotopic shift of the v_3 mode was calculated for the case of a free *X-Y-X* molecule as (Herzberg, 1945)

$$\frac{\omega'}{\omega} = \frac{M_X M_Y (M'_Y + 2M'_X \sin^2 \alpha)}{M'_X M'_Y (M_Y + 2M_X \sin^2 \alpha)} \tag{6.63}$$

where M_X and M_Y are the masses of atoms *X* and *Y*, and the primed variables represent the values following isotopic substitution. The two silicon atoms M_1 and M_2 are represented by an average mass M_X:

$$M_X = \tfrac{1}{2}\chi(M_1 + M_2) \tag{6.64}$$

where χ is an empirical factor that accounts for the vibrations of the other lattice atoms. From the experimental isotopic shifts, a bond angle of $2\alpha = 164°$ was obtained (Pajot, 1994).

These well-defined isotope shifts allow one to probe the coupling between LVMs and phonons (Section 6.4). Because the asymmetric mode frequency is more than twice that of the maximum phonon frequency, the LVM decays into three (or more) phonons. The LVM frequency for $^{17}O_i$ happens to lie on a high three-phonon density of states (Sun et al., 2004). Numerous decay channels are available, so the lifetime is short (4.5 ps). The frequency for $^{16}O_i$, in contrast, is in a spectral region where there are few three-phonon combinations, and therefore has a long lifetime (11.5 ps). Similar effects were probed with hydrostatic pressure, which shifts the silicon phonons

TABLE 6.3 Vibrational Modes of $^{16}O_i$ in Silicon

Frequency (cm^{-1})	Mode
29	Transverse (v_2)
518	Resonant
613	Symmetric stretch (v_1)
1136	Asymmetric stretch (v_3)
1203	Combination ($v_2 + v_3$)
1749	Combination ($v_1 + v_3$)

Source: After Pajot, B. 1994. Some atomic configurations of oxygen. In *Semiconductors and Semimetals* Vol. 42, ed. F. Shimura, pp. 191–249. San Diego, CA: Academic Press; McCluskey, M.D. 2000. *J. Appl. Phys.* 87: 3593–3617.

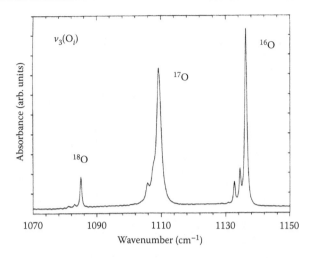

FIGURE 6.15 Infrared (IR) spectrum of silicon doped with ^{16}O, ^{17}O, and ^{18}O. The small peaks on the low-frequency sides of the main peaks correspond to ^{29}Si-O-^{28}Si and ^{30}Si-O-^{28}Si complexes. (Reprinted from Some atomic configurations of oxygen, Vol. 42, Pajot, B., *Semiconductors and Semimetals*, F. Shimura, ed., pp. 191–249, 1994. Copyright 1994, with permission from Elsevier.)

to higher frequency while decreasing the LVM frequency. The lifetime of ^{18}O$_i$ decreased abruptly as it entered the two-phonon density of states, causing the line to broaden (Hsu et al., 2003).

In addition to the asymmetric stretch mode, O$_i$ has low-frequency modes in the (111) plane, referred to as *transverse modes*. The fundamental transverse mode has a frequency of only 29 cm^{-1} (Bosomworth et al., 1970). The oxygen potential in the (111) plane was described as a harmonic (parabolic) two-dimensional (2D) potential, with a Gaussian bump in the center. This bump accounts for the off-center position of the oxygen and causes perturbations in the spectrum of vibrational energies. A similar potential was used by Yamada-Kaneta et al. (1990) to model the effect of the transverse mode on the stretch mode frequency.

As the temperature is raised, the transverse-mode excited states become thermally populated. An O$_i$ defect in its ground state can be excited to a stretch mode, which is labeled "I" in Figure 6.16. An O$_i$ defect in a thermally populated transverse mode can be excited to a stretch-plus-transverse mode, labeled "II." Due to coupling between the transverse and stretch modes, the frequencies of these transitions are different—peak II lies on the low-energy side of peak I by about 8 cm^{-1}. At higher temperatures, additional transitions (III, IV) become possible.

The *symmetric stretch mode* does not have an induced dipole moment and therefore does not absorb IR light. Its existence has been inferred from a combination mode. A weak absorption peak at 1749 cm^{-1} shows similar isotopic fine structure as the 1136 cm^{-1} peak. The weak peak was attributed to a combination of an asymmetric (v_3) and a symmetric (v_1) stretch mode (1136 + 613 = 1749 cm^{-1}). The calculated frequency for the symmetric stretch mode was 596 cm^{-1} (Artacho et al., 1995), in agreement with the experimentally determined value of 613 cm^{-1}.

Finally, the O$_i$ defect has a *resonant mode* at 518 cm^{-1}. This mode involves the vibrations of many silicon neighbors and is therefore not truly localized. Whereas the 1136 cm^{-1} line shifts to lower frequency when oxygen is replaced with a heavier isotope, the resonant mode does not exhibit such a shift. That is because the vibrational motion of the oxygen is small compared to that of the surrounding silicon atoms. One can think of the resonant mode as a perturbed phonon or "pseudolocalized" mode.

As in the case of silicon, oxygen in germanium also exists primarily as O$_i$, albeit at much lower concentrations. The asymmetric stretch mode has a frequency of 855 cm^{-1} at room temperature

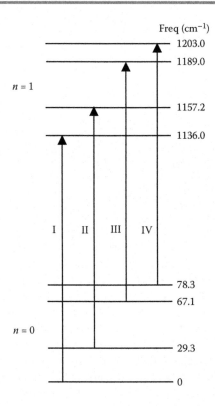

FIGURE 6.16 Vibrational energy levels for ^{28}Si:^{16}O. The *n* values correspond to the asymmetric mode. The closely spaced levels correspond to transverse modes. The transitions I, II, III, and IV can be excited by IR light. (After Bosomworth, D.R. et al. 1970. *Proc. R. Soc. London, Ser. A*. 317: 133–152.)

(Kaiser et al., 1956). At liquid-helium temperatures, a rich IR-absorption band is observed, centered at 862.5 cm^{-1} (Figure 6.17). The fine structure arises from two effects. First, the five naturally occurring germanium isotopes result in 11 average masses (Equation 6.64). Second, as in the case of Si:O, the transverse modes are thermally populated, resulting in transitions I, II, III, and IV (Figure 6.16).

To simplify the IR spectrum, isotopically pure germanium samples were grown and doped with oxygen. With increasing temperature, the appearance of three new peaks (II, III, and IV) was observed, as the transverse modes were thermally populated. Unlike Si:O$_i$, which is nearly linear, the larger Ge-O bond length pushes the oxygen atom further in the radial direction, for a Ge-O-Ge bond angle of $2\alpha = 111°$ (Mayur et al., 1994). By analyzing the temperature dependence, the transverse-mode energies were found to be in agreement with that of a rotor,

$$E_{ROT} = \frac{\hbar^2 l^2}{2mr_0^2}$$
(6.65)

where $\hbar l$ is the angular momentum about the [111] axis, m is the oxygen mass, and r_0 is the radial distance. Interestingly, under large applied pressures, O$_i$ in silicon can be forced to buckle outward (McCluskey and Haller, 1997). This outward displacement causes it to act like a rotor, similar to the case in germanium.

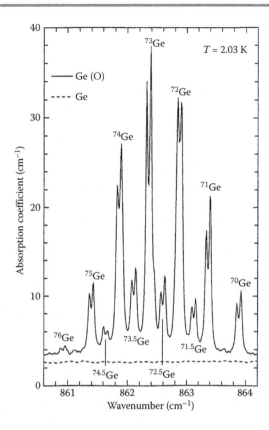

FIGURE 6.17 Infrared (IR) spectrum of Ge:O. Each doublet corresponds to the indicated *average* germanium mass in the Ge-O-Ge quasimolecule. The doublets arise from transitions I and II, similar to the case in silicon (Figure 6.16). (Reprinted with permission from Mayur, A.J. et al. 1994. Fine structure of the asymmetric stretching vibration of dispersed oxygen in monoisotopic germanium. *Phys. Rev. B* 49: 16293–16299. Copyright 1994 by the American Physical Society.)

6.8 IMPURITY VIBRATIONAL MODES IN GaAs

In this section, we discuss vibrational modes arising from carbon and silicon impurities in GaAs, and their complexes with hydrogen. Because the masses of these impurities are lower than those of the host atoms, they have LVMs that are observed in IR and Raman spectra.

High levels of *carbon* doping can be achieved by ion implantation or during MOCVD growth. While carbon in GaAs could, in principle, be an amphoteric impurity, it prefers to occupy the substitutional arsenic site and act as an acceptor. The LVM frequencies for $^{12}C_{As}$ and $^{13}C_{As}$ are 582 and 561 cm^{-1}, respectively (Newman et al., 1972). High-resolution IR spectroscopy reveals that these LVMs have fine structure due to the isotope combinations of ^{69}Ga and ^{71}Ga (Leigh et al., 1994). The possible nearest-neighbor combinations are as follows:

- All four neighbors are the same isotope. The symmetry is T_d, so there are three degenerate T_2 modes (Table 5.1).
- One neighbor is one isotope, and the other three neighbors are the other isotope. This symmetry is C_{3v}, which has two degenerate E modes and one A_1 mode (Table 5.1).
- Two neighbors are one isotope, and two neighbors are the other isotope. This symmetry, C_{2v}, has three nondegenerate modes, labeled A_1, B_1, and B_2 (Cotton, 1990).

6.8 Impurity Vibrational Modes in GaAs

The fine structure arising from these isotope combinations is shown in Figure 6.18.

Hydrogen can passivate carbon acceptors, particularly during MOCVD growth. Hydrogen binds directly to the carbon acceptor in a bond-centered location, forming a C_{As}-H pair (Figure 6.19). The stretch modes of ^{12}C-H and ^{12}C-D complexes have frequencies of 2635 and 1969 cm^{-1}, respectively, for an isotopic frequency ratio of $r = 1.3386$. The observation of the ^{13}C-H stretch mode at 2628 cm^{-1} provided direct evidence that hydrogen attaches to the carbon. The relative intensities of the ^{13}C-H and ^{12}C-H peaks match the natural abundances of ^{13}C and ^{12}C (1:90). The frequency shift is consistent with the diatomic model (Equation 6.16).

As discussed in Section 6.6, the stretch mode has A_1 symmetry. It involves out-of-phase motion of the hydrogen and carbon atoms. The C-H complex also has a lower frequency "longitudinal mode," in which the hydrogen and carbon oscillate in phase. Because the carbon and hydrogen move as a single unit, the frequencies are approximately proportional to the square root of the C-H molecular mass (Table 6.4). Like the stretch mode, the symmetry of the longitudinal mode is A_1.

The transverse modes involve motion of the hydrogen and carbon in the (111) plane. If the hydrogen and carbon oscillate out of phase, the motion is referred to as a "wag mode" (or bending mode). If they oscillate in phase, then we call it a "transverse mode." The frequencies of these modes are listed in Table 6.4. Davidson et al. (1993) used a spring-and-mass model to calculate the frequencies for the various isotopic combinations. Because the C-D wag modes involve significant motion of the carbon atom, the isotopic frequencies r do not follow a simple relation.

Silicon can be a donor (Si_{Ga}) or an acceptor (Si_{As}) in GaAs, with donors generally being dominant. The $^{28}Si_{Ga}$ donor has an LVM with a frequency of 384 cm^{-1} and a linewdith of 0.4 cm^{-1}. The narrow linewidth is due to the lack of isotopic splitting from the ^{75}As nearest neighbors, which have only one stable isotope. $^{28}Si_{As}$ acceptors have an LVM at 399 cm^{-1}, with fine structure arising

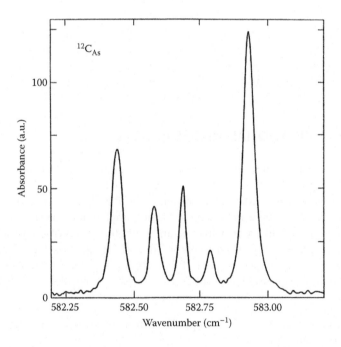

FIGURE 6.18 Infrared (IR) spectrum for carbon in GaAs. The fine structure is due to different combinations of gallium nearest-neighbor isotopes. (The source of the material Leigh, R.S. et al. Host and impurity isotope effects on local vibrational modes of GaAs:C_{As} and GaAs:B_{As}. *Semicond. Sci. Technol.* 1994, IOP Publishing is acknowledged.)

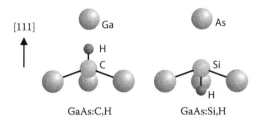

FIGURE 6.19 Ball-and-stick models for C-H and Si-H pairs in GaAs.

from various combinations of ^{69}Ga and ^{71}Ga neighbors. This isotopic splitting is similar to that observed for carbon acceptors.

Heavily doped GaAs contains Si_{Ga}–Si_{As} nearest-neighbor pairs. While the substitutional impurities have T_d symmetry, the Si_{Ga}–Si_{As} pairs have C_{3v} symmetry. This symmetry results in transverse and longitudinal modes at 464 and 393 cm^{-1}, respectively, where the two silicon atoms oscillate out of phase. In-phase oscillations have not been observed, presumably because they are resonant modes. If such modes lie in the *reststrahlen* band, in particular, then the high reflectivity of GaAs in that spectral region make IR transmission experiments impractical.

As mentioned in Section 2.4, Si_{Ga} donors form *DX* centers in GaAs under large pressures. In the *DX* configuration, the silicon atom is displaced along a <111> direction (Figure 2.6). Because the *DX* center has C_{3v} symmetry, the silicon atom has one A_1 mode and two

TABLE 6.4 Local Vibrational Modes (LVMs) of Carbon-Hydrogen Complexes in GaAs

Mode	Isotope Combination	Frequency (cm^{-1})
Stretch (A_1)	^{12}C-H	2635.2
↑	^{13}C-H	2628.5
↓	^{12}C-D	1968.6
	^{13}C-D	1958.3
Longitudinal (A_1)	^{12}C-H	452.7
↑	^{13}C-H	437.8
↑	^{12}C-D	440.2
	^{13}C-D	426.9
Wag (E)	^{12}C-H	739
→	^{13}C-H	730
←	^{12}C-D	637.2
	^{13}C-D	616.6
Transverse (E)	^{12}C-H	562.6
→	^{13}C-H	547.6
→	^{12}C-D	466.2
	^{13}C-D	463.8

Source: After Davidson, B.R. et al. 1993. *Phys. Rev. B* 48: 17106–17113; McCluskey, M.D. 2000. *J. Appl. Phys.* 87: 3593–3617.

Note: The unit vectors indicate the relative motion of the carbon and hydrogen atoms, where an "up" arrow denotes the [111] direction.

6.8 Impurity Vibrational Modes in GaAs

degenerate E modes. The frequency of the E mode, extrapolated to zero pressure, is 376 cm^{-1} (Wolk et al., 1991). The A_1 mode frequency lies within the phonon band and was not observed experimentally.

Hydrogen can passivate silicon donors, forming Si$_{Ga}$-H complexes (Figure 6.19). In these complexes, hydrogen forms a bond with the silicon, giving rise to ^{28}Si$_{Ga}$-H stretch and wag modes at 1717 and 896 cm^{-1}, respectively. The corresponding ^{28}Si$_{Ga}$-D frequencies are 1248 and 647 cm^{-1}. Small peaks, corresponding to ^{29}Si$_{Ga}$-H and ^{30}Si$_{Ga}$-H complexes, were observed on the low-frequency side of the ^{28}Si$_{Ga}$-H stretch mode (Figure 6.20). The observation of these isotopic shifts provided unambiguous evidence that hydrogen forms a bond with silicon.

Hydrogen forms complexes with many other acceptors and donors in GaAs. For group-II acceptors (Be, Zn, Cd), the hydrogen assumes a bond-centered position, forming a bond with the host arsenic atom (Chevallier et al., 1991). For group-VI donors (S, Se, Te), hydrogen attaches to

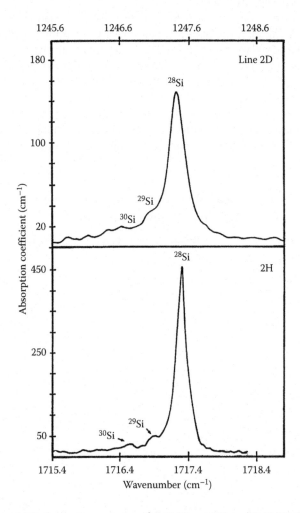

FIGURE 6.20 Infrared (IR) absorption spectra of Si-D ("line 2D") and Si-H ("2H") pairs in GaAs. Peaks due to different silicon isotopes are indicated. (Reprinted with permission from Pajot, B. et al. 1988. High-resolution infrared study of the neutralization of silicon donors in gallium arsenide. *Phys. Rev. B* 37: 4188–4195. Copyright 1988 by the American Physical Society.)

a host gallium atom in an antibonding location, similar to the case in silicon (Vetterhöffer and Weber, 1996). In all these studies, LVMs provided needed information to determine the microscopic structure. Further examples of hydrogen complexes in III-V semiconductors are discussed by McCluskey and Haller (1999).

6.9 HYDROGEN VIBRATIONAL MODES

The previous section discussed LVMs arising from complexes between hydrogen and impurities. The light mass of hydrogen, along with the large isotope effect when replacing it with deuterium, make LVM spectroscopy a natural characterization technique. In this section, we discuss forms of hydrogen that do not involve other impurities. These forms include vacancy–hydrogen complexes, interstitial hydrogen, and hydrogen molecules.

As discussed in Section 2.5, hydrogen can passivate vacancy dangling bonds. *Vacancy–hydrogen complexes* in silicon have been studied extensively with LVM spectroscopy. These studies, which measured the Si-H stretch modes, revealed that the silicon vacancy can trap up to four hydrogen atoms. To determine the number of hydrogen atoms in a center, samples were implanted with a mixture of protons and deuterons. A vacancy decorated by a single hydrogen atom, VH or VD, has a frequency of 2039 or 1495 cm^{-1} (Stallinga et al., 1998). A vacancy decorated by four hydrogen atoms, in contrast, has five possible isotopic combinations: VH_4, VH_3D, VH_2D_2, VHD_3, and VD_4. Each of these complexes has a different set of hydrogen and deuterium stretch modes, in the range of 2220–2250 and 1615–1640 cm^{-1}, respectively (Nielsen et al., 1996). In general, the frequencies increase with the number of hydrogen atoms. This is because the "crowding" of additional atoms shortens the bonds, making them stiffer.

For III-V semiconductors, the removal of a cation results in three dangling bonds. A cation vacancy is therefore fully passivated by three hydrogen atoms. A cation vacancy decorated with four hydrogen atoms has an extra electron and acts as a donor. Theorists have calculated the properties of $V_{In}H_n$ complexes in InP (Ewels et al., 1996) and $V_{Ga}H_n$ complexes in GaN (Van de Walle, 1997). IR spectroscopy performed on GaN (Weinstein et al., 1998) and InP (Zach et al., 1996) found evidence for multihydrogen centers that were consistent with the calculations. The $V_{In}H_4$ complex, in particular, is an important compensating donor in semi-insulating InP. This complex has T_d symmetry (Figure 6.21). The symmetric A_1 mode does not have an induced dipole moment and is IR inactive. The degenerate T_2 modes are IR active, resulting in a single IR peak. When InP samples were grown in a mixture of hydrogen and deuterium, additional peaks arose, similar to the case for silicon vacancies.

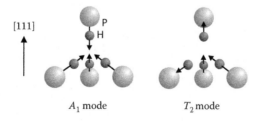

FIGURE 6.21 Hydrogen stretch modes in the InP:VH_4 complex. The arrows indicate the relative displacements of the hydrogen atoms.

156 6.9 Hydrogen Vibrational Modes

In ZnO, a II-VI semiconductor, zinc vacancies are passivated by two hydrogen atoms (Lavrov et al., 2002). As shown in Figure 6.22, one O-H bond is oriented along the c axis while the second O-H bond is nearly perpendicular to the c axis. The O-H bonds undergo stretch-mode vibrations in phase or out of phase, resulting in two IR absorption peaks. When samples contain a mixture of hydrogen and deuterium, new peaks appear. These peaks correspond to O-H or O-D stretch modes in a V_{Zn}HD complex.

In its positive charge state, *interstitial hydrogen* in silicon resides in a bond-centered location (Figure 2.16). Silicon implanted with protons, at low temperature, has a strong IR absorption peak at 1998 cm^{-1}, corresponding to the Si-H bond-stretching vibration. For temperatures above 180 K, the peak disappears as the hydrogen becomes mobile and gets trapped by vacancies and other defects. Another form of hydrogen found in proton-implanted silicon is called H_2^* (Holbech et al., 1993). The H_2^* complex consists of one hydrogen in a bond-centered location and one hydrogen in an antibonding location. The two hydrogen atoms give rise to two stretch modes at 1839 and 2062 cm^{-1}.

Interstitial *hydrogen molecules*, speculated to exist in silicon since the early 1980s, were observed by IR (Pritchard et al., 1998) and Raman (Leitch et al., 1998) spectroscopy. In free space, the H_2 molecule is not IR active, because its vibration does not induce a dipole moment. In a crystal lattice, however, the vibration induces a weak dipole moment, allowing it to be observed in the IR absorption spectrum. H_2 and D_2 dimers in silicon exhibit sharp vibrational peaks at 3618.4 cm^{-1} and 2642.6 cm^{-1}, respectively. In its interstitial T site, the hydrogen molecule is free to rotate, where J denotes the rotational quantum number. Owing to the T_d symmetry of the

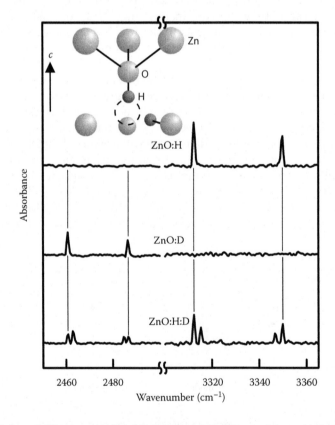

FIGURE 6.22 Infrared (IR) spectra of $V_{Zn}H_2$, $V_{Zn}D_2$, and $V_{Zn}HD$ complexes in ZnO. (Reprinted with permission from Lavrov, E.V. et al. 2002. Hydrogen-related defects in ZnO studied by infrared absorption spectroscopy. *Phys. Rev. B* 66: 165205 [7 pages]. Copyright 2002 by the American Physical Society.)

interstitial site, dimers with $J = 0$ are IR inactive (Shi et al., 2005). The peaks at 3618.4 cm^{-1} and 2642.6 cm^{-1} therefore correspond to vibrational transitions of dimers in the $J = 1$ state. In contrast to the dimers, which must exist in either an *ortho* or *para* state, HD molecules obey different selection rules. The HD absorption line, observed at 3265 cm^{-1}, corresponds to a vibrational transition with $J = 0$ in the ground state and $J = 1$ in the excited state.

PROBLEMS

6.1 The interaction between two atoms is modeled by $U(R) = -2a^2R^2 + b^2R^4$.
  a. Sketch $U(R)$.
  b. Calculate the equilibrium bond length R_0.
  c. Calculate the force constant C.

6.2 A one-dimensional linear chain is composed of identical masses M, connected by springs with alternating force constants $3C$ and $5C$. Find the phonon frequencies at $K = 0$ (see Appendix C).

6.3 In this problem, we model Si as a linear chain. From Equation 6.8, the maximum phonon frequency is $\omega_{TO} = \sqrt{4C/M}$. Experimentally, $\omega_{TO} = 520$ cm^{-1}. A mass defect, ^{10}B, is introduced. Assume that the LVM involves the oscillation of ^{10}B with no motion of the neighboring Si atoms. (This approximation allows you to solve the problem without doing the full linear chain calculation.) Calculate the LVM frequency (cm^{-1}).

6.4 Would you expect Bi in GaAs to have a local vibrational mode? Explain in 1–2 sentences.

6.5 The ^{14}N-H complex in ZnO has a fundamental bond-stretching mode at 3151 cm^{-1} (Jokela and McCluskey, 2010). The ^{14}N-H overtone is at 6133 cm^{-1}.
  a. Use the diatomic model to find the fundamental mode for ^{15}N-H, assuming $\chi = 1$.
  b. Find the Morse parameters ω_e (cm^{-1}) and x_e.

6.6 Sketch the IR absorption peak due to the LVM of an impurity, at low and high temperatures. Briefly explain the temperature dependence.

6.7 An impurity of mass m and charge q is attached to a host crystal by a spring C. It is initially at rest. A constant electric field E is applied for a very short time t. (This is an example of impulsive excitation. Assume the displacement of the mass is negligible during the excitation).
  a. How much energy was given to the mass?
  b. What is the amplitude of its oscillation after the excitation?

6.8 The displacement of a vibrating impurity is given by the real part of Equation 6.43. Assume x_0 is real and $\gamma \ll \omega_0$. Find an expression for the potential energy, kinetic energy, and total energy versus time.

6.9 A substitutional impurity with T_d symmetry is excited from the ground state to a $n_x = 1, n_y = 0, n_z = 1$ vibrational state. The fundamental frequency is ω.
  a. What is the energy of the photon that excited the impurity? Ignore anharmonic shifts.
  b. Write the wave function of the excited state (up to a proportionality constant) and indicate its symmetry.
  c. Write two wave functions that are degenerate to the excited state.

6.10 Consider the wag mode of a hydrogen complex with C_{3v} symmetry. The second harmonic of this mode has $n_x + n_y = 2$ and $n_z = 0$. Show that linear combinations of harmonic oscillator wave functions yield functions that are proportional to $x^2 + y^2 - 1$, $x^2 - y^2$, and xy. Give the symmetry of each mode and indicate which modes are degenerate.

REFERENCES

Artacho, E., A. Lizón-Nordström, and F. Ynduráin. 1995. Geometry and quantum delocalization of interstitial oxygen in silicon. *Phys. Rev. B* 51: 7862–7865.

Barker, A.S. Jr. and A.J. Sievers. 1975. Optical studies of the vibrational properties of disordered solids. *Rev. Mod. Phys.* 47: S1–S179.

Bosomworth, D.R., W. Hayes, A.R.L. Spray, and G.D. Watkins. 1970. Absorption of oxygen in silicon in the near and the far infrared. *Proc. R. Soc. London, Ser. A.* 317: 133–152.

Cardona, M. 1982. Resonance phenomena. In *Light Scattering in Solids II*, eds. M. Cardona and G. Guntherodt, pp. 19–98. Berlin: Springer.

Chevallier, J., B. Clerjaud, and B. Pajot. 1991. Neutralization of defects and dopants in III-V semiconductors. In *Semiconductors and Semimetals*, Vol. 34, eds. J.I. Pankove and N.M. Johnson, pp. 447–510. San Diego, CA: Academic Press.

Cotton, F.A. 1990. *Chemical Applications of Group Theory*, 3rd ed. New York, NY: John Wiley & Sons.

Davidson, B.R., R.C. Newman, T.J. Bullough, and T.B. Joyce. 1993. Dynamics of the H-C_{As} complex in GaAs. *Phys. Rev. B* 48: 17106–17113.

Davidson, B.R., R.C. Newman, T.B. Joyce, and T.J. Bullough. 1996. A calibration of the H-C_{As} stretch mode in GaAs. *Semicond. Sci. Tech.* 11: 455–457.

Elliot, R.J., W. Hayes, G.D. Jones, H.F. MacDonald, and C.T. Sennet. 1965. Localized vibrations of H^- and D^- ions in the alkaline earth fluorides. *Proc. R. Soc. London, Ser. A* 289: 1–33.

Ewels, C.P., S. Öberg, R. Jones, B. Pajot, and P.R. Briddon. 1996. Vacancy- and acceptor-H complexes in InP. *Semicond. Sci. Technol.* 11: 502–507.

Grosche, E.G., M.J. Ashwin, R.C. Newman et al. 1995. Nearest-neighbor isotopic fine structure of the As_P gap mode in GaP. *Phys. Rev. B* 51: 14758–14761.

Hayes, W., M.C.K. Wiltshire, and P.J. Dean. 1970. Local vibrational modes of carbon in GaP. *J. Phys. C* 3: 1762–1766.

Herzberg, G. 1945. *Molecular Spectra and Molecular Structure Vol. II—IR and Raman Spectra of Polyatomic Molecules*. New York, NY: Van Nostrand.

Holbech, J.D., B. Bech Nielsen, R. Jones, P. Sitch, and S. Öberg. 1993. H_2^* defect in crystalline silicon. *Phys. Rev. Lett.* 71: 875–8.

Hrostowski, H.J. and R.H. Kaiser. 1957. Infrared absorption of oxygen in silicon. *Phys. Rev.* 107: 966–972.

Hsu, L., M.D. McCluskey, and J.L. Lindstrom. 2003. Resonant interaction between localized and extended vibrational modes in Si:^{18}O under pressure. *Phys. Rev. Lett.* 90: 095505 (4 pages).

Jokela, S.J. and M.D. McCluskey. 2010. Structure and stability of N-H complexes in single-crystal ZnO. *J. Appl. Phys.* 107: 113536 (5 pages).

Kaiser, W., P.H. Keck, and C.F. Lange. 1956. Infrared absorption and oxygen content in silicon and germanium. *Phys. Rev.* 101: 1264–1268.

Keating, P.N. 1966. Effect of invariance requirements on the elastic strain energy of crystals with application to the diamond structure. *Phys. Rev.* 145: 637–645.

Kittel, C. 2005. *Introduction to Solid State Physics*, 8th ed. New York, NY: John Wiley & Sons.

Lavrov, E.V., J. Weber, F. Börrnert, C.G. Van de Walle, and R. Helbig. 2002. Hydrogen-related defects in ZnO studied by infrared absorption spectroscopy. *Phys. Rev. B* 66: 165205 (7 pages).

Leigh, R.S., R.C. Newman, M.J.L. Sangster, B.R. Davidson, M.J. Ashwin, and D.A. Robbie. 1994. Host and impurity isotope effects on local vibrational modes of GaAs:C_{As} and GaAs:BA_s. *Semicond. Sci. Technol.* 9: 1054–1061.

Leitch, A.W.R., V. Alex, and J. Weber. 1998. Raman spectroscopy of hydrogen molecules in crystalline silicon. *Phys. Rev. Lett.* 81: 421–424.

Lüpke, G., X. Zhang, B. Sun, A. Fraser, N.H. Tolk, and L.C. Feldman. 2002. Structure-dependent vibrational lifetimes of hydrogen in silicon. *Phys. Rev. Lett.* 88: 135501 (4 pages).

Lüpke, G., N.H. Tolk, and L.C. Feldman. 2003. Vibrational lifetimes of hydrogen in silicon. *J. Appl. Phys.* 93: 2317–2336.

Mayur, A.J., M. Dean Sciacca, M.K. Udo et al. 1994. Fine structure of the asymmetric stretching vibration of dispersed oxygen in monoisotopic germanium. *Phys. Rev. B* 49: 16293–16299.

McCluskey, M.D. 2000. Local vibrational modes of impurities in semiconductors. *J. Appl. Phys.* 87: 3593–3617.

McCluskey, M.D. 2009. Resonant interaction between hydrogen vibrational modes in AlSb:Se. *Phys. Rev. Lett.* 102: 135502 (4 pages).

McCluskey, M.D. and E.E. Haller. 1997. Interstitial oxygen in silicon under hydrostatic pressure. *Phys. Rev. B* 56: 9520–9523.

McCluskey, M.D. and E.E. Haller. 1999. Hydrogen in III-V and II-V semiconductors. In *Semiconductors and Semimetals*, Vol. 61, ed. N.H. Nickel, pp. 373–440. San Diego, CA: Academic Press.

McCluskey, M.D., S.J. Jokela, and W.M. Hlaing Oo. 2006. Hydrogen in bulk and nanoscale ZnO. *Physica B* 376–377: 690–693.

Morse, P.M. 1929. Diatomic molecules according to wave mechanics. II. Vibrational levels. *Phys. Rev.* 34: 57–64.

Newman, R.C. 1993. Local vibrational mode spectroscopy of defects in III/V compounds. In *Semiconductors and Semimetals*, Vol. 38, ed. E. Weber, pp. 117–187. San Diego, CA: Academic Press.

Newman, R.C., F. Thompson, M. Hyliands, and R.F. Peart. 1972. Boron and carbon impurities in gallium arsenide. *Solid State Commun.* 10: 505–507.

Nielsen, B.B., L. Hoffman, and M. Budde. 1996. Si-H stretch modes of hydrogen-vacancy defects in silicon. *Mater. Sci. Eng. B* 36: 259–263.

Pajot, B. 1994. Some atomic configurations of oxygen. In *Semiconductors and Semimetals*, Vol. 42, ed. F. Shimura, pp. 191–249. San Diego, CA: Academic Press.

Pajot, B., R.C. Newman, R. Murray, A. Jalil, J. Chevallier, and R. Azoulay. 1988. High-resolution infrared study of the neutralization of silicon donors in gallium arsenide. *Phys. Rev. B* 37: 4188–4195.

Persson, B.N.J. and R. Ryberg. 1985. Brownian motion and vibrational phase relaxation at surfaces: CO on Ni(111). *Phys. Rev. B* 32: 3586–3596.

Pritchard, R.E., M.J. Ashwin, J.H. Tucker, and R.C. Newman. 1998. Isolated interstitial hydrogen molecules in hydrogenated crystalline silicon. *Phys. Rev. B* 57: R15048–R15051.

Raman, C.V. 1928. A change of wavelength in light scattering. *Nature* 121: 619–620.

Shi, G.A., M. Stavola, W. Beall Fowler, and E. Chen. 2005. Rotational-vibrational transitions of interstitial HD in Si. *Phys. Rev. B* 72: 085207 (6 pages).

Stallinga, P., P. Johannesen, S. Herstrøm, K. Bonde Nielsen, B. Bech Nielsen, and J.R. Byberg. 1998. Electron paramagnetic study of hydrogen-vacancy defects in crystalline silicon. *Phys. Rev. B* 58: 3842–3852.

Sun, B., Q. Yang, R.C. Newman et al. 2004. Vibrational lifetimes and isotope effects of interstitial oxygen in silicon and germanium. *Phys. Rev. Lett.* 92: 185503 (4 pages).

Van de Walle, C.G. 1997. Interactions of hydrogen with native defects in GaN. *Phys. Rev. B* 56: R10020–R10023.

Vetterhöffer, J. and J. Weber. 1996. Hydrogen passivation of shallow donors S, Se, and Te in GaAs. *Phys. Rev. B* 53: 12835–12844.

Watkins, G.D., W.B. Fowler, M. Stavola et al. 1990. Identification of a Fermi resonance for a defect in silicon: Deuterium-boron pair. *Phys. Rev. Lett.* 64: 467–470.

Weinstein, M.G., C.Y. Song, M. Stavola et al. 1998. Hydrogen-decorated lattice defects in proton implanted GaN. *Appl. Phys. Lett.* 72: 1703–1705.

Wilson, E.B., J.C. Decius, and P.C. Cross. 1980. *Molecular Vibrations*. New York, NY: Dover.

Wolk, J.A., M.B. Kruger, J.N. Heyman, W. Walukiewicz, R. Jeanloz, and E.E. Haller. 1991. Local-vibrational-mode spectroscopy of DX centers in Si-doped GaAs under hydrostatic pressure. *Phys. Rev. Lett.* 66: 774–777.

Yamada-Kaneta, H., C. Kaneta, and T. Ogawa. 1990. Theory of local-phonon-coupled low-energy anharmonic excitation of the interstitial oxygen in silicon. *Phys. Rev. B* 42: 9650–9656.

Zach, F.X., E.E. Haller, D. Gabbe, G. Iseler, G.G. Bryant, and D.F. Bliss. 1996. Electrical properties of the hydrogen defect in InP and the microscopic structure of the 2316 cm^{-1} hydrogen related line. *J. Electron. Mater.* 25: 331–335.

Zheng, J.-F. and M. Stavola. 1996. Correct assignment of the hydrogen vibrations of the donor-hydrogen complexes in Si: A new example of Fermi resonance. *Phys. Rev. Lett.* 76: 1154–1157.

Optical Properties

7

The optical properties of semiconductors are inextricably linked to the electronic and vibrational properties. As we discussed in Chapter 1, the band gaps of semiconductors span a wide spectral range that can be exploited for the emission or detection of light. Along with intrinsic optical properties, defects give rise to various optical phenomena. Electronic transitions from hydrogenic impurities (Chapter 5) and vibrational modes of mass defects (Chapter 6) lead to well-defined peaks in the infrared (IR) spectrum that provide information about the symmetry and chemical composition of the defects. Optical spectra also provide researchers with a nondestructive method for determining the concentration of specific impurities in a sample.

In this chapter, we continue the discussion of optical properties of defects in semiconductors. The first two sections deal with optical changes due to free carriers and their interaction with the lattice. The following sections cover the absorption and emission of photons with energies near the band gap. These processes involve the creation and destruction of electron–hole pairs. Finally, we discuss the optical properties of specific defect systems: isoelectronic impurities, large-relaxation defects such as *DX* centers, and transition metals.

7.1 FREE-CARRIER ABSORPTION AND REFLECTION

The presence of free electrons and holes causes changes in the IR reflection and absorption properties of a semiconductor. In general, an electric field $\mathbf{E}$ induces a polarization $\mathbf{P}$ (dipole moment per m^3) according to

$$(7.1) \qquad \mathbf{P} = \varepsilon_0 \chi \mathbf{E}$$

where χ is the electric susceptibility of the medium (Jackson, 1999). From the definition of the electric displacement $\mathbf{D}$,

$$(7.2) \qquad \mathbf{D} = \varepsilon_0 \mathbf{E} + \mathbf{P} = \varepsilon_0 (1 + \chi) \mathbf{E} = \varepsilon_0 \varepsilon \mathbf{E}$$

We can therefore define the dielectric function ε as

$$(7.3) \qquad \varepsilon = 1 + \chi$$

162 7.1 Free-Carrier Absorption and Reflection

This is the "relative" dielectric function (i.e., $\varepsilon = 1$ for vacuum). For convenience, we break the dielectric function into three parts:

$$\varepsilon = \varepsilon_\infty + \chi_{FC} + \chi_L \tag{7.4}$$

where $\varepsilon_\infty = 1 + \chi_\infty$ is the dielectric constant at high frequencies due to core electrons, χ_{FC} is the contribution from free carriers, and χ_L is from the lattice.

To calculate reflection or absorption, we use a complex dielectric function:

$$\varepsilon = \varepsilon_R + i\varepsilon_I \tag{7.5}$$

The index of refraction is given by

$$n + i\kappa = \sqrt{\varepsilon} \tag{7.6}$$

where n and κ are the real and imaginary parts of the index, respectively (Kittel, 2005). Squaring both sides of Equation 7.6 yields

$$\begin{aligned} n^2 - \kappa^2 &= \varepsilon_R \\ 2n\kappa &= \varepsilon_I \end{aligned} \tag{7.7}$$

which can be solved to obtain n and κ individually.

The *reflectance* is the ratio of reflected to incident light intensity. At normal incidence, the reflectance from the air–sample interface is given by

$$R = \left| \frac{\sqrt{\varepsilon} - 1}{\sqrt{\varepsilon} + 1} \right|^2 = \frac{(n-1)^2 + \kappa^2}{(n+1)^2 + \kappa^2} \tag{7.8}$$

Light that is not reflected is either transmitted or absorbed. Consider a light wave that propagates in vacuum along the z direction,

$$E = E_0 e^{i(kz - \omega t)} \tag{7.9}$$

where E_0 is the electric field amplitude, $k = 2\pi/\lambda$, and ω is the angular frequency. In a medium, k is multiplied by the index (Equation 7.6)

$$E = E_0 e^{i(knz - \omega t)} e^{-k\kappa z} \tag{7.10}$$

The imaginary part of the index, κ, causes the light wave to decay exponentially. The time-averaged light intensity is proportional to $|E|^2$:

$$I = I_0 e^{-2k\kappa z} \tag{7.11}$$

Comparing this expression to the definition for the absorption coefficient α (Equation 1.25), we see that

$$\alpha = 2k\kappa = \frac{2\omega\kappa}{c} = \frac{4\pi\kappa}{\lambda} \tag{7.12}$$

Chapter 7—Optical Properties **163**

We now use this formalism to calculate the optical properties of free electrons or holes in a semiconductor, using the *Drude model*. Consider a single free carrier with a charge e and effective mass m^*. In the presence of an electric field $E = E_0 e^{-i\omega t}$, the equation of motion is given by (Appendix A)

$$(7.13) \qquad \ddot{x} + \gamma \dot{x} = eE/m^*$$

where x is the displacement from equilibrium and γ is a damping constant. Inserting the trial solution

$$(7.14) \qquad x = x_0 e^{-i\omega t}$$

into Equation 7.13 yields

$$(7.15) \qquad x = -\frac{eE/m^*}{\omega^2 + i\omega\gamma}$$

For a free-carrier density N, the polarization is given by

$$(7.16) \qquad P = Nex$$

From this expression, the susceptibility (Equation 7.1) is

$$(7.17) \qquad \chi = \frac{P}{\varepsilon_0 E} = -\frac{Ne^2}{\varepsilon_0 m^*} \frac{1}{\omega^2 + i\omega\gamma}$$

Using Equation 7.4, we obtain the dielectric function:

$$(7.18) \qquad \varepsilon(\omega) = \varepsilon_\infty - \varepsilon_\infty \frac{\omega_p^2}{\omega^2 + i\gamma\omega}$$

where ω_p is the *plasma frequency* (rad/s)

$$(7.19) \qquad \omega_p = \sqrt{\frac{Ne^2}{\varepsilon_0 \varepsilon_\infty m^*}}$$

For a small damping factor (high-mobility carriers), we can approximate Equation 7.18 as

$$(7.20) \qquad \varepsilon(\omega) \approx \varepsilon_\infty \left(1 - \frac{\omega_p^2}{\omega^2} \right)$$

For frequencies below the plasma frequency, the dielectric function is a negative real number. The index is therefore purely imaginary and, from Equation 7.8, the reflectivity is one. In that low-frequency regime, electromagnetic waves are reflected and do not penetrate into the bulk. Instead, the oscillating free charges radiate light that propagates away from the sample. This *plasma reflection* is seen in metals and doped semiconductors. Above the plasma frequency, the dielectric function approaches ε_∞, and the sample reflects only a fraction of the incident light. An illustration of plasma reflection is shown in Figure 7.1.

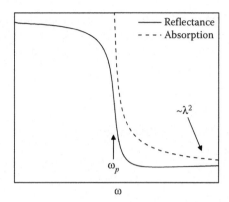

FIGURE 7.1 Plots of free-carrier reflectance (R) and absorption coefficient (α) as a function of frequency (ω), for $\omega_p/\gamma = 20$.

Light that is not reflected may be absorbed. The *free-carrier absorption* is calculated from the imaginary part of Equation 7.18:

$$\varepsilon_I = \varepsilon_\infty \frac{\omega_p^2(\gamma\omega)}{\omega^4 + (\gamma\omega)^2} \tag{7.21}$$

Let us assume that we are in the high-frequency limit such that $\gamma \ll \omega$ and $\varepsilon_\infty \approx n^2$. In that limit, using Equation 7.7, we obtain

$$\kappa = \frac{n}{2}\frac{\omega_p^2 \gamma}{\omega^3} \tag{7.22}$$

From Equation 7.12, the absorption coefficient is given by

$$\alpha = \frac{n}{c}\frac{\omega_p^2 \gamma}{\omega^2} \tag{7.23}$$

Because $\omega = 2\pi c/\lambda$, the free-carrier absorption is proportional to λ^2.

The Drude model assumes that a free carrier experiences a damping or friction term that is proportional to the velocity (Equation 7.13). More sophisticated calculations show that the free-carrier absorption is proportional to $\lambda^{3/2}$ when the carriers are scattered by acoustical phonons and to $\lambda^{5/2}$ when they are scattered by optical phonons. Scattering by ionized impurities leads to a λ^3 to $\lambda^{3.5}$ dependence. The total absorption is obtained by summing these three terms (Pankove, 1971). Empirically, one can model free-carrier absorption as being proportional to λ^p, where p is an adjustable parameter. The value of p tends to increase with the number of ionized dopants.

7.2 LATTICE VIBRATIONS

The preceding discussion of free-carrier reflection and absorption neglected the optical properties of phonons. For a polar crystal with no free carriers, the dielectric function is given by (Yu and Cardona, 1996)

$$\varepsilon(\omega) = \varepsilon_\infty \left(1 + \frac{\omega_{LO}^2 - \omega_{TO}^2}{\omega_{TO}^2 - \omega^2 - i\Gamma\omega}\right) \tag{7.24}$$

where ε_∞ is the dielectric constant at high frequencies, due to core electrons; ω_{LO} and ω_{TO} are the longitudinal optical (LO) and transverse optical (TO) zone-center phonon frequencies ($\omega_{LO} > \omega_{TO}$); and Γ is the phonon damping factor. In the limit of low damping, Equation 7.24 is approximated by

$$\varepsilon(\omega) \approx \varepsilon_\infty \left(1 - \frac{\omega_{LO}^2 - \omega_{TO}^2}{\omega^2 - \omega_{TO}^2}\right) \tag{7.25}$$

From this expression, it can be seen that $\varepsilon(\omega)$ is negative when ω lies between the TO and LO frequencies. This results in a reflectivity of one (Equation 7.8) in the *reststrahlen* band.

For nonpolar semiconductors such as silicon and germanium, $\omega_{LO} = \omega_{TO}$ at the zone center, and there is no *reststrahlen* band. These crystals do, however, exhibit *multiphonon absorption* (Spitzer, 1967). The reason can be understood qualitatively for the case of two-phonon absorption (Lax and Burstein, 1955). The first phonon excitation distorts the lattice, lowering the symmetry of the unit cell. This results in a slight positive and negative charge on neighboring atoms, allowing them to absorb a second, optical phonon. All semiconductors, polar and nonpolar, exhibit multiphonon absorption. Because photons carry no momentum, the momenta of the phonons must sum to zero. One example of such a combination is a zone-edge TO phonon with momentum **K** plus a TA phonon with momentum −**K**. Figure 7.2 shows multiphonon absorption features in silicon.

Now, let us consider the optical properties of polar semiconductors with free carriers. By combining Equations 7.24 and 7.18, the dielectric function can be written

$$\varepsilon(\omega) = \varepsilon_\infty \left(1 + \frac{\omega_{LO}^2 - \omega_{TO}^2}{\omega_{TO}^2 - \omega^2 - i\Gamma\omega} - \frac{\omega_p^2}{\omega^2 + i\gamma\omega}\right) \tag{7.26}$$

This is called the *Lorentz–Drude model* (Mitra, 1985). For a polar semiconductor with a low free-carrier concentration, the plasma frequency ω_p lies far below the optical phonon frequencies. Hence, the contribution from lattice vibrations dominates the dielectric response. When the carrier concentration is high enough to move the plasma frequency into or above the *reststrahlen* band, then there will be changes in the reflectivity profile in that spectral region. Specifically, the minimum in the reflectivity on the high-energy side is sensitive to the free-carrier concentration.

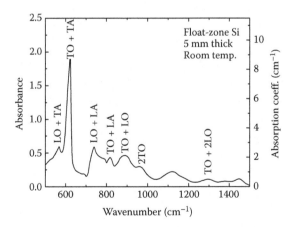

FIGURE 7.2 Multiphonon absorption spectrum of a float-zone silicon sample at room temperature. (Spitzer, W.G. 1967. In *Semiconductors and Semimetals*, Vol. 3, eds. R.K. Willardson and A.C. Beer, pp. 17–69. New York, UK: Academic Press. Spectrum taken by S. Teklemichael.)

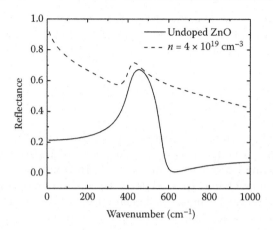

FIGURE 7.3 Reflectance spectra for ZnO, from Equation 7.26. Parameters are $\varepsilon_\infty = 3.75$, $\omega_{TO} = 410$ cm^{-1}, and $\omega_{LO} = 575$ cm^{-1}. For the undoped sample, $\gamma = 150$ cm^{-1} and $\Gamma = 50$ cm^{-1}. For the heavily doped sample, $\gamma = 1350$ cm^{-1} and $\Gamma = 60$ cm^{-1}. (After Hlaing Oo, W.M. et al. 2007. *J. Appl. Phys.* 102: 043529.)

Calculated reflectance spectra for bulk ZnO samples with different free-carrier concentrations are shown in Figure 7.3.

The Lorentz–Drude model can also be used to model *plasmon–LO coupling* observed in Raman spectra. For an uncoupled system, the plasmon (Equation 7.20) and LO (Equation 7.25) modes correspond to frequencies where $\varepsilon(\omega) = 0$. To calculate the coupled modes, we set Equation 7.26 equal to zero, neglecting the damping coefficients. That procedure results in two longitudinal modes, commonly labeled L_- and L_+ (Mooradian and McWhorter, 1967). Physically, these two modes arise from an interaction between the LO phonon and the free carriers. When an LO vibration is excited, the motion of the ions causes the carriers to oscillate, and vice versa.

The modes L_- and L_+ are linear combinations of phonons and plasmons. The carrier damping γ is typically much larger than the phonon damping Γ. Therefore, the plasmon–LO coupling causes the Raman peaks to be broader than a pure LO mode. For a heavily doped semiconductor, this broadening can cause the LO peak to disappear into the noise. This effect was used to investigate oxygen *DX* centers in GaN (Wetzel et al., 1996). At ambient pressures, a GaN:O sample with a free-electron concentration of 10^{19} cm^{-3} showed no LO mode, due to plasmon–LO coupling. At pressures above 20 GPa, however, the oxygen transformed from a shallow donor to a deep *DX* center, resulting in a drop in carrier concentration. This led to a dramatic decrease in plasmon–LO coupling and the appearance of the LO Raman peak (Figure 7.4).

A second type of interaction occurs when a local vibrational mode (LVM) lies on a strong free-carrier absorption profile. The interaction between the localized vibration and free carriers results in a *Fano resonance*. In general, a Fano resonance occurs whenever a localized state is superimposed on a continuum of states (Fano, 1961). The interaction produces an asymmetric line shape:

$$(7.27) \qquad f(z) = \frac{(q+z)^2}{1+z^2}$$

The variable z is defined as the reduced frequency

$$(7.28) \qquad z \equiv 2(\omega - \omega_0)/\Gamma$$

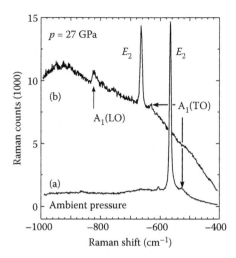

FIGURE 7.4 Raman spectrum of GaN:O under pressure. The appearance of a longitudinal optical (LO) line at high pressure is due to the reduction of free carriers, as oxygen transforms from a shallow donor to a *DX* center. (Reprinted with permission from Wetzel, C. et al. 1996. Carrier localization of as-grown *n*-type gallium nitride under large hydrostatic pressure. *Phys. Rev. B* 53: 1322–1326. Copyright 1996 by the American Physical Society.)

where ω_0 is the resonance frequency and Γ is the linewidth. The variable q is the asymmetry parameter; small values of q indicate strong coupling and a highly asymmetric line shape. The Fano function (Equation 7.27) approaches a symmetric Lorentzian line shape (Equation 6.36) as $q \to \infty$. Plots of the Fano function with different q values are shown in Figure 7.5.

An example of a Fano resonance can be seen in a heavily doped GaAs:C epilayer grown by MBE (Woodhouse et al., 1991). The layer had a hole concentration of 2×10^{20} cm^{-3}. Figure 7.6 shows the C_{As} LVM peak, which lies on top of strong free-carrier absorption. The interaction between the local mode and the free-carrier continuum leads to the asymmetric Fano line shape. After a rapid thermal annealing (RTA) treatment in AsH$_3$, the hole concentration decreased to 4×10^{19} cm^{-3}, due to hydrogen passivation of the carbon acceptors. The drop in free-carrier absorption led to a reduction in the asymmetry of the line shape (i.e., a higher q value).

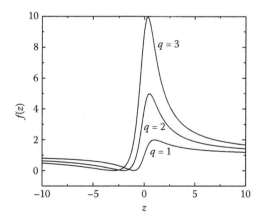

FIGURE 7.5 Fano function *f(z)* for three different asymmetry parameters (*q*).

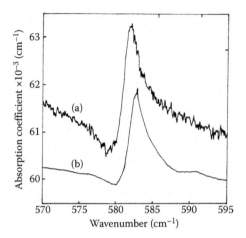

FIGURE 7.6 Infrared (IR) spectra of heavily doped GaAs:C, (a) before and (b) after rapid thermal annealing (RTA) treatment in AsH_3. The asymmetry of the local vibrational mode (LVM) is due to a Fano resonance. The RTA treatment caused the hole concentration to decrease, leading to a decrease in the Fano interaction. (The source of material Woodhouse, K. et al. LVM spectroscopy of carbon and carbon-hydrogen pairs in GaAs grown by MOMBE. *Semicond. Sci. Technol.*, 1991. IOP Publishing Ltd. is acknowledged.)

7.3 DIPOLE TRANSITIONS

In order for a material to absorb or emit light, certain selection rules must be obeyed. Energy must be conserved. For an intrinsic semiconductor, crystal momentum must be conserved, but this requirement is relaxed in the presence of a defect. In the following analysis, we use the electric dipole approximation, in which the wavelength of light (λ) is assumed to be much longer than the scale of the defect. Within that approximation, transitions that are caused by the absorption or emission of light are *dipole-allowed*, whereas transitions with zero probability are *dipole-forbidden*. For vibrational or hydrogenic spectra, such transitions are often called "IR-allowed" or "IR-forbidden."

As discussed in Chapter 6, a vibrational mode can absorb IR light if the mode causes a change in the dipole moment. Consider the diatomic model, where the positions of the nuclei are given by the normal-mode coordinate Q (Equation 6.18), and the atoms have charges of $\pm|e|$. For light polarized along the normal-mode direction, the *oscillator strength* is defined (Sakurai, 1985):

$$(7.29) \qquad f \equiv \frac{2\mu}{\hbar}\omega|\langle n|Q|0\rangle|^2$$

where μ is the reduced mass and ω is the vibrational frequency. Here, the oscillator strength is a dimensionless quantity that is proportional to the probability that the oscillator will transition from the ground state (0) to an excited state (n), when driven by light of frequency ω. The transition matrix element is given by

$$(7.30) \qquad \langle n|Q|0\rangle = \int \psi_n^*(Q) Q \psi_0(Q) dQ$$

where the wave function ψ_n is given by Equation 6.56. It turns out that this integral is only non-zero for $n = 1$. In general, for a simple harmonic oscillator, the dipole-allowed transitions are

Chapter 7—Optical Properties **169**

those that change the n value by ±1. In real crystals, the potential is anharmonic (Section 6.2), and this condition is relaxed. Transitions from $n = 0$ to 2, for example, are weakly allowed.

The oscillator strength can also be defined for an electronic transition. For light polarized along the x axis, the oscillator strength is given by

$$(7.31) \qquad f \equiv \frac{2m^*}{\hbar^2} E \left| \langle f | x | i \rangle \right|^2$$

where E is the energy difference between the final (f) and initial (i) states, and the origin is set at the impurity nucleus. Ignoring spin, the transition matrix element is given by

$$(7.32) \qquad \langle f | x | i \rangle = \int \psi_f^*(\mathbf{r}) x \psi_i(\mathbf{r}) dV$$

where ψ_f and ψ_i are spatial wave functions. Light polarized along the y or z axes have similar expressions for the transition probabilities, with x replaced by y or z. One can see that if both the initial and final wave functions are even with respect to inversion, then the oscillator strength equals zero. That is why, for hydrogenic donors, $1s \rightarrow 2p$ absorption peaks are IR-allowed whereas the $1s \rightarrow 2s$ transition is IR-forbidden. Similarly, $2p \rightarrow 1s$ transitions emit light while $2s \rightarrow 1s$ transitions are "dark."

By considering the symmetry of the initial and final states, one can determine which transitions are dipole-forbidden, without the need to calculate the transition matrix explicitly. While group theory does not tell us which transitions are strong or weak, it does tell us which transitions are forbidden by symmetry. Table 7.1 shows the allowed transitions for several point groups. Transitions that are not on this list are dipole-forbidden. For example, consider a defect of C_{3v} symmetry that is in a ground state A_1. The system can be promoted to an A_1 excited state (electronic or vibrational) by the absorption of light polarized along the z axis, or to an E excited state by light polarized perpendicular to the z axis. The transition from A_1 to A_2 is forbidden. Conversely, the decay from the A_1 or E excited state to the A_1 ground state can *emit* light polarized parallel or perpendicular to the z axis, respectively.

Initial and final states that are spatially separated from each other will lead to small, but non-zero, transition probabilities. To illustrate this, consider a donor electron in its $1s$ ground state. Suppose an unoccupied, deep acceptor is a distance R away. The transition of the electron from the donor to the acceptor is an example of quantum-mechanical *tunneling*. The transition matrix element is given by

$$(7.33) \qquad \langle f | x | i \rangle \sim \int \psi_A^*(\mathbf{r} - \mathbf{R}) \, x \, \psi_D(\mathbf{r}) dV$$

where ψ_A and ψ_D are the acceptor and donor wave functions, respectively (Williams, 1968). The magnitude of the integral (Equation 7.33) depends on the overlap between the donor and acceptor wave functions. As R increases, the overlap, and hence the tunneling rate, decreases. This behavior is important in explaining emission spectra due to donor–acceptor pairs (Section 7.6).

7.4 BAND-GAP ABSORPTION

As discussed in Chapter 1, semiconductors can absorb light via the excitation of an electron from the valence band into the conduction band. For direct-gap semiconductors, such a transition is vertical in **k** space. For indirect-gap semiconductors, the conduction-band minima occur

170 7.4 Band-Gap Absorption

TABLE 7.1   Dipole-Allowed Transitions for Several Point-Group Symmetries

Point Group	Polarization	Dipole-Allowed Transitions
T_d	x, y, or z	$A_1 \leftrightarrow T_2$
		$A_2 \leftrightarrow T_1$
		$E \leftrightarrow T_1$
		$E \leftrightarrow T_2$
		$T_1 \leftrightarrow T_1$
		$T_1 \leftrightarrow T_2$
		$T_2 \leftrightarrow T_2$
C_{3v}	x or y	$A_1 \leftrightarrow E$
		$A_2 \leftrightarrow E$
		$E \leftrightarrow E$
	z	$A_1 \leftrightarrow A_1$
		$A_2 \leftrightarrow A_2$
		$E \leftrightarrow E$
C_s	y or z	$A' \leftrightarrow A'$
		$A'' \leftrightarrow A''$
	x	$A' \leftrightarrow A''$
C_1	(All transitions allowed)	
D_{2d}	x or y	$A_1 \leftrightarrow E$
		$A_2 \leftrightarrow E$
		$B_1 \leftrightarrow E$
		$B_2 \leftrightarrow E$
	z	$A_1 \leftrightarrow B_2$
		$A_2 \leftrightarrow B_1$
		$E \leftrightarrow E$
D_{3d}	x or y	$A_{1g} \leftrightarrow E_u$
		$A_{2g} \leftrightarrow E_u$
		$E_g \leftrightarrow E_u$
		$A_{1u} \leftrightarrow E_g$
		$A_{2u} \leftrightarrow E_g$
	z	$A_{1g} \leftrightarrow A_{2u}$
		$E_g \leftrightarrow E_u$
		$A_{1u} \leftrightarrow A_{2g}$

Note: These transitions can lead to the absorption or emission of a photon. The polarization of the electric field is indicated.

at nonzero **k** values. Because photons carry essentially no momentum, a phonon is needed to conserve momentum. This requirement of an additional particle makes indirect-gap absorption coefficients smaller than those for direct gaps.

For a direct-gap semiconductor with parabolic bands, the energy of the conduction band is given by

(7.34)
$$E = E_g + \frac{\hbar^2 k^2}{2m_e}$$

while the energy of the valence band is

$$(7.35) \qquad E = -\frac{\hbar^2 k^2}{2m_h}$$

For a vertical electronic transition, conservation of energy requires that

$$(7.36) \qquad h\nu = E_g + \frac{\hbar^2 k^2}{2\mu}$$

where $h\nu$ is the photon energy and μ is the reduced electron-hole mass:

$$(7.37) \qquad \frac{1}{\mu} \equiv \frac{1}{m_e} + \frac{1}{m_h}$$

Equation 7.36 is similar to the equation for the energy of the conduction band (Equation 7.34), with m_e replaced by μ. From Equation 5.41, the transition density of states is given by

$$(7.38) \qquad D(h\nu) \sim (h\nu - E_g)^{1/2} \qquad (h\nu > E_g)$$

This expression is proportional to the number of vertical transitions that have an energy $h\nu$. Assuming that the oscillator strength (Equation 7.31) for all such transitions is constant, the absorption coefficient has the same dependence as Equation 7.38:

$$(7.39) \qquad \alpha(h\nu) = A(h\nu - E_g)^{1/2} \qquad (h\nu > E_g)$$

Indirect transitions involve the absorption or emission of a phonon of energy $\hbar\Omega$. As discussed by Pankove (1971), the absorption coefficient is given by

$$(7.40) \qquad \alpha(h\nu) = \alpha_a(h\nu) + \alpha_e(h\nu)$$

where α_a and α_e are transitions accompanied by the absorption and emission, respectively, of a phonon. Assuming parabolic bands and a constant phonon density of states, these terms are

$$(7.41) \qquad \alpha_a(h\nu) = B\frac{(h\nu + \hbar\Omega - E_g)^2}{e^{\hbar\Omega/k_B T} - 1} \qquad (h\nu + \hbar\Omega > E_g)$$

and

$$(7.42) \qquad \alpha_e(h\nu) = B\frac{(h\nu - \hbar\Omega - E_g)^2}{1 - e^{-\hbar\Omega/k_B T}} \qquad (h\nu - \hbar\Omega > E_g)$$

Note that the phonon-absorption term (Equation 7.41) vanishes as T approaches zero. That term disappears because, at low temperatures, there are no phonons around to be absorbed.

Equations 7.39 through 7.42 describe direct and indirect *band-to-band transitions*. The electron and hole created by the photon are free carriers in the conduction and valence band, respectively. An alternate possibility is that the electron and hole will experience Coulomb attraction and form a bound pair called an *exciton*. An exciton can be treated analogously to the hydrogen atom, with a hole instead of a proton. The hole, unlike a proton, has a mass that is similar to that of the electron. Therefore, the center of mass is roughly midway between the electron and hole.

As in the case of hydrogenic donors, the exciton Bohr radius is much greater than that of a hydrogen atom, due to dielectric screening. Such spatially extended pairs are called "Wannier excitons." From the Bohr model (Equation 5.10), the exciton binding energy is given by

$$E_X = 13.6\,\mathrm{eV}\,\frac{\mu/m}{\varepsilon^2} \qquad (7.43)$$

where μ is the reduced electron-hole mass (Equation 7.37), m is the fundamental electron mass, and ε is the static (low frequency) dielectric constant. In the absence of defects, the electron–hole pair is a *free exciton* and is able to move throughout the crystal.

The formation of excitons leads to sub-band-gap absorption features. The photon energy required to form an exciton is

$$h\nu = E_g - E_X/n^2 \qquad (7.44)$$

where n is the principal quantum number ($n = 1$ is the ground state, which typically dominates the spectrum). In GaAs at low temperatures, free exciton formation leads to a peak slightly below the band-to-band absorption threshold (Figure 7.7). At room temperature, the free exciton is unstable and the absorption peak is not distinct. Indirect-gap semiconductors also have exciton absorption features below the band gap. Indirect absorption processes involve phonons, which have a range of energies. This leads to a broad absorption threshold rather than a sharp peak. Discontinuities in the slope of the absorption occur at energies corresponding to specific phonons and phonon combinations (Gershenzon et al., 1962).

Dopants affect band-gap absorption in several ways. Consider a semiconductor with a high concentration of donors and compensating acceptors. Due to the random distribution of ionized dopants, certain regions will contain excess positive or negative electric charge. It is possible to excite an electron–hole pair such that the electron resides in a region of positive charge while the hole is in a negatively charged region. The energy of such a pair is below the intrinsic band gap.

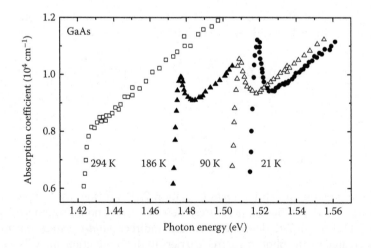

FIGURE 7.7 Band-gap absorption spectra of GaAs at several temperatures. At low temperatures, the absorption peak is due to free-exciton formation. (After Sturge, M.D. 1962. *Phys. Rev.* 121, 359–362.)

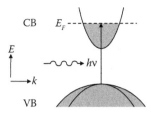

FIGURE 7.8 Burstein–Moss shift. Electrons excited from the valence band must go into states that are above the Fermi level. This effect leads to an increase in the optical band gap.

The randomly fluctuating potential results in exponentially decaying *band tails*. Empirically, the band-tail absorption follows *Urbach's rule* (Urbach, 1953):

$$\alpha(h\nu) = \alpha_0 e^{(h\nu - E_g)/\Delta} \tag{7.45}$$

where Δ is a broadening parameter that increases with doping or temperature. The Urbach tail is also found in amorphous semiconductors or samples with inhomogeneous strain.

In addition to band tails, high doping levels cause the Fermi level to enter the conduction band or valence band, affecting the band-gap absorption. In Figure 5.19, we saw a schematic of a degenerate (metallic) *n*-type semiconductor. Electrons that are excited from the valence band cannot go into the filled states. Instead, they must go into states that are higher than the Fermi level E_F (Figure 7.8). This increase in the optical band gap is called the *Burstein–Moss shift* (Burstein, 1954; Moss, 1954). At very high doping levels ($n > 10^{19}$ cm^{-3}), the Burstein–Moss shift is reduced due to *band renormalization*. This reduction is caused by the Coulomb attraction between electrons and positively ionized donors, as well as many-body correlation effects (Berggren and Sernelius, 1981).

The Burstein–Moss shift is greatest in semiconductors with low electron effective masses. That is because a low value of m^* means that the curvature of the band is high (Equation 1.17) and the density of states is low (Equation 5.41). With a low density of states, E_F rises significantly as electrons are added to the conduction band. An example of this is seen in InN, which at low doping levels has an electron effective mass of $m_e \approx 0.07m$ and room-temperature band gap of $E_g \approx 0.7$ eV (Wu et al., 2002). For *n*-type InN with $n = 5 \times 10^{19}$ cm^{-3}, the Burstein–Moss effect causes the optical band gap to increase by 0.2 eV. Figure 7.9 shows optical absorption spectra, taken at liquid-helium temperatures, for InN samples with $n = 4 \times 10^{17}$ and 1×10^{19} cm^{-3} (Wu et al., 2003). These spectra show the increase in band tailing with doping level, as well as the Burstein–Moss shift.

The final examples of near-band-gap transitions are transitions to ionized donors or from ionized acceptors. In the first case, an electron from the valence band is excited into an unoccupied, positive donor level. In the second case, an electron from an occupied, negative acceptor is excited into the conduction band. Both of these transitions result in an absorption threshold below the band gap (Pankove, 1971). Transitions from neutral hydrogenic donors and acceptors, discussed in Section 5.3, occur at much lower (IR) energies. Figure 7.10 illustrates several types of electronic absorption processes discussed so far. Optical absorption from deep levels, which involve significant lattice relaxation, will be discussed in Section 7.8.

7.5 CARRIER DYNAMICS

As discussed in the previous section, free carriers can be generated by exposing a semiconductor to light. Devices that utilize optical *generation* of carriers include photodiodes, charge-coupled

7.5 Carrier Dynamics

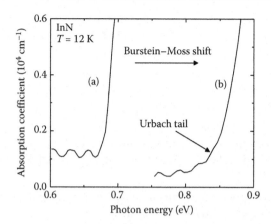

FIGURE 7.9 Band-gap absorption spectra of InN, for (a) $n = 4 \times 10^{17}$ cm^{-3} and (b) $n = 1 \times 10^{19}$ cm^{-3}. The Burstein–Moss shift causes an increase in the optical band gap. An increase in band tailing can also be seen. (After Wu, J. et al. 2003. *J. Appl. Phys.* 94: 4457–4460.)

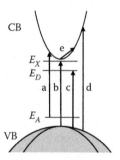

FIGURE 7.10 Several absorption processes in semiconductors: (a) acceptor-to-band, (b) free exciton formation, (c) band-to-donor, (d) band-to-band, and (e) free-carrier absorption. (Free-carrier absorption requires phonon emission/absorption to conserve momentum.)

device (CCD) arrays, and photoconductor detectors. A photon with energy near the band gap can generate an electron–hole pair. High-energy photons such as x-rays and γ-rays, or energetic particles such as electrons, generate many electrons and holes. In light-emitting diodes (LEDs) and laser diodes, high densities of electrons and holes are injected electrically, resulting in the emission of light.

Free-carrier *recombination* is the inverse process of free-carrier generation. In most cases, a specific carrier generation path is not retraced inversely by the subsequent recombination process. For example, a photon-generated electron–hole pair in silicon recombines without emission of a photon ~10^6 times more frequently than by photon emission. This is due, in part, to the indirect band gap of silicon. In addition, there always are some deep levels in silicon which act as effective recombination centers even at very low impurity concentrations. Even in the purest, structurally most perfect silicon single crystals grown today, the recombination of excess electrons or holes is dominated by deep-level imperfections.

In *nonradiative* processes, the recombination energy is transformed, through one or more steps, into heat. In *radiative* recombination, a photon is generated. The recombination energy that is freed when an electron and hole recombine across the band gap leaves the crystal in the form of electromagnetic radiation. All solid-state lasers and LEDs work on the basis of this process. A direct band gap is most favorable for radiative recombination because no phonons are required for the conservation of momentum. Several III-V compound and alloy semiconductors (e.g., AlGaAs and InGaN) have direct band gaps and are widely used for the fabrication of efficient light-emitting sources. Indirect-gap semiconductors such as silicon and germanium also exhibit radiative recombination, but the probability for such processes is significantly reduced.

Radiative recombination leads to the emission of light, or luminescence, which is classified according to how the electron–hole pairs are generated. When electrons and holes are produced by optical excitation, the emitted light is referred to as *photoluminescence* (PL). If an electron beam produces the electron–hole pairs, then the light emission is *cathodoluminescence*. Light that is produced by electrical injection of carriers, as in a laser diode or LED, is called *electroluminescence*. In the following sections, we restrict the discussion to PL. In general, the emission phenomena occur with the other techniques as well.

Electrons that are created high in the conduction band lose energy by phonon emission and accumulate near the conduction-band minimum. Similarly, holes accumulate near the top of the valence band. The populations of electrons and holes are commonly described by the quasi-Fermi levels μ_e and μ_h, respectively (Figure 7.11). When an electron in the conduction band falls into an empty state (hole) in the valence band, we have radiative band-to-band recombination. The band-tail states, which are low in energy, tend to dominate the spectrum. This results in a PL peak that is *Stokes-shifted* (lower in energy) relative to the absorption threshold. With high doping levels, the PL peak shifts even lower, due to band renormalization (Section 7.4).

Another radiative decay path can occur when a carrier transitions from a band to an impurity. In a *p*-type semiconductor, a photogenerated electron can fall from the conduction band to a neutral acceptor. The energy of the emitted photon is $E_g - E_a$, where E_a is the acceptor binding energy (Yu and Cardona, 1996). Similarly, a hole can transition from the valence band to a neutral donor. These types of emission are essentially the inverse of the absorption processes shown in Figure 7.10.

In a PL experiment, a light source with photon energy above the band gap creates concentrations of electrons (n) and holes (p) that are higher than the equilibrium values n_0 and p_0. Now, consider what happens when the light source is turned off. The product np is greater than that found at thermal equilibrium (Equation 5.48). For radiative recombination, the decay in excess electrons is given by (Orton and Blood, 1990)

$$\frac{d\Delta n}{dt} = -B(np - n_0 p_0) \tag{7.46}$$

where $n = n_0 + \Delta n$ and B is the radiative recombination rate. Because electrons and holes annihilate each other, the concentration of excess holes is equal to Δn. Table 7.2 lists the band-to-band recombination coefficients for four semiconductors. The recombination rates for

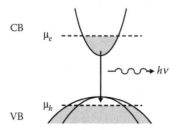

FIGURE 7.11 Population of electrons and holes during a photoluminescence (PL) experiment, as described by the quasi-Fermi levels μ_e and μ_h. The transition of an electron from the conduction band to the valence band results in the emission of a photon.

176   7.5 Carrier Dynamics

TABLE 7.2   Recombination Coefficients (B) for Radiative Band-to-Band Recombination

Semiconductor	$T(K)$	B (cm^3 s^{-1})
Si	90	1.3×10^{-15}
	290	1.8×10^{-15}
Ge	77	4.1×10^{-13}
	300	5.3×10^{-14}
GaAs	90	1.8×10^{-8}
	294	7.2×10^{-10}
GaSb	80	2.8×10^{-8}
	300	2.4×10^{-10}

Source: After Varshni, Y.P. 1967. *Phys. Stat. Sol. (b)* 19: 459–514 and 20: 9–36.

indirect-gap semiconductors are several orders of magnitude lower than those for the direct-gap semiconductors.

For a weak light source, the concentration of excess carriers is low compared to $n_0 + p_0$. In that limit, $(\Delta n)^2$ can be neglected and Equation 7.46 can be approximated as

$$(7.47) \qquad \frac{d\Delta n}{dt} = -B(n_0 + p_0)\Delta n$$

This equation is solved by a decaying exponential:

$$(7.48) \qquad \Delta n = \Delta n(0)e^{-t/\tau_m}$$

where

$$(7.49) \qquad \tau_m = \frac{1}{B(n_0 + p_0)}$$

is called the *minority carrier recombination lifetime.* Consider the example of a doped, p-type silicon sample with $p_0 = 10^{16}$ cm^{-3} at 290 K. Because the concentration of minority electrons n_0 is low, the lifetime is determined by the hole concentration p_0. From the B value given in Table 7.2, Equation 7.49 yields a radiative lifetime of $\tau_m = 56$ ms. Now consider intrinsic silicon, which has $n_0 = p_0 = 7 \times 10^9$ cm^{-3} at 290 K (Varshni, 1967). The intrinsic concentrations result in a lifetime of $\tau_m = 11$ hr. These long lifetimes mean that defects are guaranteed to play a dominant role in carrier recombination.

Defects enhance the nonradiative recombination of electrons and holes through the *Shockley–Read–Hall* (SRH) mechanism (Hall, 1952; Shockley and Read, 1952). Suppose we generate electron–hole pairs in a p-type sample. Along with shallow dopants, the sample contains a concentration N_T of deep electron traps. These deep levels can be unoccupied donors or acceptors. The SRH process occurs by two steps (Figure 7.12). First, a trap captures an electron, with a *capture cross section* σ_n. The capture rate is given by

$$(7.50) \qquad A \equiv 1/\tau_n = N_T \sigma_n v_{th}$$

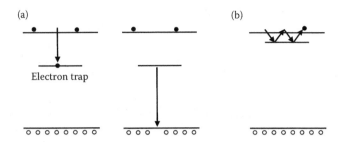

FIGURE 7.12 (a) Shockley–Read–Hall recombination. First, a conduction-band electron is captured by an electron trap. Second, a hole is captured from the valence band. (b) An electron trap near the conduction-band minimum. At room temperature, the probability is high that a captured electron will be emitted back into the conduction band.

where N_T is the concentration of unfilled traps and $v_{th} = (3k_B T/m^*)^{1/2}$ is the average thermal velocity of electrons (Equation 5.63). Second, the filled trap captures a hole. Because $p \gg n$, the hole capture occurs much faster than the electron capture so that the rate-limiting step is given by Equation 7.50. The energy for both steps is released in the form of phonons (i.e., heat). Therefore, SRH is a nonradiative recombination process.

For an n-type sample, τ_p is obtained by replacing σ_n in Equation 7.50 by σ_p, the capture cross section for holes. Intrinsic samples involve a more complex expression (Sze and Ng, 2007), but the process is qualitatively the same. Table 7.3 contains values of capture cross sections for some deep-level impurities in silicon.

The SRH mechanism is most efficient when the trap level is near the middle of the band gap. If the trap level is near either band extremum, then the SRH mechanism becomes less efficient and the recombination lifetime increases. The reason for this can be understood qualitatively with the example of an electron trap near the conduction band. Such a trap can capture an electron, with the emission of one or more phonons. However, there is a high probability of emitting the electron back into the conduction band (Figure 7.12). Therefore, the electrons and holes do not recombine at a high rate. A deep trap, in contrast, will hold on to the electron until a hole comes along.

In addition to SRH recombination, a second nonradiative process is the *Auger* (pronounced like "OJ") effect. Band-to-band Auger recombination occurs when an electron and hole recombine,

TABLE 7.3 Electron and Hole Capture Cross Sections of Deep-Level Impurities in Silicon, at Room Temperature ($E_g = 1.12$ eV)

Dopant	Level (eV)	Type	Measured Cross Section (cm²)
Au	0.57	Acceptor (0/−)	$\sigma_n = 10^{-14}, \sigma_p = 2 \times 10^{-15}$
	0.35	Donor (0/+)	$\sigma_n = 5 \times 10^{-15}, \sigma_p = 4 \times 10^{-15}$
Fe	0.39	Donor (0/+)	$\sigma_n = 10^{-15}$
Pt	0.89	Acceptor (0/−)	$\sigma_n = 5 \times 10^{-15}$
	0.32	Donor (0/+)	$\sigma_n = 5 \times 10^{-15}, \sigma_p = 10^{-15}$
Zn	0.55	Acceptor (−/2−)	$\sigma_n = 5 \times 10^{-19}, \sigma_p = 5 \times 10^{-15}$
	0.26	Acceptor (0/−)	$\sigma_n = 5 \times 10^{-15}, \sigma_p = 5 \times 10^{-15}$

Source: Madelung, O., U. Rössler, and M. Schulz, eds. 2002. *The Landolt-Börnstein Database*, Group III, Vol. 41: A2a, Berlin: Springer-Verlag.

Note: Donor and acceptor levels are relative to the valence-band maximum. Since cross section values from the literature can vary by an order of magnitude, only approximate values are listed.

but instead of emitting a photon, the energy goes into exciting a second electron higher into the conduction band. Another version of the Auger effect is trap-assisted Auger recombination. In that process, when an electron is trapped by a defect, the energy goes into exciting a second electron. The hole capture then proceeds normally, by the emission of phonons. The Auger effect is used, perhaps too often, to explain why radiative recombination does not occur under circumstances where light emission is expected.

Equations 7.47 through 7.49 apply for low minority carrier concentrations. For a light-emitting device, however, np can be much larger than $n_0 p_0$ during operation. This condition also applies for a PL experiment with high laser intensity. Assuming $n = p$ and large n, the carrier decay is given by the *ABC* model (Schubert, 2006),

$$(7.51) \qquad \frac{dn}{dt} = -An - Bn^2 - Cn^3$$

where A, B, and C are the SRH, radiative, and Auger recombination coefficients, respectively.

7.6 EXCITON AND DONOR–ACCEPTOR EMISSION

Section 7.5 discussed the recombination of free electrons with free holes. In addition to such band-to-band recombination, free excitons can form (Section 7.4). These excitons are mobile and can move throughout the crystal. When an exciton recombines, it may emit light. From Equation 7.44, an exciton in its ground state emits a photon of energy:

$$(7.52) \qquad h\upsilon = E_g - E_X$$

Alternatively, the exciton-hole recombination may occur with the emission of a photon plus n LO phonons of energy $\hbar\Omega$. In that case, the photon energy is

$$(7.53) \qquad h\upsilon = E_g - E_X - n\hbar\Omega$$

In PL spectroscopy, the $n = 0$ emission gives rise to the *zero-phonon line*, and the $n > 0$ emissions are referred to as *phonon replicas*.

In a pure semiconductor at low temperatures, the PL signal is dominated by free-exciton emission. As the temperature is raised, the excitons dissociate into free electrons and holes, and band-to-band emission dominates. For most semiconductors, the free exciton binding energy (Equation 7.43) is low such that excitons are unstable at room temperature. One exception is ZnO, which has a relatively high free exciton binding energy of 60 meV (Meyer et al., 2004; Özgür et al., 2005).

Free excitons can get trapped at many different kinds of impurities and defects, forming *bound excitons*. In indirect semiconductors at low temperatures, bound excitons can have lifetimes of several ms. Consider the case of a neutral donor, which consists of an electron orbiting around a positively charged impurity atom. The neutral donor and free exciton are both analogous to hydrogen atoms. They can form a bound complex, analogous to a H_2 molecule. This complex is referred to as a *donor bound exciton*. Similarly, an exciton can become trapped by an acceptor, forming an *acceptor bound exciton*. The binding energy for a bound exciton, E_{BX}, is defined as the energy required to remove the exciton from the impurity. Empirically, for many semiconductors, this energy is $\sim 1/10$ that of the donor/acceptor binding energy (Haynes, 1960). An exciton can also bind to a positively charged donor, forming a

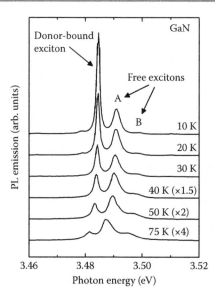

FIGURE 7.13 Photoluminescence (PL) spectra of donor-bound excitons and free excitons in GaN. (After Shan et al. 1995. *Appl. Phys. Lett.* 67, 2512–2514.)

weakly bound complex that is analogous to H_2^+. For any bound exciton, the zero-phonon line is given by

$$h\nu = E_g - E_X - E_{BX} \tag{7.54}$$

Figure 7.13 shows PL spectra of free and bound excitons in GaN over a range of temperatures (Shan et al., 1995). At low temperatures, the donor bound exciton line dominates (Monemar, 2001). As the temperature increases, the PL lines shift to lower energy, due to the lowering of the band gap. Bound excitons are liberated from the donors, and the free exciton lines dominate the spectrum. Because GaN has the wurtzite structure, the degeneracy of the valence band at the Γ point is lifted, resulting in two kinds of valence-band maxima, A and B (Jain et al., 2000). This splitting results in two free excitons, depending on whether the hole wave function comes from the A or B valence band. Time-resolved measurements showed that the excitons have lifetimes of 30–60 ps at 10 K. Nonradiative processes such as the SRH effect play an important role in the decay of the PL signal.

At higher temperatures, the PL peaks broaden, and it is difficult to determine whether the emission is excitonic or band-to-band. To elucidate the nature of the emission, researchers examine the dependence of the PL intensity on excitation power. The luminescence intensity is given by

$$I_{PL} = \eta I_{laser}^\alpha \tag{7.55}$$

where η is the emission efficiency and the exponent α depends on the nature of the emission (Fouquet and Siegman, 1985). For exciton emission, the emission intensity is proportional to the number of excitons, so $\alpha = 1$. For band-to-band emission, the intensity is proportional to np, leading to $\alpha = 2$. In general, a sample that contains a mixture of free carriers and excitons will have $1 < \alpha < 2$. For the case of GaN, the value of α jumps from 1.2 to 1.6 at a temperature of 280 K (Chen et al., 2006). This behavior was attributed to the breakup of free excitons, which

7.6 Exciton and Donor–Acceptor Emission

have a binding energy of ~25 meV, into free electrons and holes. For ZnO, with its free-exciton binding energy of 60 meV, the dissociation of free excitons occurs at 700 K (Chen et al., 2007).

Due to their long lifetimes, bound excitons in indirect-gap semiconductors give rise to extremely sharp peaks in the PL spectrum. In chemically pure silicon, the sharpness of the PL lines is limited by the random distribution of silicon isotopes. This is due to perturbation of the bands by zero-point vibrational motion of the atoms (Haller, 1995). Regions of heavier or lighter isotopes give rise to slightly different local band gaps, causing inhomogeneous broadening of the emission peak. The ultimate limit in narrow peaks was achieved for phosphorus donor bound excitons in isotopically enriched ^{28}Si (Figure 7.14). The width of the zero-phonon PL line was limited only by the instrumental resolution of ~1 μeV (Karaiskaj et al., 2001). Later experiments, using a tunable laser as an excitation source, were able to resolve the hyperfine splitting (~0.5 μeV) due to the coupling between the ^{31}P nuclear spin and donor electron (Thewalt et al., 2007).

As shown in Figure 7.14, the acceptor bound exciton gives rise to several sharp peaks. This splitting is due to the complicated structure of the bound exciton, which consists of two holes and one electron. The ground state of the electron splits into three levels due to valley-orbit splitting (Section 5.3). The hole-hole interaction results in three levels. When these interactions are taken into account, along with the hole-electron interaction and the anisotropy of the conduction-band minima, it is possible to explain the rich fine structure of the PL spectrum (Karasyuk et al., 1992). Donor bound excitons, which consist of two electrons and one hole, also show splitting. Only one phosphorus donor bound exciton line is shown in Figure 7.14.

The discussion so far has focused on low exciton concentrations. At high concentrations, free excitons can combine to form *biexcitons*, like H_2 molecules. Because the binding energy for such quasi-molecules is low, biexcitons are only stable at low temperatures. At sufficiently low temperatures and high concentrations, excitons condense into a liquid phase known as an *electron-hole drop* (Wolfe and Jeffries, 1983). Figure 7.15 shows an image of PL emission from a single electron-hole drop in germanium, at a temperature of 1.8 K (Wolfe et al., 1975). The drop in this image had a radius of 0.3 mm and was confined by the nonuniform strain caused by a nylon screw pressing against the sample, which was mounted in a plastic ring. The laser generating the electron–hole pairs was incident on the back of the sample. The density of electron–hole pairs was on the order

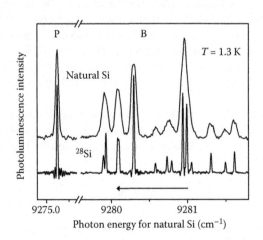

FIGURE 7.14 Photoluminescence (PL) spectra of donor (P) and acceptor (B) bound excitons, in natural and isotopically enriched silicon. The arrow indicates that the bottom spectrum was shifted to lower frequency, in order to line up with the top spectrum. (Reprinted with permission from Karaiskaj, D. et al. 2001. Photoluminescence of isotopically purified silicon: How sharp are the bound exciton transitions? *Phys. Rev. Lett.* 86: 6010–6013. Copyright 2001 by the American Physical Society.)

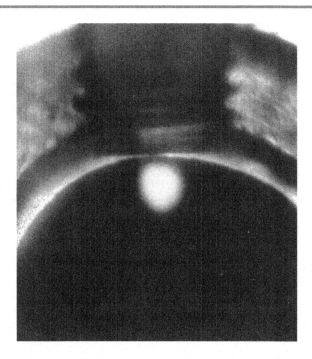

FIGURE 7.15 Electron-hole drop in germanium. The bright spot is the drop, which had a radius of 0.3 mm. A nylon screw is seen pressing on the sample. Note that the maximum stress is not at the surface between the nylon screw and the Ge disk but at a little distance toward the center of the disk. Russian railroad engineers discovered this effect in the 18th century. (Reprinted with permission from Wolfe, J.P. et al. 1975. Photograph of an electron-hole drop in germanium. *Phys. Rev. Lett.* 34: 1292–1293. Copyright 1975 by the American Physical Society.)

of 10^{17} cm^{-3}. The electrons fell into the lowest conduction-band minima generated by the strain. In germanium, this is one of the four *X*-band valleys.

Along with exciton emission, samples may exhibit *donor–acceptor pair* transitions. Consider a compensated, *p*-type sample at low temperature. The uncompensated acceptors are neutral, while all the donors are ionized. When electron–hole pairs are generated in a PL experiment, some of the electrons get trapped by the ionized donors to form neutral donors. The neutral donors are in a high-energy, or excited, state. An electron can then tunnel from a neutral donor to a neutral acceptor. The products of this reaction are a positively charged donor and a negatively charged acceptor. The reduction in energy results in the emission of a photon.

The donor and acceptor ions, separated by a distance *R*, experience a Coulomb attraction that lowers the energy of the final state. By conservation of energy, the zero-phonon line is given by

$$h\nu = E_g - E_a - E_d + \frac{1}{4\pi\varepsilon\varepsilon_0} \frac{e^2}{R} \tag{7.56}$$

where E_a and E_d are the acceptor and donor binding energies, respectively. Because this is a tunneling process, the transition probability decreases with *R* (Equation 7.33). However, in the crystal lattice, the number of atoms that are on a "shell" of radius *R* increases with *R*. This balance, along with phonon replicas, leads to complex fine structure in the emission spectrum. Figure 7.16 shows a PL spectrum of GaP doped with Si$_P$ acceptors and S$_P$ donors (Thomas et al., 1964). Donor–acceptor pairs with low *R* values give rise to sharp peaks on the high-energy side of the

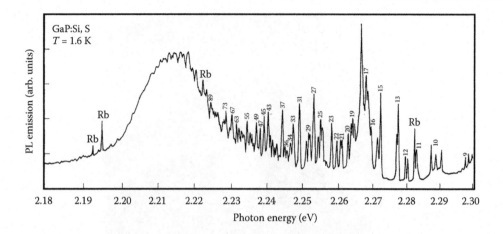

FIGURE 7.16 Photoluminescence spectrum of GaP doped with Si_p acceptors and S_p donors. The peaks are labeled by the "shell number," where small numbers correspond to close donor–acceptor pairs. Peaks labeled "Rb" are from a rubidium calibration source. (Reprinted with permission from Thomas, D.G., M. Gershenzon, and F.A. Trumbore. 1964. Pair spectra and "edge" emission in gallium phosphide. *Phys. Rev.* 133: A269–A279. Copyright (1964) by the American Physical Society.)

spectrum. The broad peak at ~2.21 eV is due to distant pairs. This broad peak has phonon replicas at lower energies not shown in the figure.

If a short excitation pulse is used, one can monitor the temporal evolution of the PL spectrum. Because the tunneling rate is highest for the closest pairs, the PL spectrum is initially dominated by these high-energy emissions. As time progresses, more distant pairs contribute to the PL spectrum. This leads to a red shift in the emission peak as a function of time (Thomas et al., 1965). The spectrum also shows an interesting dependence on excitation intensity. For a weak light source, only a few donors will be in their excited state, and distant pairs will dominate the PL spectrum. At high photoexcitation intensities, all the donors will be excited, and the close pairs will dominate. Hence, the emission energies increase (blue shift) with light intensity (Yu and Cardona, 1996).

7.7 ISOELECTRONIC IMPURITIES

Isoelectronic impurities have the same number of valence electrons as the atoms they replace. In GaP, nitrogen substitutes for a phosphorus atom (N_p). Due to the high electronegativity of nitrogen, it acts as an electron trap, with an energy level $E_T = 9$ meV below the conduction-band minimum (Thomas and Hopfield, 1966). The electronic wave function is highly localized. In a PL experiment, a nitrogen impurity captures an electron, forming N^-. The negatively charged center can then attract a hole. The hole orbits around the N^-, similar to a hydrogenic acceptor, with a binding energy of $E_a = 11$ meV. The electron and hole can be thought of as an exciton that is bound to a neutral impurity. The bound-exciton energy is obtained by adding the electron and hole energies:

(7.57) $$E = E_g - E_T - E_a$$

When the electron and hole recombine, a photon with $h\nu = E$ is emitted.

Because GaP is an indirect-gap semiconductor, momentum conservation requires the absorption or emission of a phonon during band-to-band recombination. Nitrogen, however, has an electronic

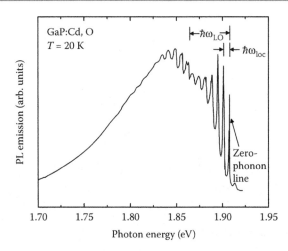

FIGURE 7.17 Photoluminescence (PL) spectrum of Cd-O pairs in GaP. The peaks are separated by $\hbar\omega_{loc}$, where ω_{loc} is the localized vibrational frequency of the Cd-O pair. A phonon replica can also be observed. (After Dean, P.J. 1986. In *Deep Centers in Semiconductors*, ed. S.T. Pantelides, pp. 185–347. New York, NY: Gordon and Breach.)

wave function that is localized in real space and spread out in **k**-space (Equation 5.14). This feature means that momentum conservation is relaxed so that a phonon is no longer required. In the early days of LED development, isoelectronic impurities provided the primary way to enhance the luminescence efficiency. GaP:N is still used for low-brightness green LEDs, with InGaN alloys preferred for high-brightness green LEDs (Schubert, 2006). In GaAs, the nitrogen electron trap level lies in the conduction band. This means that an electron will not form a bound state.

With increasing nitrogen concentration, a series of deep traps appears due to nitrogen atoms located at neighboring anion sites, the "NN pairs." These pairs, along with LO phonon replicas, can be observed with PL spectroscopy. As the concentration of nitrogen increases to ~1%, the material enters the alloy regime (Vurgaftman and Meyer, 2003). These *dilute nitride semiconductors* have potential applications in solar cells or infrared devices. $GaP_{1-x}N_x$ and $GaAs_{1-x}N_x$ have large, composition-dependent bowing parameters (Bellaiche et al., 1996) such that the band gap decreases significantly for small concentrations of nitrogen (x). For example, PL emission energies in $GaAs_{1-x}N_x$ epilayers and quantum wells exhibited shifts of ~0.1 eV for $x = 0.01$ (Buyanova et al., 1999). The large reduction in the band gap was explained by an interaction between a narrow band of N-states and the GaAs conduction-band minimum (Shan et al., 1999).

In GaP, nearest-neighbor donor–acceptor pairs such as Cd_{Ga}-O_P and Zn_{Ga}-O_P give rise to red emission (Dean, 1986). In these pairs, a shallow acceptor sits next to a deep oxygen donor, forming a kind of "isoelectronic molecule." The high electronegativity of oxygen attracts an electron. The negatively charged defect then attracts a hole, and the electron–hole pair is considered to be a bound exciton. In a room-temperature PL experiment, Cd-O and Zn-O pairs result in broad bound-exciton emission bands, centered at ~1.8 eV. At liquid-helium temperatures, vibrational replicas of the Cd-O pair are clearly resolved (Figure 7.17). These peaks arise from the emission of one photon plus n localized vibrations, similar to phonon replicas (Equation 7.53).

Similar to nitrogen, the Cd-O electronic wave function is highly localized. This results in a slight shift in the vibrational frequency depending on whether a bound exciton is present or not. For the zero-phonon line, the emission energy is given by

$$h\nu = E + \tfrac{1}{2}\hbar\omega^* - \tfrac{1}{2}\hbar\omega \qquad (7.58)$$

where E is the electronic part of the energy; and ω^* and ω are the vibrational frequencies with and without an exciton, respectively. Here we consider only a single, nondegenerate vibrational mode. In the harmonic approximation, Equation 7.58 becomes

$$(7.59) \qquad h\nu = E + \frac{1}{2}\hbar\frac{\sqrt{C^*} - \sqrt{C}}{\sqrt{M}}$$

where the C values are force constants and M is the defect mass. This mass dependence gives researchers a powerful tool to obtain chemical information about the defect. In the present example, replacing ^{16}O with ^{18}O led to a shift of 0.65 meV, unambiguously establishing the presence of oxygen in this defect. A smaller shift between ^{110}Cd and ^{114}Cd was also observed.

7.8 LATTICE RELAXATION

As we saw in Section 5.3, the optical absorption spectra of shallow impurities were like that of the hydrogen atom, with transitions from the ground state to the p states and the conduction-band continuum. Implicit in that discussion was the lack of relaxation upon excitation. Because the hydrogenic wave functions are delocalized, their effect on the local environment of the impurity is minimal. When the electron or hole is promoted to a higher state, it is reasonable to assume that the impurity-host bond lengths do not change appreciably.

Deep-level defects, in contrast, have localized wave functions and may exhibit significant lattice relaxation (Lang and Logan, 1977; Alkauskas et al., 2016). One can simulate the effect of relaxation using the diatomic model (Section 6.2), where a deep donor impurity m is attached to a host atom M. In its neutral (occupied) charge state, the equilibrium bond length is R_0. Let $Q = R - R_0$ denote the normal-mode coordinate (Equation 6.18). In the harmonic approximation, the electronic energy is given by

$$(7.60) \qquad U = \frac{1}{2}CQ^2$$

where the minimum is arbitrarily set at zero.

Suppose a photon promotes the donor electron to the conduction-band minimum. The system now consists of a free electron and a positively charged donor. In its positive charge state, the equilibrium bond length is not R_0. Instead, the minimum in the potential energy curve shifts to a new value. The energy of the positive donor plus free electron is given by

$$(7.61) \qquad U^* = \frac{1}{2}C(Q - \Delta Q)^2 + E_0$$

where we assume, for simplicity, that the force constant does not change.

Consider the vertical transition caused by a photon of energy E_{opt} (Figure 7.18):

$$(7.62) \qquad E_{opt} = \frac{1}{2}C(\Delta Q)^2 + E_0$$

In this process, the positions of the nuclei do not change during the absorption. After the electronic transition, the nuclei will relax to the energy minimum by the emission of phonons. The *Huang–Rhys factor*,

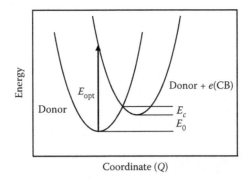

FIGURE 7.18 Configuration-coordinate diagram for a deep donor. The curve on the left is for a neutral donor. The curve on the right is for a positive donor plus an electron at the conduction-band minimum.

$$S = \frac{\frac{1}{2}C(\Delta Q)^2}{\hbar \omega} \tag{7.63}$$

is essentially the number of vibrational quanta of frequency ω that are emitted during the relaxation (Huang and Rhys, 1950). For $S = 0$, there is no lattice relaxation. For $S \gg 1$, lattice relaxation lowers the energy significantly. The *Franck–Condon shift* is defined as the amount of energy lost during relaxation. In the excited electronic state, this is

$$E_{FC} = S\hbar\omega = E_{opt} - E_0 \tag{7.64}$$

Quantum mechanically, we describe the initial and final states as products of electronic and nuclear (vibrational) wave functions. Assuming the defect is initially in its ground state, we have

$$\Psi_i = \psi_i(\mathbf{r})\psi_0(Q) \tag{7.65}$$

where $\psi_i(\mathbf{r})$ is the electronic wave function, and $\psi_0(Q)$ is the ground-state vibrational wave function. The final state is

$$\Psi_f = \psi_f(\mathbf{r})\psi_n(Q) \tag{7.66}$$

where n denotes the vibrational level. The transition from the initial to the final state can be excited by a photon of energy $E_0 + n\hbar\omega$. From Equation 7.31, the electronic transition probability is proportional to

$$\left|\langle f|x|i\rangle\right|^2 = \left|\int \psi_f^*(\mathbf{r}) x \psi_i(\mathbf{r}) dV\right|^2 \left|\int \psi_n^*(Q)\psi_0(Q) dQ\right|^2 \tag{7.67}$$

The vibrational overlap term is given by

$$\left|\int \psi_n^*(Q)\psi_0(Q) dQ\right|^2 = \frac{e^{-S}S^n}{n!} \tag{7.68}$$

This overlap term is maximized when the photon energy equals E_{opt}. Plots of Equation 7.68 are shown in Figure 7.19. For $S = 0$, there is just one sharp peak, corresponding to $n = 0$. For $S = 10$, the peak is vibrationally broadened.

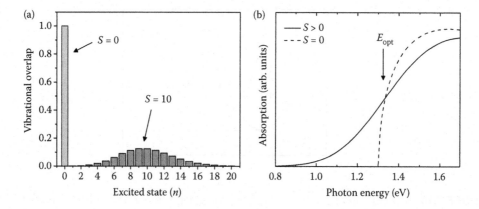

FIGURE 7.19 (a) Overlap between the ground state and a vibrationally excited state, for $S = 0$ and $S = 10$. (b) Absorption spectrum for a *DX* center. The dashed line shows the absorption spectrum if S is set to zero. The nonzero value of S causes a vibrational broadening of the absorption onset. (After McCluskey, M.D. et al. 1998. *Phys. Rev. Lett.* 80: 4008–4011.)

The preceding discussion concerned the specific transition of an electron from the ground state to the conduction-band minimum. In reality, the electron can also be excited to the continuum of states above the conduction-band minimum. This leads to a vibrationally broadened absorption profile. Examples of absorption profiles are shown in Figure 7.19, for $S = 0$ and $S = 10$, at 100 K. At higher temperatures, vibrationally excited states become thermally populated, leading to more broadening. An explicit form for the absorption profile is given in Alkauskas et al. (2016). Additional details about optical absorption of deep-level centers can be found in Pajot and Clerjaud (2012).

At thermal equilibrium, the energy difference between the two charge states is E_0. The proportion of thermally ionized donors is related to the Boltzmann factor, $\exp(-E_0/k_BT)$. For a defect with large relaxation, E_{opt} is much larger than E_0 (Equation 7.64). This difference between the optical and thermal energies, the Franck–Condon shift, is a hallmark of large-relaxation deep levels. For these systems, optical spectroscopy and variable-temperature electrical measurements will lead to much different defect energies.

A second important phenomenon is called *persistent photoconductivity*. More accurately, it is conductivity that persists after the light source has been turned off. When a neutral donor is photoionized, it becomes positively charged and relaxes to its new energy minimum. The free electron increases the electrical conductivity of the sample. In order for the defect to return to the neutral charge state, two obstacles must be overcome. First, the defect must capture a free electron (Equation 7.50). Second, the defect must have enough thermal energy to surmount the capture barrier E_c (Figure 7.18). After the defect surmounts the barrier, it relaxes to the neutral ground state by the emission of phonons. Assuming the free electron is at the conduction-band minimum, the capture cross section is given by

$$\sigma = \sigma_\infty e^{-E_c/k_BT} \tag{7.69}$$

where the prefactor σ_∞ is proportional to $T^{1/2}$ (Lang, 1986). Because the exponential dominates the temperature dependence, σ_∞ is often treated as a constant. At liquid-nitrogen temperatures, the capture cross section (Equation 7.69) is small such that the photo-induced conductivity persists for hours or days. Such a long-lived state is called *metastable*.

The quintessential example of a large-relaxation donor is the DX center (Mooney, 1990). As discussed in Sections 2.4 and 6.8, Si_{Ga} donors form DX centers in $Al_xGa_{1-x}As$ ($x > 0.22$) or in GaAs under large pressures. In the DX configuration, the change in the normal-mode coordinate Q corresponds to displacement of the silicon atom along the [111] direction (Figure 2.6). The value of E_{opt} lies in the 1.3–1.6 eV range (Mooney et al., 1988). Unlike the neutral donor model presented in this section, the charge state of the DX center is actually −1. The defect reaction that produces DX centers is given by

$$(7.70) \qquad\qquad 2Si^0 \rightarrow Si^+ + DX^-$$

In an $Al_xGa_{1-x}As{:}Si$ ($x > 0.22$) sample with no other defects, at low temperature, half of the silicon impurities will be positively charged donors while half will be DX centers. Despite the complication of the charge state, the metastability and Franck–Condon shift can be explained reasonably well with the simple configuration coordinate diagram in Figure 7.18.

Optical measurements on GaAs:Si under large pressures were important in establishing the negative charge state of the DX center. As discussed in Section 2.4, the Chadi and Chang model was supported by the results of infrared spectroscopy experiments on GaAs:Si under hydrostatic pressure (Wolk et al., 1991). The local vibrational mode frequency of the silicon DX center was found to be ~2% lower than that of the substitutional silicon donor, due to the softening of the bonds. This frequency shift agrees with first-principles calculations (Saito et al., 1992), which found the negatively charged DX center to be the lowest-energy configuration. In addition, the estimated concentrations of Si^+ donors and DX centers were in agreement with the defect reaction given in Equation 7.70.

DX centers are also found in other III-V semiconductors. For example, oxygen forms DX centers in $Al_xGa_{1-x}N$ (McCluskey et al., 1998) and GaN under pressure. For pressures greater than 20 GPa, the transformation of oxygen from a shallow donor to a deep DX center causes a reduction in free-carrier concentration. This reduction was observed via free-carrier absorption (Perlin et al., 1995) and Raman spectroscopy (Section 7.2). Along with optical techniques, deep-level transient spectroscopy (DLTS) has been crucial to elucidate the properties of DX centers. DLTS is discussed in Chapter 9.

7.9 TRANSITION METALS

Transition metals are common contaminants that introduce deep levels into the band gap (Clerjaud, 1985; Hennel, 1993). These levels provide efficient recombination pathways for electrons and holes, reducing the minority carrier lifetime. The electronic states are due to unfilled d orbitals in the atomic core. As discussed in Section 5.2, the five d orbitals may be written as functions proportional to xy, yz, xz, $x^2 - y^2$, and $2z^2 - x^2 - y^2$. In T_d symmetry, these states split into two energy levels, a twofold degenerate E state and a threefold degenerate T_2 state. The transition of an electron from E and T_2 leads to absorption peaks in the mid-IR (typically 2–3 μm). Conversely, the transition from T_2 to E produces PL peaks in the same spectral region. The IR emission competes with band-edge emission, reducing the performance of LEDs and laser diodes.

A well-studied example is InP:Fe, where the deep Fe acceptor is used to make semi-insulating substrates (Bishop, 1986). An isolated Fe atom has six $3d$ and two $4s$ electrons. Substitutional Fe^{3+} gives up three electrons to form bonds with P atoms, leaving five $3d$ electrons. It is electrically neutral with respect to the host. The negatively charged acceptor, Fe^{2+}, has six $3d$ electrons. In the Fe^{2+} ground state, four electrons occupy the E levels while two electrons occupy T_2 levels. IR

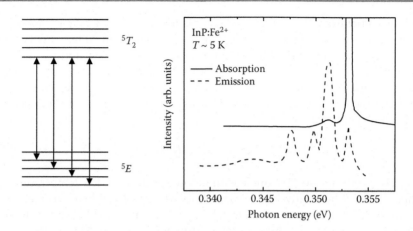

FIGURE 7.20 Energy levels and infrared (IR) spectra for intra-*d* transitions in InP:Fe^{2+}. The four peaks in the photoluminescence (PL) spectrum arise from transitions to the ^5E states. At low temperatures, the IR absorption spectrum is dominated by the transition from the lowest ^5E state. (After Bishop, S.G. 1986. *Deep Centers in Semiconductors*, ed. S.T. Pantelides, pp. 541–626. New York, NY: Gordon and Breach.)

light can promote an electron from E to T_2, producing an excited state. In the multiple-electron picture, the ground state is denoted ^5E and the excited state is $^5T_2$. Fine structure in the IR spectrum arises from the splitting of the ground state (Figure 7.20).

IR spectral lines provide unique "fingerprints" that are used to identify transition metals in semiconductor samples. Because they arise from internal *d-d* transitions, however, these peaks do not yield information about where the energy levels are relative to the band edges. Optical absorption experiments can be performed to estimate the position of these levels. In the case of InP:Fe, an electron is promoted from the valence band to the (0/−) acceptor level, changing the oxidation state from Fe^{3+} to Fe^{2+}. The onset for the optical absorption is the difference between the energy of the Fe^{2+} ground state and the energy of a valence-band electron plus Fe^{3+}. A configuration-coordinate diagram is shown in Figure 7.21. Although the lattice relaxation is not as great as for *DX* centers, there is nonetheless a small difference between the optical and thermal energies (Franck–Condon shift). For the InP:Fe acceptor at low temperatures, the optical absorption threshold is ∼0.7 eV, whereas electrical measurements indicate an acceptor level ∼0.6 eV above the valence band (Look, 1979).

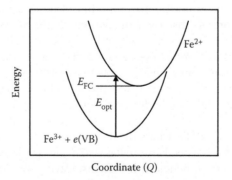

FIGURE 7.21 Configuration-coordinate diagram for InP:Fe. The bottom curve is for Fe^{3+} plus an electron in the valence band. The top curve is for Fe^{2+} in its ground state (^5E).

Chapter 7—Optical Properties **189**

PROBLEMS

7.1 A free electron in a solid experiences a static electric field and a damping force that is proportional to velocity. Consider the steady-state situation; i.e., constant velocity.
   a. Derive an expression for the power absorbed by the electron.
   b. How does the electron-mobility dependence of your answer for (a) compare with the high-frequency limit (Equation 7.23)?

7.2 A GaN sample has a free-electron density of 3×10^{19} cm^{-3}. Let $\varepsilon_\infty = 5$ and $m^* = 0.2m$.
   a. Calculate the plasma frequency (rad/s).
   b. Convert your answer in (a) to cm^{-1}.
   c. Sketch reflectivity versus frequency (cm^{-1}).

7.3 The density of states in the conduction band is given by Equation 5.41. Consider a degenerate semiconductor (band gap E_g) with E_F in the conduction band at 0 K.
   a. Find the number of free electrons (N_{el}) as a function of E_F.
   b. Find the optical band gap versus free-electron density (n).
   c. Qualitatively, how would your answer to (b) change if band renormalization were taken into account?

7.4 Sketch the band gap absorption spectrum for a direct-gap semiconductor, including exciton absorption and Urbach tail.

7.5 A laser of wavelength 325 nm shines on a semiconductor with a direct band gap of 3.10 eV and LO phonon energy of 35 meV. The free exciton binding energy is 50 meV. The donor bound exciton has a binding energy $E_{BX} = 10$ meV. Indicate the peaks in the PL spectrum (eV) due to the following. (Assume excitons are in their ground state.)
   a. Band-to-band transition
   b. Free exciton recombination
   c. Donor bound exciton recombination
   d. First LO phonon replica of (c)

7.6 Donors in ZnO (band gap 3.4 eV) have a donor binding energy of 50 meV. A PL peak at 3.0 eV is attributed to a donor–acceptor pair transition (distant pairs). What is the acceptor binding energy?

7.7 GaP has the zincblende structure with $a = 5.45$ Å, band gap 2.34 eV, and static dielectric constant $\varepsilon = 10.7$. Sulfur (S$_P$) has a donor binding energy of 80 meV. Silicon (Si$_P$) has an acceptor binding energy of 60 meV.
   a. Calculate the donor–acceptor pair (DAP) transition energy for distant pairs (S and Si infinitely far away from each other).
   b. What is the DAP transition energy if the distance between S and Si is $10a$?

7.8 A donor and acceptor are located at $(-x_0,0,0)$ and $(x_0,0,0)$. Let their electron/hole wave functions be

$$\psi_D = B\exp\{-[(x+x_0)^2 + y^2 + z^2]/2a^2\}$$

$$\psi_A = B\exp\{-[(x-x_0)^2 + y^2 + z^2]/2a^2\}$$

Calculate the transition matrix element,

$$\langle f|x|i\rangle = \iiint \psi_A{}^* x\psi_D dxdydz$$

Note:

$$\int_{-\infty}^{\infty} \exp(-x^2/a^2)dx = \sqrt{\pi}\,a$$

7.9 A silicon sample contains $[B] = 5 \times 10^{16}$ cm^{-3} and $[Au] = 1 \times 10^{14}$ cm^{-3}. The sample is at room temperature; assume all the boron acceptors are ionized.
   a. At equilibrium (in the dark), what is the charge state of the Au impurities?
   b. A light bulb shines on the sample. At what rate (s^{-1}) do the Au impurities capture minority electrons? Approximate the electron effective mass as $m^* = m$.
7.10 The optical absorption threshold for a deep donor is 1.1 eV. An electrical measurement determines the donor level to be 0.4 eV below the conduction-band minimum.
   a. What is the Franck–Condon shift for the excited electronic state?
   b. What is the Huang–Rhys factor? Assume the vibrational energy is $\hbar\omega = 50$ meV.

REFERENCES

Alkauskas, A., M.D. McCluskey, and C.G. Van de Walle. 2016. Tutorial: Defects in semiconductors–Combining experiment and theory. *J. Appl. Phys.* 119: 181101 (11 pages).

Bellaiche, L., S.-H. Wei, and A. Zunger. 1996. Localization and percolation in semiconductor alloys: GaAsN vs. GaAsP. *Phys. Rev. B* 54: 17568–17576.

Berggren, K.-F. and B.E. Sernelius. 1981. Band-gap narrowing in heavily doped many-valley semiconductors. *Phys. Rev. B* 24: 1971–1986.

Bishop, S.G. 1986. Iron impurity centers in III-V semiconductors. In *Deep Centers in Semiconductors*, ed. S.T. Pantelides, pp. 541–626. New York, NY: Gordon and Breach.

Burstein, E. 1954. Anomalous optical absorption limit in InSb. *Phys. Rev.* 93: 632–633.

Buyanova, I.A., W.M. Chen, G. Pozina et al. 1999. Mechanism for low-temperature photoluminescence in GaNAs/GaAs structures grown by molecular-beam epitaxy. *Appl. Phys. Lett.* 75: 501–503.

Chen, X.-B., J. Huso, J.L. Morrison, and L. Bergman. 2006. Dynamics of GaN band edge photoluminescence at near-room-temperature regime. *J. Appl. Phys.* 99: 046105 (3 pages).

Chen, X.-B., J. Huso, J.L. Morrison, and L. Bergman. 2007. The properties of ZnO photoluminescence at and above room temperature. *J. Appl. Phys.* 102: 116105 (3 pages).

Clerjaud, B. 1985. Transition-metal impurities in III-V compounds. *J. Phys. C* 18: 3615–3661.

Dean, P.J. 1986. Oxygen and oxygen associates in gallium phosphide and related semiconductors. In *Deep Centers in Semiconductors*, ed. S.T. Pantelides, pp. 185–347. New York, NY: Gordon and Breach.

Fano, U. 1961. Effects of configuration interaction on intensities and phase shifts. *Phys. Rev.* 124: 1866–1878.

Fouquet, J.E. and A.E. Siegman. 1985. Room-temperature photoluminescence times in a GaAs/Al$_x$Ga$_{1-x}$As molecular beam epitaxy multiple quantum well structure. *Appl. Phys. Lett.* 46: 280–282.

Gershenzon, M., D.G. Thomas, and R.E. Dietz. 1962. Radiative transitions near the band edge of GaP. In *Proc. Intl. Conf. on the Phys. Semicond.*, ed. A.C. Stickland, pp. 752–759. London, UK: IOP.

Hall, R.N. 1952. Electron-hole recombination in germanium. *Phys. Rev.* 87: 387.

Haller, E.E. 1995. Isotopically engineered semiconductors. *J. Appl. Phys.* 77: 2857–2878.

Haynes, J.R. 1960. Experimental proof of the existence of a new electronic complex in silicon. *Phys. Rev. Lett.* 4: 361–363.

Hennel, A.M. 1993. Transition metals in III/V compounds. In *Semiconductors and Semimetals*, Vol. 38, ed. E. Weber, pp. 189–234. San Diego, CA: Academic Press.

Hlaing Oo, W.M., M.D. McCluskey, J. Huso, and L. Bergman. 2007. Infrared and Raman spectroscopy of ZnO nanoparticles annealed in hydrogen. *J. Appl. Phys.* 102: 043529 (5 pages).

Huang, K. and A. Rhys. 1950. Theory of light absorption and non-radiative transitions in F-centres. *Proc. Roy. Soc. A (London)* 204: 406–423.

Jackson, J.D. 1999. *Classical Electrodynamics*, 3rd ed. New York, NY: Wiley.

Jain, S.C., M. Willander, J. Narayan, and R. van Overstraeten. 2000. III-nitrides: Growth, characterization, and properties. *J. Appl. Phys.* 87: 965–1006.

Karaiskaj, D., M.L.W. Thewalt, T. Ruf et al. 2001. Photoluminescence of isotopically purified silicon: How sharp are the bound exciton transitions? *Phys. Rev. Lett.* 86: 6010–6013.

Karasyuk, V.A., A.G. Steele, A. Mainwood et al. 1992. Ultrahigh-resolution photoluminescence studies of excitons bound to boron in silicon under uniaxial stress. *Phys. Rev. B* 45: 11736–11743.

Kittel, C. 2005. *Introduction to Solid State Physics*, 8th ed. New York, NY: John Wiley & Sons.

Lang, D.V. 1986. DX centers in III-V alloys. In *Deep Centers in Semiconductors*, ed. S.T. Pantelides, pp. 489–539. New York, NY: Gordon and Breach.

Lang, D.V. and R.A. Logan. 1977. Large-lattice-relaxation model for persistent photoconductivity in compound semiconductors. *Phys. Rev. Lett.* 39: 635–639.

Lax, M. and E. Burstein. 1955. Infrared lattice absorption in ionic and homopolar crystals. *Phys. Rev.* 97: 39–52.

Look, D.C. 1979. Model for Fe^{2+} intracenter-induced photoconductivity in InP:Fe. *Phys. Rev. B* 20: 4160–4166.

Madelung, O., U. Rössler, and M. Schulz, eds. 2002. *The Landolt-Börnstein Database*, Group III, 41: A2a. Berlin: Springer-Verlag.

McCluskey, M.D., N.M. Johnson, C.G. Van de Walle, D.P. Bour, M. Kneissl, and W. Walukiewicz. 1998. Metastability of oxygen donors in AlGaN. *Phys. Rev. Lett.* 80: 4008–4011.

Meyer, B.K., H. Alves, D.M. Hofmann et al. 2004. Bound exciton and donor-acceptor pair recombinations in ZnO. *Phys. Stat. Solidi B* 241: 231–260.

Mitra, S.S. 1985. Optical properties of nonmetallic solids. In *Handbook of Optical Constants of Solids*, ed. E.D. Palik, pp. 263–267. Orlando, FL: Academic Press.

Monemar, B. 2001. Bound excitons in GaN. *J. Phys.: Condens. Matter* 13: 7011–7026.

Mooney, P. 1990. Deep donor levels (*DX* centers) in III-V semiconductors. *J. Appl. Phys.* 67: R1–R26.

Mooney, P., G.A. Northrop, T.N. Morgan, and H.G. Grimmeiss. 1988. Evidence for large lattice relaxation at the *DX* center in Si-doped $Al_xGa_{1-x}As$. *Phys. Rev. B* 37: 8298–8307.

Mooradian, A. and A.L. McWhorter. 1967. Polarization and intensity of Raman scattering from plasmons and phonons in gallium arsenide. *Phys. Rev. Lett.* 19: 849–852.

Moss, T.S. 1954. The interpretation of the properties of indium antimonide. *Proc. Phys. Soc. (London).* B76: 775–782.

Orton, J.W. and P. Blood. 1990. *The Electrical Characterization of Semiconductors: Measurement of Minority Carrier Properties*. London, UK: Academic Press.

Özgür, Ü., Y.I. Alivov, C. Liu et al. 2005. A comprehensive review of ZnO materials and devices. *J. Appl. Phys.* 98: 041301 (103 pages).

Pajot, B. and B. Clerjaud. 2012. *Optical Properties of Impurities and Defects in Semiconductors II: Electronic Absorption of Deep Centres and Vibrational Spectra*. Berlin: Springer-Verlag.

Pankove, J.I. 1971. *Optical Processes in Semiconductors*. New York, NY: Dover.

Perlin, P., T. Suski, H. Teisseyre et al. 1995. Toward the identification of the dominant donor in GaN. *Phys. Rev. Lett.* 75: 296–299.

Saito, M., A. Oshiyama, and O. Sugino. 1992. Validity of the broken-bond model for the DX center in GaAs. *Phys. Rev. B* 45: 13745–13748.

Sakurai, J.J. 1985. *Modern Quantum Mechanics*, 337. Reading, MA: Addison-Wesley.

Schubert, E.F. 2006. *Light-Emitting Diodes*, 2nd ed. Cambridge, UK: Cambridge University Press.

Shan, W., X.C. Xie, J.J. Song, and B. Goldenberg. 1995. Time-resolved exciton luminescence in GaN grown by metalorganic chemical vapor deposition. *Appl. Phys. Lett.* 67: 2512–2514.

Shan, W., W. Walukiewicz, J.W. Ager III et al. 1999. Band anticrossing in GaInNAs alloys. *Phys. Rev. Lett.* 82: 1221–1224.

Shockley, W. and W.T. Read. 1952. Statistics of the recombinations of holes and electrons. *Phys. Rev.* 87: 835–842.

Spitzer, W.G. 1967. Multiphonon lattice absorption. In *Semiconductors and Semimetals*, Vol. 3, eds. R.K. Willardson and A.C. Beer, pp. 17–69. New York, UK: Academic Press.

Sturge, M.D. 1962. Optical absorption of gallium arsenide between 0.6 and 2.75 eV. *Phys. Rev.* 121: 359–362.

Sze, S.M. and K.K. Ng. 2007. *Physics of Semiconductor Devices*, 3rd ed., pp. 40–45. New York, UK: John Wiley & Sons.

Thewalt, M.L.W., A. Yang, M. Steger et al. 2007. Direct observation of the donor nuclear spin in a near-gap bound exciton transition: ^{31}P in highly enriched ^{28}Si. *J. Appl. Phys.* 101: 081724 (5 pages).

Thomas, D.G. and J.J. Hopfield. 1966. Isoelectronic traps due to nitrogen in GaP. *Phys. Rev.* 150: 680–703.

Thomas, D.G., M. Gershenzon, and F.A. Trumbore. 1964. Pair spectra and "edge" emission in gallium phosphide. *Phys. Rev.* 133: A269–A279.

Thomas, D.G., J.J. Hopfield, and W.N. Augustyniak. 1965. Kinetics of radiative recombination of randomly distributed donors and acceptors. *Phys. Rev.* 140: A202–A220.

Urbach, F. 1953. The long-wavelength edge of photographic sensitivity and of the electronic absorption of solids. *Phys. Rev.* 92: 1324.

Varshni, Y.P. 1967. Band-to-band radiative recombination in groups IV, VI, and III-V semiconductors. *Phys. Stat. Sol. (b)* 19: 459–514 and 20: 9–36.

Vurgaftman, I. and J.R. Meyer. 2003. Band parameters for nitrogen-containing semiconductors. *J. Appl. Phys.* 94: 3675–3696.

Wetzel, C., W. Walukiewicz, E.E. Haller et al. 1996. Carrier localization of as-grown n-type gallium nitride under large hydrostatic pressure. *Phys. Rev. B* 53: 1322–1326.

Williams, F. 1968. Donor-acceptor pairs in semiconductors. *Phys. Stat. Solid.* 25: 493–512.

Wolfe, J.P. and C.D. Jeffries. 1983. Strain-confined excitons and electron-hole liquid. In *Electron-Hole Droplets in Semiconductors*, eds. C.D. Jeffries and L.V. Keldysh. Amsterdam, UK: North-Holland.

Wolfe, J.P., W.L. Hansen, E.E. Haller, R.S. Markiewicz, C. Kittel, and C.D. Jeffries. 1975. Photograph of an electron-hole drop in germanium. *Phys. Rev. Lett.* 34: 1292–1293.

Wolk, J.A., M.B. Kruger, J.N. Heyman, W. Walukiewicz, R. Jeanloz, and E.E. Haller. 1991. Local-vibrational-mode spectroscopy of DX centers in Si-doped GaAs under hydrostatic pressure. *Phys. Rev. Lett.* 66: 774–777.

Woodhouse, K., R.C. Newman, T.J. de Lyon, J.M. Woodall, G.J. Scilla, and F. Cardone. 1991. LVM spectroscopy of carbon and carbon-hydrogen pairs in GaAs grown by MOMBE. *Semicond. Sci. Technol.* 6: 330–334.

Wu, J., W. Walukiewicz, W. Shan et al. 2002. Effects of the narrow band gap on the properties of InN. *Phys. Rev. B* 66: 201403 (4 pages).

Wu, J., W. Walukiewicz, W. Shan et al. 2003. Temperature dependence of the fundamental band gap of InN. *J. Appl. Phys.* 94: 4457–4460.

Yu, P.Y. and M. Cardona. 1996. *Fundamentals of Semiconductors*. Berlin: Springer-Verlag.

Thermal Properties

The electronic, vibrational, and optical properties of defects in semiconductors are all affected by temperature. Within the broad subject of thermal properties, this chapter focuses on defect formation and diffusion, phenomena that play important roles during crystal growth and processing of semiconductors. We first discuss the thermodynamics of defect formation, including the effect of the Fermi level and chemical potentials. We then review diffusion, from the continuum to the atomic scale. Examples of self-diffusion, dopant diffusion, and superlattice disordering are given.

8.1 DEFECT FORMATION

At elevated temperatures, defects will always form in semiconductors. A system in thermodynamic equilibrium, at a pressure P and temperature T, will minimize the Gibbs free energy:

(8.1) $$G = H - TS$$

where H is the enthalpy and S is the entropy. The enthalpy is defined as

(8.2) $$H = E + PV$$

where E is the total energy and V is the volume of the sample. In semiconductors at ambient pressures, the PV term plays a minor role. With the exception of high-pressure environments, such as those produced by diamond-anvil cells, energy and enthalpy are essentially interchangeable in the solid phase.

A simple derivation of the dependence of the vacancy concentration on temperature has been given by Kittel (2005). A monatomic crystal with vacancies is considered. The enthalpy required to remove one atom from the interior of the crystal and place it on the surface is H_V. In order to create n vacancies, an amount of enthalpy nH_V must be invested. It is assumed that the defect concentration is small such that the vacancies are isolated from each other and do not interact.

194   8.1 Defect Formation

We now determine the configurational entropy S of the crystal with vacancies. The number of possibilities to form n vacancies in an N atom crystal is

$$(8.3) \qquad \Omega = \frac{N!}{(N-n)!n!}$$

We divide by $n!$ because it does not matter in which sequence we remove the n identical atoms (Reif, 1965). The entropy is proportional to the logarithm of the number of possible arrangements:

$$(8.4) \qquad S = k_B \ln \Omega$$

where k_B is the Boltzmann constant. In addition, there are some subtle changes in entropy that arise primarily from vibrational modes. The term nS_V denotes the change in vibrational entropy that is caused by the introduction of n vacancies.

From Equations 8.1 through 8.4, the change in the Gibbs free energy that results from the formation of n vacancies is given by

$$(8.5) \qquad \Delta G = nH_V - k_B T \{\ln(N!) - \ln[(N-n)!] - \ln(n!)\} - nTS_V$$

Using Stirling's approximation, $\ln(x!) \approx x \ln(x) - x$ for large x, Equation 8.5 becomes

$$(8.6) \qquad \Delta G = nH_V - k_B T[N \ln N - (N-n)\ln(N-n) - n\ln n] - nTS_V$$

We now find the value of n that minimizes G, by setting the derivative equal to zero:

$$(8.7) \qquad \frac{dG}{dn} = H_V - k_B T[\ln(N-n) - \ln n] - TS_V = 0$$

Solving this equation yields

$$(8.8) \qquad \frac{n}{N-n} = e^{S_V/k_B} e^{-H_V/k_B T}$$

The number of vacancies is small compared to the number of atoms in the crystal. With the approximation $n \ll N$, Equation 8.8 becomes

$$(8.9) \qquad n/N = e^{S_V/k_B} e^{-H_V/k_B T}$$

The vibrational entropy S_V is typically a few k_B. In defect calculations, it is often neglected. This simplification leads to

$$(8.10) \qquad n = Ne^{-H_V/k_B T}$$

As an example, suppose that it costs $H_V = 2$ eV to form a vacancy thermodynamically (in contrast to formation of defects with energetic electrons or ion beams). Silicon has $N = 5 \times 10^{22}$ atoms/cm³. At the melting point of silicon ($T_m = 1415°C = 1688$ K), Equation 8.10 yields a vacancy concentration of 5×10^{16} cm⁻³. Because of the exponential in Equation 8.10, the computed vacancy concentration is very sensitive to H_V. If the value of H_V is increased to 3 eV, then n drops to 5×10^{13} cm⁻³.

Chapter 8—Thermal Properties **195**

Although we performed this exercise for vacancies in silicon, it should be made clear that in silicon, self-interstitials (Si_i) are also important native defects at high temperatures (Seeger and Chik, 1968). The experimental proof for this conclusion came mainly from detailed electron microscopy studies that showed extrinsic stacking faults in dislocation-free silicon crystals. Extrinsic stacking faults are small sections of extra planes that form when interstitial silicon atoms condense during cooling. These extended defects result in A-swirls, as shown in Figure 2.5. Because self-interstitials are not important in metals, it took a long time for researchers to accept the role of interstitials in semiconductors (Gösele, 2000).

The concentration of interstitials can be calculated by Equation 8.10, with H_V replaced by H_I, the enthalpy required to take an atom from the surface and place it into an interstitial site. The presence of native defects at elevated temperatures affects a range of phenomena, including diffusion, annealing of radiation-induced defects (e.g., during ion implantation), dislocation formation and motion, and gettering. During cooling, the vacancies become supersaturated and recombine with interstitials and dislocations, or they migrate to the crystal surface. They can also condense into an intrinsic stacking fault, a part of a crystal plane that is missing (Section 2.3).

Contrary to silicon, germanium is believed to be dominated by vacancies up to the melting point $T_m = 936°C = 1209$ K. In general, the situation for germanium and most other semiconductors is not nearly as well understood as for silicon.

8.2 CHARGE STATE

The previous section showed that, given knowledge of the defect *formation enthalpy* H^f, one can calculate the defect concentration n:

$$(8.11) \qquad n = Ne^{-H^f/k_B T}$$

where N is the concentration of sites where the defect can reside. Here, the formation entropy is neglected (Equation 8.10). Also, under normal conditions of pressure, H^f is equivalent to the formation energy E^f. Therefore, formation "enthalpy" and "energy" are practically interchangeable.

One factor that affects the formation enthalpy is the *charge state* of the defect. If the defect has a charge of $+1$, then it has given an electron to the semiconductor. The energy of that electron is equal to the Fermi energy E_F. This energy must be added to the defect formation enthalpy. If the defect has a charge of -1, then it has *taken* an electron from the Fermi level. In that case, E_F must be subtracted from the formation enthalpy. In general, we add the term qE_F, where q is the charge state ($-1, 0, 1$, etc.).

Consider a deep single donor (Figure 8.1). When the Fermi level is at the valence band ($E_F = 0$), the neutral state ($q = 0$) has a higher enthalpy than the positively charged state ($q = 1$). This means that it is advantageous for the donor to give up its electron and compensate an acceptor. As the Fermi level rises, the neutral level has zero slope but the positive charge state has a slope of $+1$. They cross at the (0/+) level. When E_F is above the (0/+) level, the neutral charge state is preferred.

The effect of charge state is important because it can lead to *self-compensation*. For example, in p-type semiconductors, donor defects will form spontaneously and compensate the acceptor dopants. Because the Fermi energy is low in p-type material, the formation enthalpy for such donor defects is reduced. In highly n-type semiconductors, compensating acceptors form.

An important type of self-compensation is described by the *amphoteric defect model* (Walukiewicz, 1989). In compound semiconductors such as GaAs, vacancies and antisites form and may act as donors or acceptors. In heavily doped n-type GaAs, the gallium vacancy V_{Ga} is a triple acceptor, with a charge state of $q = -3$. When E_F is high, the formation enthalpy for V_{Ga} is low. As the concentration of compensating V_{Ga} acceptors increases, though, E_F is lowered. When

8.2 Charge State

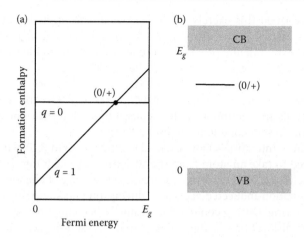

FIGURE 8.1 (a) Formation enthalpy for a deep donor. (b) Band diagram showing the position of the donor level.

the Fermi level is ~0.6 eV above the valence band, V_{Ga} becomes unstable. One of the neighboring arsenic atoms moves into the gallium vacancy, forming an arsenic antisite (As_{Ga}). This defect reaction is given by

$$V_{Ga} \rightarrow V_{As} + As_{Ga} \tag{8.12}$$

The vacancy–antisite pair ($V_{As} + As_{Ga}$) is a triple donor. The transformation of the native acceptor into a donor prevents the Fermi level from decreasing any further. The energy $E_F = 0.6$ eV is called the *Fermi stabilization energy* (E_{FS}).

The formation enthalpies for these defects are shown in Figure 8.2. In this figure, only the lowest-enthalpy configurations are plotted. The slopes of the lines are equal to the charge states ($q = -2$,

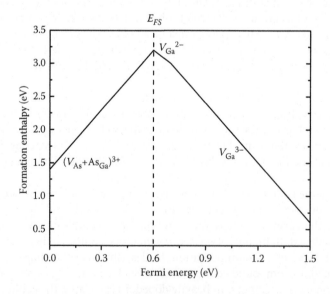

FIGURE 8.2 Formation enthalpies for native defects in GaAs. The Fermi stabilization energy E_{FS} is ~0.6 eV above the valence-band maximum. (After Walukiewicz, W. 1989. *Appl. Phys. Lett.* 54: 2094–2096.)

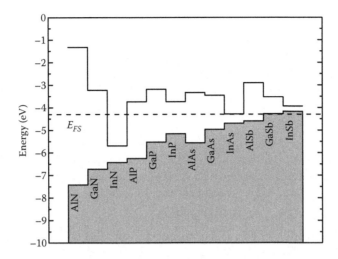

FIGURE 8.3 Approximate Fermi stabilization energy level E_{FS}, for various III-V semiconductors. For consistency, the band alignments of Figure 5.11 were used. (After Walukiewicz, W. 2001. *Physica B* 302–303: 123–134.)

−3, or +3). The maximum enthalpy occurs at $E_F = E_{FS}$. In a similar manner, arsenic vacancy donors can transform into acceptor defects, with a comparable value for E_{FS}. The Fermi level of a sample that is exposed to particle irradiation will approach E_{FS} as the concentration of native defects increases. When $E_F = E_{FS}$, the Fermi level is stabilized and is insensitive to further damage.

Remarkably, the absolute value of E_{FS} appears to be independent of the particular semiconductor (Walukiewicz, 2001). For most III-V semiconductors, E_{FS} lies in the band gap (Figure 8.3). In those cases, n- or p-type samples that are exposed to particle irradiation will experience a decrease in carrier concentration as the Fermi level is pushed toward the middle of the gap. For semiconductors with low conduction-band minima, however, E_{FS} actually lies in the conduction band. InN, for example, has $E_{FS} \sim 1$ eV above the conduction-band minimum (Li et al., 2005). When InN is irradiated with energetic electrons, protons, or alpha particles (^{4}He$^+$), the free-electron concentration increases as E_F is pushed upward toward E_{FS}.

One way to influence the formation of defects during growth is to shine above-gap light on the sample. The optically generated electrons and holes lead to quasi-Fermi levels (Section 7.5), which can suppress the formation of compensating centers (Alberi and Scarpulla, 2016).

8.3 CHEMICAL POTENTIAL

Along with the charge state, the *chemical potentials* of the various molecular species affect formation enthalpies. The chemical potential provides information about the environment in which the semiconductor resides. In GaAs, for example, if the arsenic chemical potential (μ_{As}) is high, then it will be more likely for arsenic antisites to form (Zhang and Northrup, 1991). A thermodynamic definition is given by the Gibbs free energy:

$$(8.13) \qquad G = \sum_i n_i \mu_i$$

where n_i and μ_i are the number and chemical potential, respectively, of the *i*th molecular species. At $T = 0$, $G = H$ (Equation 8.1), and the chemical potential is just the enthalpy of a single atom or molecule. This low-temperature approximation is commonly used in DFT calculations.

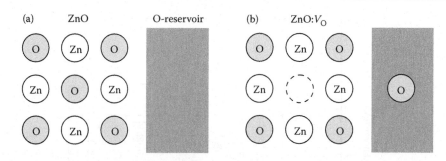

FIGURE 8.4 Formation of an oxygen vacancy. (a) Pure ZnO. (b) To form a vacancy, the oxygen atom moves from the ZnO crystal to an oxygen reservoir.

Consider the formation of an oxygen vacancy in ZnO (Section 2.3). Initially, we have bulk ZnO, the enthalpy of which is $H_{tot}(ZnO)$:

$$H_{\text{before}} = H_{tot}(ZnO) \tag{8.14}$$

An oxygen vacancy is formed, with a charge state q. To form the vacancy, an oxygen atom leaves the pure ZnO and enters a reservoir (Figure 8.4). The chemical potential μ_O is essentially the enthalpy of an oxygen atom in the reservoir. The enthalpy of the system is now

$$H_{\text{after}} = H_{tot}(ZnO:V_O) + \mu_O + qE_F \tag{8.15}$$

The defect formation enthalpy is given by $H_{\text{after}} - H_{\text{before}}$, or

$$H^f = H_{tot}(ZnO:V_O) - H_{tot}(ZnO) + \mu_O + qE_F \tag{8.16}$$

The chemical potential μ_O is a parameter that depends on the experimental conditions. In practice, there are limits on its value. If μ_O is very low, for example, then the loss of oxygen from the sample will result in the growth of zinc precipitates. Using rules from equilibrium thermodynamics, one can place bounds on the chemical potentials (Janotti and Van de Walle, 2007). Under chemical equilibrium, Zn and O atoms can react according to

$$Zn + O \leftrightarrow ZnO \tag{8.17}$$

Given this reaction, the chemical potentials are constrained by

$$\mu_{Zn} + \mu_O = \mu_{ZnO} \tag{8.18}$$

where μ_{ZnO} is the chemical potential for a Zn–O pair in bulk ZnO. The formation enthalpy for bulk ZnO is given by

$$\Delta H = \mu_{ZnO} - \mu_{Zn(\text{bulk})} - \mu_{O(\text{gas})} \tag{8.19}$$

where $\mu_{Zn(\text{bulk})}$ is the chemical potential for a zinc atom in a bulk zinc crystal and $\mu_{O(\text{gas})}$ is for an oxygen atom in an O_2 molecule. Empirically, $\Delta H = -3.6$ eV for ZnO (Pauling, 1988).

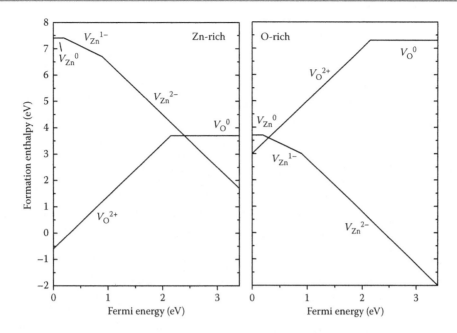

FIGURE 8.5 Formation enthalpies for oxygen and zinc vacancies in ZnO, for zinc-rich and oxygen-rich conditions. The Fermi energy is relative to the valence-band maximum. (After Janotti, A. and C.G. Van de Walle. 2007. *Phys. Rev. B* 76: 165202 (22 pages).)

We now look at the two bounding cases. For a zinc-rich environment, the chemical potential is given by $\mu_{Zn} = \mu_{Zn(bulk)}$. This condition would exist if zinc metal were in contact with the ZnO sample. From Equations 8.18 and 8.19, this leads to $\mu_O = \mu_{O(gas)} + \Delta H$. In an oxygen-rich environment, the chemical potentials are given by $\mu_O = \mu_{O(gas)}$ and $\mu_{Zn} = \mu_{Zn(bulk)} + \Delta H$. For this oxygen-rich extreme, μ_O is at its maximum, which raises the oxygen vacancy formation energy (Equation 8.16). This increase in formation energy leads to a reduction in the concentration of oxygen vacancies (Equation 8.11).

In DFT calculations, $\mu_{Zn(bulk)}$ and $\mu_{O(gas)}$ are often estimated by their values at low pressures and temperatures. In that regime, $G = E$ (Equations 8.1 and 8.2), $\mu_{Zn(bulk)}$ is the energy of a zinc atom in a bulk zinc crystal, and $\mu_{O(gas)}$ is ½ the energy of an isolated O_2 molecule.

The formation enthalpy for zinc vacancies is similar to Equation 8.16, with μ_O replaced by μ_{Zn}. Figure 8.5 shows a plot of the formation enthalpy as a function of Fermi level, for zinc and oxygen vacancies, calculated using DFT. The formation enthalpy for oxygen vacancies increases in oxygen-rich conditions, while the formation enthalpy for zinc vacancies decreases. Different slopes correspond to different charge states. For oxygen vacancies, the 1− charge state is always energetically unfavorable. The point where the 0 and 2+ lines intersect is the (0/2+) level. For zinc vacancies, the acceptor levels (0/−) and (−/2−) occur at 0.2 and 0.9 eV, respectively.

8.4 DIFFUSION

As discussed in Section 4.8, impurity diffusion is one of the most important technologies for the production of well-defined doped layers in semiconductors (Shaw, 1973). Diffusion is not driven by concentration gradients, as is sometimes claimed. Rather, diffusion is a thermally activated process that tends to reduce concentration differences in a system, through the random walk of individual dopant atoms and defects. Imagine two sections of a sample that have high and low

200 8.4 Diffusion

concentrations of impurities. At an elevated temperature, the random motion of the impurities will cause them to cross the boundary between the two sections. On average, more impurities will move from the high-concentration side to the low-concentration side than the reverse. The net flow of impurities is therefore in a direction opposite to the concentration gradient.

Fick's laws quantitatively describe matter transport by diffusion (Ghez, 2001). *Fick's first law* is

$$(8.20) \qquad \mathbf{J} = -D\vec{\nabla}C$$

where $\mathbf{J}$ is the concentration flux ($cm^{-2}s^{-1}$), D is the *diffusion coefficient* or *diffusivity* (cm^2/s), and $C(\mathbf{r})$ is the concentration (cm^{-3}). This is a mathematical statement that the net flow of matter is opposite to the concentration gradient. The time dependence of the concentration is given by

$$(8.21) \qquad \frac{\partial C}{\partial t} = -\vec{\nabla} \cdot \mathbf{J}$$

Equation 8.21 states that, if there is a gradient in the flux, the concentration of impurities at a given location will change over time. For example, if there is high flux into a region but low flux out of the region, then the impurity concentration in that region will grow.

These equations can be combined to produce *Fick's second law*. First, we take the divergence of both sides of Equation 8.20:

$$(8.22) \qquad \vec{\nabla} \cdot \mathbf{J} = -D\nabla^2 C$$

where we have assumed that D is constant. Equation 8.22 is then substituted into Equation 8.21, resulting in

$$(8.23) \qquad \frac{\partial C}{\partial t} = D\nabla^2 C$$

The one-dimensional form of Fick's second law is expressed as

$$(8.24) \qquad \frac{\partial C}{\partial t} = D\frac{\partial^2 C}{\partial x^2}$$

Two types of boundary conditions are of general importance for many technical applications (Runyan, 1965). Both deal with the case of an impurity that enters a "pure" single crystal from or near the surface. First, we consider an *inexhaustible source* on the surface at $x = 0$. We assume that there are no impurities in the crystal at $t = 0$ and that D is constant at a given temperature. For these conditions, the solution to Fick's second law (Equation 8.24) is a complementary error function:

$$(8.25) \qquad C(x,t) = C_0 \mathrm{erfc}\left(\frac{x}{2\sqrt{Dt}}\right)$$

where C_0 is the surface concentration (constant). The error function is defined by an integral:

$$(8.26) \qquad \mathrm{erfc}(y) = 1 - \mathrm{erf}(y) = 1 - \frac{2}{\sqrt{\pi}} \int_0^y e^{-z^2} dz$$

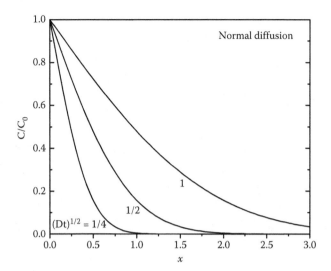

FIGURE 8.6 Normal diffusion, according to Equation 8.25, for three values of $\sqrt{Dt}$.

The dimensionless parameter

$$y = \frac{x}{2\sqrt{Dt}} \tag{8.27}$$

is the "normalized" diffusion depth. Figure 8.6 shows plots of Equation 8.25 for three values of $\sqrt{Dt}$. When $x = \sqrt{Dt}$ ($y = 0.5$), the concentration drops to approximately half the value at the surface. For $y = 1.82$, the value of erfc(y) has dropped to 1%, and for $y = 4$ it is 1.5×10^{-8}, a very rapid drop-off.

The second boundary condition is diffusion from a *finite source* into a semi-infinite space. At $t = 0$, there is a fixed supply of S diffusing impurities per cm², located at $x = 0$. With that initial condition, the solution of Fick's second law (Equation 8.24) is

$$C(x,t) = \frac{S}{\sqrt{\pi Dt}} e^{-\left(x/2\sqrt{Dt}\right)^2} \tag{8.28}$$

This solution is for a perfectly "reflecting" surface (i.e., a surface where diffusing atoms do not get trapped but are reflected back into the crystal). The surface concentration, which is assumed to be infinite at $t = 0$, decreases over time:

$$C(0,t) = S/\sqrt{\pi Dt} \tag{8.29}$$

During in-diffusion of a dopant impurity, from a semi-infinite or finite source, we observe *out-diffusion* of the bulk dopants. Fick's second law is used to model out-diffusion from a semi-infinite space. In this case, the surface acts as a *sink*, opposite to the assumption made for diffusion from a finite source. The boundary conditions are given by

$$\begin{array}{l} C(x \geq 0, t = 0) = C_{\text{bulk}} \\ C(x = 0, t > 0) = 0 \end{array} \tag{8.30}$$

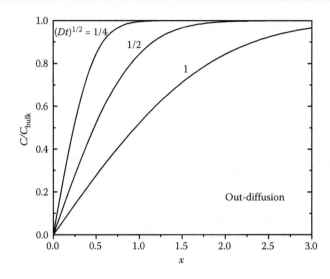

FIGURE 8.7 Out-diffusion, according to Equation 8.31, for three values of $\sqrt{Dt}$.

Given these conditions, the solution to Equation 8.24 is an error function:

$$C(x,t) = C_{bulk} \text{erf}\left(\frac{x}{2\sqrt{Dt}}\right) \tag{8.31}$$

and is plotted in Figure 8.7. Out-diffusion leads to a reduction of the bulk dopant concentration near the surface. This has important consequences for devices. If a *p-n* junction is produced by diffusion of a donor impurity into *p*-type silicon, the depth of the junction is deeper with out-diffusion occurring than when it does not take place.

The final boundary effect to be discussed involves the influence of an oxide layer on diffusion profiles. When a doped semiconductor is oxidized, the impurities prefer to either reside in the semiconductor or in the oxide. A segregation coefficient k, analogous to the one used in crystal growing (Section 4.2), is defined:

$$k = \frac{\text{impurity concentration in semiconductor}}{\text{impurity concentration in oxide}} \tag{8.32}$$

$k > 1$ leads to a "pile-up" of impurities in the bulk, while $k < 1$ leads to a depletion or "pile-down" (Figure 8.8). The segregation coefficient is temperature dependent and is empirically modeled as $k = k_0 \exp(-E/k_B T)$. For the Si/SiO$_2$ system, indium and boron acceptors have $k < 1$ (Kizilyalli et al., 1996), while the substitutional donor impurities have $k > 1$ (Steen et al., 2008).

A thin *delta-doped* layer at $x = x'$, far from the surface, can be described by a Gaussian function, similar to Equation 8.28:

$$C(x,t) = \frac{S}{2\sqrt{\pi Dt}} e^{-(x-x')^2/4Dt} \tag{8.33}$$

where S is the concentration of impurities per cm^2. Over time, the width of the Gaussian peak increases from zero, as the atoms diffuse away from their initial location. We can use this result

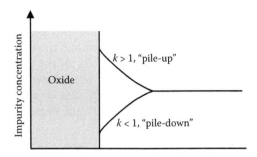

FIGURE 8.8 Impurity segregation near an oxide-semiconductor interface.

to model the diffusion of *any* doping profile that is far from the surface. Given an initial doping concentration $C_0(x)$, the concentration at later times is given by

(8.34) $$C(x,t) = \int \frac{C_0(x')dx'}{2\sqrt{\pi Dt}} e^{-(x-x')^2/4Dt}$$

A special case, important for quantum wells, is given by

(8.35) $$C_0(x) = \begin{cases} C_0 & -x_0 < x < x_0 \\ 0 & \text{otherwise} \end{cases}$$

This concentration profile describes a "square well" of width $2x_0$. As diffusion occurs, the sharp interfaces of this concentration profile begin to broaden (Figure 8.9). Using Equation 8.34, the concentration is given by (Crank, 1975)

(8.36) $$C(x,t) = \frac{C_0}{2}\left[\text{erf}\left(\frac{x+x_0}{\sqrt{4Dt}}\right) - \text{erf}\left(\frac{x-x_0}{\sqrt{4Dt}}\right)\right]$$

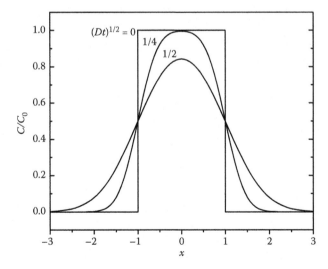

FIGURE 8.9 Diffusion of atoms from a quantum well or delta-doped layer, described by Equation 8.36.

Examples of quantum-well intermixing are discussed in Section 8.8.

As discussed in Section 4.8, dopants can be introduced into a semiconductor by a *two-step diffusion* process. During the first step, predeposition, the dopant is diffused into a near-surface region. The source is effectively inexhaustible, so the impurity profile is given by a complementary error function (Equation 8.25). The second step, drive-in, is diffusion from a finite source. After a prolonged drive-in, a Gaussian profile is approached (Equation 8.28). For an arbitrary predeposition time t and drive-in time t', the impurity concentration is given by (Smith, 1953):

$$(8.37) \qquad C(x) = C_0 \frac{2}{\sqrt{\pi}} \int_{\sqrt{\beta}}^{\infty} e^{-z^2} \mathrm{erf}(\alpha z)\, dz$$

where

$$(8.38) \qquad \begin{aligned} \alpha &= \sqrt{Dt/D't'} \\ \beta &= \frac{x^2}{4(Dt + D't')} \end{aligned}$$

C_0 is the surface concentration during predeposition; D and D' are the diffusion coefficients during predeposition and drive-in, respectively.

The previous cases assumed a constant diffusion coefficient D. As discussed in Section 8.7, there are instances where D depends on the local dopant concentration. In that case, numerical methods are employed to calculate the diffusion profile.

8.5 MICROSCOPIC MECHANISMS OF DIFFUSION

In general, diffusion is divided into direct and indirect mechanisms. A direct mechanism does not require native defects. For example, consider an impurity that diffuses interstitially in a hypothetical lattice. The impurity performs a series of N random steps, hopping from one equivalent interstitial site to another. Let δx_i denote the displacement along the x direction for the ith step. The total displacement is given by

$$(8.39) \qquad x = \delta x_1 + \delta x_2 + \cdots + \delta x_N$$

The mean value of x^2 is

$$(8.40) \qquad \langle x^2 \rangle = \langle \delta x_1{}^2 \rangle + \langle \delta x_2{}^2 \rangle + \cdots + \langle \delta x_N{}^2 \rangle = N d^2$$

where d is an average (root mean squared) hopping distance. Here, we used the assumption of a random walk, in which the cross terms (e.g., $\langle \delta x_1 \delta x_2 \rangle$) are equal to zero.

The rate at which steps are taken can be estimated from first-order kinetics. The impurity atom must surmount a barrier to get to the next interstitial site. When it is in the barrier, its enthalpy and entropy increase by H^M and S^M, respectively. These values are referred to as the *migration enthalpy* and *migration entropy*. The probability that the dopant will have sufficient energy to overcome the barrier is given by the Boltzmann factor, similar to Equation 8.9:

(8.41) $$e^{-G^M/k_BT} = e^{S^M/k_B} e^{-H^M/k_BT}$$

The hopping rate is the probability (Equation 8.41) times the attempt frequency v_0:

(8.42) $$v = v_0 e^{S^M/k_B} e^{-H^M/k_BT}$$

The number of steps taken in a time t is $N = vt$. Inserting this into Equation 8.40 yields

(8.43) $$\langle x^2 \rangle = v_0 e^{S^M/k_B} e^{-H^M/k_BT} d^2 \cdot t$$

From Section 8.4, the diffusion of impurities proceeds as $\langle x^2 \rangle \sim Dt$. Comparing this to Equation 8.43, we can write the diffusion coefficient as

(8.44) $$D = D_0 e^{-H^M/k_BT}$$

where the pre-exponential factor is

(8.45) $$D_0 = d^2 v_0 e^{S^M/k_B}$$

In semiconductors, the mechanisms of diffusion are more complicated than the simple random-walk model. We distinguish between *self-diffusion*, the migration of host atoms through the crystal, and dopant or *impurity diffusion*. Focusing on the latter type, we can again form two groups: interstitial impurities (e.g., lithium in silicon and germanium), which diffuse rapidly through a crystalline solid on their own, and substitutional impurities, which must leave their host lattice site in order to move through the lattice. *Indirect* mechanisms, which involve native defects such as vacancies and self-interstitials, dominate the diffusion of substitutional impurities.

With vacancy-assisted diffusion, one assumes that there is a strong tendency for a substitutional impurity to move into a neighboring vacancy whenever possible (Frank et al., 1984). In the *vacancy mechanism* (Figure 8.10) the impurity atom forms a complex with a vacancy:

(8.46) $$A_S + V \leftrightarrow A_S V$$

where A_S is the substitutional impurity, V is the vacancy, and $A_S V$ is the complex. Once the complex is formed, it is easy for the impurity to step into the vacancy. The vacancy may then leave and diffuse through the crystal. Diffusion of substitutional impurities in germanium occurs primarily by the vacancy mechanism (Chroneos et al., 2008).

FIGURE 8.10 Vacancy mechanism for diffusion. (a) A substitutional impurity and a vacancy ($A_S + V$), where the vacancy is the dashed circle. (b) The impurity and vacancy form a complex ($A_S V$). (c) The impurity moves into the vacancy.

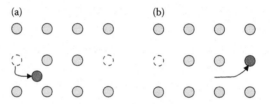

FIGURE 8.11 Dissociative (Frank–Turnbull) mechanism for diffusion. (a) An impurity atom leaves its substitutional site, resulting in a vacancy (dashed circle). (b) After diffusing interstitially, the impurity occupies a vacancy.

Another diffusion process based on vacancies is termed the *Frank–Turnbull* or *dissociative* mechanism (Frank and Turnbull, 1956). In this process, the substitutional impurity leaves its lattice site, becoming an interstitial impurity. After it diffuses interstitially, it encounters a vacancy and occupies it, becoming a substitutional impurity again (Figure 8.11). The defect reaction is given by

$$A_I + V \leftrightarrow A_S \tag{8.47}$$

where A_I is the interstitial impurity. The dissociative mechanism dominates the diffusion of copper, silver, and gold in germanium (Bracht et al., 1991).

Another way that substitutional impurities diffuse is through the *kick-out* mechanism (Gösele et al., 1980). In this process, an impurity leaves its host lattice site because a self-interstitial arrives and "kicks out" the impurity (Figure 8.12). The defect reaction is

$$A_S + I \leftrightarrow A_I \tag{8.48}$$

where I is the self-interstitial (e.g., Si_i). In an interstitial position, the impurity can move much more rapidly than in a substitutional position. After diffusing some distance, the impurity kicks out a silicon host atom and assumes a new substitutional position in the lattice. In silicon, boron acceptors and phosphorus donors diffuse primarily by the kick-out mechanism, while arsenic donors experience contributions from kick-out and vacancy-assisted diffusion (Uematsu, 1997). The "interstitialcy" mechanism is essentially a subset of the kick-out mechanism, in which the diffusing atom moves from one substitutional site to a nearest-neighbor substitutional site.

Native defects play a crucial role in the diffusion of the standard dopant impurities, which normally occupy substitutional sites. Dopants and native defects are inseparable when we are interested in their random walks through the lattice. In the simple random-walk model, the steps are uncorrelated (Equation 8.40). When native defects are involved, however, there is a correlation between successive steps of an impurity. In a vacancy-assisted diffusion step, for example,

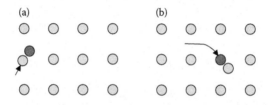

FIGURE 8.12 Kick-out mechanism for diffusion. (a) A self-interstitial atom kicks an impurity atom out of its substitutional site. (b) After diffusing interstitially, the impurity kicks out a host atom to occupy a substitutional site again.

Chapter 8—Thermal Properties **207**

the impurity moves into a neighboring vacancy (Figure 8.10). The next step could be the impurity simply moving back into the vacancy, resulting in no net displacement. Or, the vacancy could diffuse to another site and then the impurity could move into the vacancy. Adding together all of the possibilities yields the *correlation factor f* (Compaan and Haven, 1955). The correlation factor modifies Equation 8.40 such that

$$(8.49) \qquad \langle x^2 \rangle = f \cdot Nd^2$$

For uncorrelated steps, $f = 1$. Usually, $f < 1$ (i.e., the effect of correlations is to reduce the diffusion coefficient).

8.6 SELF-DIFFUSION

Self-diffusion is the diffusion of host atoms in the lattice. To track the motion of a host atom, we consider the atom to be "marked" and treat it similarly to an impurity. Experimentally, a host atom is marked by its mass or by using a radioactive isotope. As in the case of impurity diffusion, self-diffusion is strongly affected by the concentration of native defects.

In an elemental semiconductor, self-interstitials and vacancies contribute to self-diffusion. The self-diffusion coefficient, D^{SD}, is given by

$$(8.50) \qquad D^{SD} = \frac{1}{C_0}(f_I D_I C_I^{eq} + f_V D_V C_V^{eq})$$

where C_0 is the concentration of host atoms; C^{eq} denotes equilibrium concentration; I and V refer to interstitials and vacancies, respectively; D_I and D_V are the corresponding diffusivities; and f_I and f_V are correlation factors. The concentrations and diffusion coefficients are proportional to Boltzmann factors of the form $\exp(-H/k_B T)$. The diffusion coefficients depend on migration enthalpies of the native defects (H^M) and the concentrations are described by formation enthalpies (H^f). From Equations 8.9 and 8.44, the self-diffusion terms are given by

$$(8.51) \qquad \frac{1}{C_0}(DC^{eq}) \sim e^{(S^M + S^f)/k_B} e^{-(H^M + H^f)/k_B T}$$

Therefore, the self-diffusion activation enthalpy is the *sum* of the migration and formation enthalpies.

Zinc, a double acceptor in silicon, diffuses by the kick-out mechanism (Equation 8.48). The diffusion rate is sensitive to the silicon self-interstitial concentration, making it a useful probe of the interstitial contribution to self-diffusion. A detailed study of zinc diffusion (Bracht et al., 1995) revealed

$$(8.52) \qquad \frac{1}{C_0} C_I^{eq} D_I = 2980 \, e^{-(4.95 \text{eV}/k_B T)} \text{ cm}^2\text{s}^{-1}$$

for temperatures ranging from 870 to 1208°C. A later study used silicon isotope heterostructures and secondary ion mass spectrometry (SIMS) (Chapter 11) to measure the diffusion of silicon atoms (Bracht and Haller, 1998). These investigations determined the silicon self-diffusion coefficient to be

$$(8.53) \qquad D^{SD} = 560 \, e^{-(4.76 \text{eV}/k_B T)} \text{ cm}^2\text{s}^{-1}$$

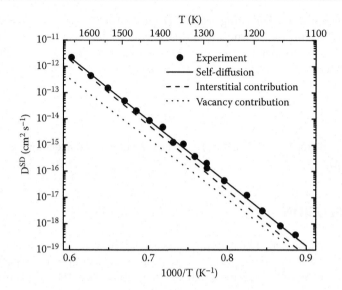

FIGURE 8.13 Arrhenius plot of self-diffusion in silicon, determined from experiments on isotope heterostructures. The solid line is a fit to the data (Equation 8.53). The interstitial and vacancy contributions to the self-diffusion, given by $f_I D_I C_I^{eq}/C_0$ and $f_V D_V C_V^{eq}/C_0$ in Equation 8.50, are shown by the dashed and dotted lines.

over a similar temperature range as Equation 8.52 (Bracht, 2006). From Equations 8.50, 8.52, and 8.53, the vacancy term was estimated as

$$\frac{1}{C_0} C_V^{eq} D_V = 43\, e^{-(4.56\,\text{eV}/k_B T)}\, \text{cm}^2\text{s}^{-1} \tag{8.54}$$

where correlation factors of $f_I = 0.56$ and $f_V = 0.5$ were used (Bracht et al., 2007). The result represented by Equation 8.54 is similar to that obtained from the analysis of metal-diffusion experiments (Tan and Gösele, 1985). As shown in Figure 8.13, the silicon interstitial contribution to self-diffusion dominates for temperatures above ~900°C.

Bracht et al. (2007) also estimated the equilibrium concentrations of interstitials and vacancies. The interstitial concentration was determined from the zinc-diffusion studies, while the vacancy concentration was estimated from a variety of experimental studies, including radiation-enhanced diffusion. The following estimates were obtained:

$$C_I^{eq} = 2.9 \times 10^{24}\, e^{-(3.18\,\text{eV}/k_B T)}\, \text{cm}^{-3}$$
$$C_V^{eq} = 1.4 \times 10^{24}\, e^{-(2.44\,\text{eV}/k_B T)}\, \text{cm}^{-3} \tag{8.55}$$

Expressions for the diffusion coefficients, D_I and D_V, can be obtained from Equations 8.52, 8.54, and 8.55.

To model diffusion in *compound* semiconductors and their alloys, one must consider self-interstitials and vacancies of each host lattice species, as well as antisites. A study of Ga self-diffusion in GaAs was conducted using isotopically pure layers of GaAs on natural GaAs substrates (Wang et al., 1996). SIMS profiles show the concentrations of ^{69}Ga and ^{71}Ga in the substrate and in the two MBE layers grown from isotopically pure Ga effusion sources (Figure 8.14). Moving from

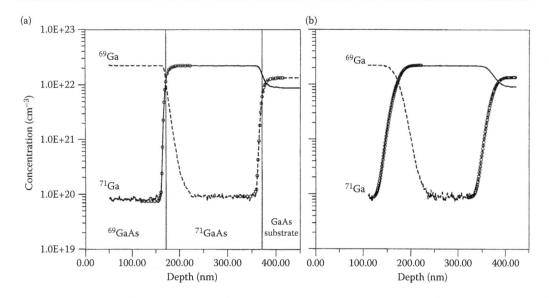

FIGURE 8.14 Concentration profiles for Ga isotopes in a GaAs isotope heterostructure, measured by secondary ion mass spectrometry (SIMS). (a) As grown. (b) After annealing for 3321 s at 974°C. (After Wang, L. et al. 1996. *Phys. Rev. Lett.* 76: 2342–2345.)

the surface on the left side of the figure toward the substrate on the right side, the ^{71}Ga concentration profile rises more sharply than the ^{69}Ga concentration profile drops at the ^{69}Ga–^{71}Ga interface. The inverse situation occurs at the ^{71}Ga–natural Ga interface, where the ^{69}Ga concentration increases more sharply than the ^{71}Ga concentration drops. The reason for this difference stems from the knock-on of the SIMS primary ions (Cs$^+$), which push the Ga atoms deeper into the crystal. Going from a high concentration to a low concentration leads to a noticeable profile change, while in the opposite direction the effect is practically undetectable.

In Figure 8.14, we see the SIMS profile after Ga self-diffusion for 3321 s at 974°C. An arsenic overpressure of ~1 atm was maintained during the heat treatment. Fitting error functions (Equation 8.25) over two orders of magnitude leads to precise values of the diffusion coefficients. Data from numerous samples are shown in Figure 8.15. The straight line is accurately described by

$$D^{SD} = 43\, e^{-(4.24\,\mathrm{eV}/k_B T)}\ \mathrm{cm^2 s^{-1}} \tag{8.56}$$

Using Equation 8.45, we can estimate the migration entropy. Using the lattice constant for GaAs ($d = 5.4$ Å) and a cut-off phonon frequency of $\nu_0 = 8 \times 10^{12}$ Hz, we find $S^{SD} = 8k_B$. This value is comparable to the self-diffusion entropies found for silicon and germanium (Frank et al., 1984). The Ga atoms are believed to diffuse by a vacancy-assisted process.

An inherent drawback to the GaAs study was that because As has only one stable isotope, the self-diffusion of As could not be measured. A similar limitation is present in GaP (Wang et al., 1997). In GaSb, on the other hand, Ga and Sb each have two stable isotopes, allowing one to measure simultaneously the self-diffusion of both host sublattices. The results of these measurements are surprising: near the melting temperature, the self-diffusion of Ga (10^{-15} to 10^{-14} cm^{2}s^{-1}) is over *three orders of magnitude* greater than that of Sb (10^{-18} cm^{2}s^{-1}). While it takes more than an hour for Sb to diffuse one lattice constant (6.1 Å), Ga atoms require only a few seconds to diffuse the same distance. This asymmetry indicates that the Ga and Sb atoms diffuse independently, on their own sublattices.

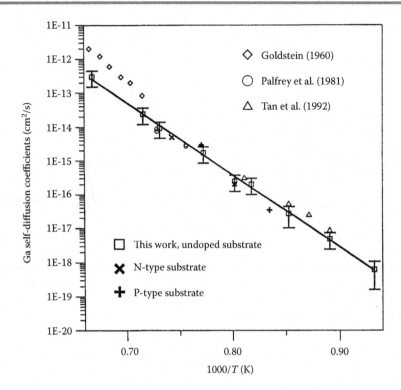

FIGURE 8.15 Arrhenius plot of Ga self-diffusion in GaAs, determined by isotope heterostructures (Tan et al. 1992. Wang et al. 1996 where the latter is labeled "this work") and radioactive ^{72}Ga diffusion experiments (Goldstein 1960; Palfrey et al. 1981). (After Wang, L. et al. 1996. *Phys. Rev. Lett.* 76: 2342–2345.)

The extremely slow Sb self-diffusion suggests that the concentration of Sb vacancies is low. The authors proposed a defect reaction, similar to Equation 8.12, in which a neighboring Ga atom fills the Sb vacancy:

$$V_{Sb} \rightarrow V_{Ga} + Ga_{Sb} \tag{8.57}$$

This reaction transforms the V_{Sb} donor into a vacancy–antisite pair ($V_{Ga} + Ga_{Sb}$), which is an acceptor defect. In GaSb, the Fermi stabilization energy is near the valence-band maximum (Figure 8.3). The tendency toward *p*-type conductivity favors the formation of acceptors over donors, pushing the defect reaction (Equation 8.57) toward the right. The lack of Sb vacancies deprives Sb host atoms of vacancy-assisted diffusion mechanisms, resulting in a small diffusion coefficient.

The foregoing discussion has dealt with equilibrium concentrations of vacancies and interstitials. In many instances, C_I and C_V deviate by orders of magnitude from C_I^{eq} and C_V^{eq}. This in turn will change D^{SD}. We will discuss these effects in the following section.

8.7 DOPANT DIFFUSION

In this section, we focus on impurity diffusion in silicon, for two reasons. First, in silicon, one knows more than in other semiconductors about diffusion mechanisms, although this knowledge

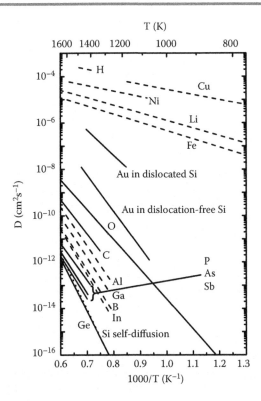

FIGURE 8.16 Diffusivities of impurities in silicon. Silicon self-diffusion is shown by the dotted line (Equation 8.53). (After Gösele, U. and T.Y. Tan. 1985. *Mater. Res. Soc. Symp. Proc.* 36: 105–116.)

is still incomplete. Second, silicon is and will remain the most important semiconductor from an applications point of view.

Figure 8.16 shows a compilation of the experimentally determined diffusion coefficients of various dopants in silicon. The data fall into three major groups: (1) fast diffusers such as lithium, copper, and hydrogen; (2) slow diffusers including silicon, germanium, and shallow substitutional dopants; and (3) a few impurities such as carbon, oxygen, and gold that lie between the two extremes. All the rapid diffusers occupy interstitial sites, where they are not strongly bound to the lattice.

In the case of lithium, Li$^+$ ions reside on interstitial sites and act as shallow donors. The Li$^+$ ions are very mobile down to temperatures as low as ~100°C and readily form complexes with negative acceptor ions. This kind of dopant passivation (Section 2.5) is used to form *p-i-n* junction devices with very wide *i*, or "intrinsic" regions. (In fact, "intrinsic" is a misnomer because the *i* region is not pure.) The passivation can be extremely effective, resulting in net dopant concentrations as low as 10^8 cm^{-3} in the *i* regions. Reverse-biased *p-i-n* junctions are used widely as detectors in nuclear and x-ray spectrometers with high energy resolution.

Like lithium, hydrogen also diffuses interstitially, although it is not a shallow donor in silicon. Hydrogen diffusion was studied by Van Wieringen and Warmholtz (1956) at high temperatures (1092 to 1200°C). At a hydrogen pressure of 1 atm, the solubility is

$$N = 5\times 10^{21} e^{-1.86\text{eV}/k_B T} \text{ atoms/cm}^3 \tag{8.58}$$

and the diffusion coefficient is

$$D = 0.01\, e^{-(0.48\,\text{eV}/k_B T)} \text{ cm}^2\text{s}^{-1} \tag{8.59}$$

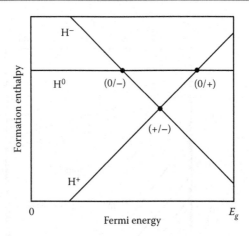

FIGURE 8.17 Formation enthalpy for atomic hydrogen in silicon, for the negative, neutral, and positive charge states. When the Fermi level is below the (+/−) level, hydrogen prefers to be a donor (H⁺). When the Fermi level is above (+/−), hydrogen preferentially acts as an acceptor (H⁻). (After Johnson, N.M. and C.G. Van de Walle. 1999. *Semiconductors and Semimetals*, Vol. 61, pp. 113–137. San Diego, CA: Academic Press.)

At lower temperatures, hydrogen becomes "trapped" as it forms complexes with defects. In the low-temperature regime, the diffusion and solubility are determined by the defect concentrations. For example, in boron-doped silicon exposed to a hydrogen plasma at 150°C, the hydrogen concentration nearly matches the boron concentration due to the formation of boron–hydrogen pairs (Johnson, 1985).

As discussed in Section 2.5, atomic hydrogen can exist in positive, negative, or neutral charge states. The formation enthalpies for these different charge states depend on the Fermi level. As shown in Figure 8.17, H⁰ is never the lowest-enthalpy state. As the Fermi level increases, the energy of H⁺ increases while the energy of H⁻ decreases (Johnson and Van de Walle, 1999). The acceptor (0/−) and donor (0/+) levels occur 0.6 and 0.2 eV below the conduction-band minimum, respectively. The (+/−) level is midway between these two levels, at $E_g - 0.4$ eV. Ignoring vibrational contributions to entropy, the relative concentrations of the charge states are given by

(8.60)
$$n_-/n_0 = (N_{\text{sites},-}/N_{\text{sites},0})e^{(E_F-E_a)/k_BT}$$
$$n_+/n_0 = (N_{\text{sites},+}/N_{\text{sites},0})e^{(E_d-E_F)/k_BT}$$

where E_a and E_d are the acceptor and donor levels, respectively. N_{sites} is the number of equivalent sites where a charge state can reside; from Herring and Johnson (1991), $N_{\text{sites},-}/N_{\text{sites},0} = 1/4$ and $N_{\text{sites},+}/N_{\text{sites},0} = 1/2$. As E_F goes from below the (+/−) level to above it, the dominant charge state switches abruptly from positive to negative.

The impurities iron and gold have been known for a long time because of their strong effects on minority carrier lifetimes even at very low impurity concentrations. With the exception of gold doping used in very fast switching diodes, these two impurities are diligently kept away from silicon devices. Elaborate gettering mechanisms have been designed to remove iron and gold from the active regions in silicon wafers and chips. SiO₂ precipitates, for example, are often desirable because they getter metal impurities (Section 2.4).

The substitutional shallow dopants diffuse assisted by vacancies and interstitials. The relative contributions of these two native defects depend on temperature, Fermi-level position, and external disturbances such as oxidation, nitridation, and implantation. From a large number of studies

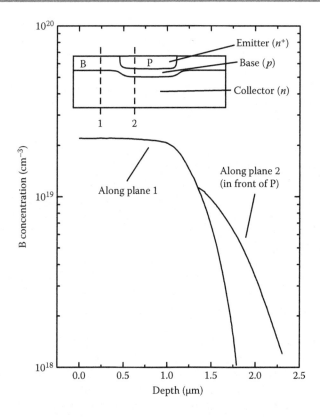

FIGURE 8.18 The emitter push effect in silicon. In an *n-p-n* transistor, the emitter is a phosphorus-doped layer. Boron is diffused in to create the base layer. Boron that is in front of diffusing phosphorus atoms experiences enhanced diffusion due to excess interstitials.

(Uematsu, 1997), there is strong evidence that boron and phosphorus diffuse primarily via the kick-out mechanism (Equation 8.48). The diffusion is enhanced by the presence of self-interstitials. An injection of vacancies, on the other hand, decreases the concentration of self-interstitials and slows down the diffusion. Arsenic diffusion is enhanced by the injection of interstitials or vacancies, indicating that it diffuses by both kick-out and vacancy-assisted mechanisms (Fahey et al., 1989). Antimony diffuses primarily by vacancy-assisted mechanisms.

Various effects can disturb the vacancy and interstitial populations. One of the earliest and most detrimental deviations from normal diffusion was the *emitter push effect*. Figure 8.18 shows what happens to a boron diffusion profile in front of a second, high-concentration phosphorus diffusion front. The high phosphorus concentration increases the silicon self-interstitial concentration, producing an *interstitial wind*. The kick-out diffusion of boron increases in front of the diffusing phosphorus but not outside the range of the interstitial wind. This enhanced diffusion of boron led to much broader base layers in *n-p-n* bipolar silicon transistors than originally designed. Such transistors exhibited much lower cut-off operating frequencies. Advanced modeling and processing techniques eventually overcame this problem.

A second source for interstitial wind originates from silicon *oxidation*. Exposure to O_2 gas at elevated temperatures results in the growth of a SiO_2 layer on the silicon surface. When the silicon is oxidized, the volume expansion of 125% is mostly accommodated by viscoelastic flow. Vacancies from the bulk and injection of interstitials from the SiO_2-Si interface both help in reducing the surface strains. Hence, oxidation leads to an increase in self-interstitials and a decrease in vacancies. Dopants that diffuse mostly via the kick-out mechanism (essentially all except antimony) exhibit increased diffusivities, while antimony diffusion is retarded.

The opposite effect occurs upon *nitridation*, which can be achieved by exposing a silicon sample to a nitrogen-containing gas such as ammonia (Mizuo et al., 1983). Nitridation causes the injection of vacancies, at the expense of interstitials. Under nitridation, antimony diffuses at an accelerated rate while all the other group-III and group-V dopants diffuse at a slower pace.

A third source of excess interstitials is related to rodlike defects caused by ion implantation. These defects are oriented along <110> directions and lie on {131} planes. The defects are believed to emit interstitials during the early phase of radiation damage annealing, leading to *transient enhanced diffusion*. SIMS measurements showed the enhanced diffusivity of the first three or four near-surface boron delta layers grown by MBE (Eaglesham et al., 1994). The enhanced diffusion was caused by implantation of silicon ions, which produced primary damage at a depth of less than 1000 Å.

The same group of authors showed that carbon in silicon is a very effective silicon-interstitial sink (Stolk et al., 1996). The proposed mechanism involves two steps. First, a silicon interstitial kicks out a substitutional carbon, according to Equation 8.48. Second, the interstitial carbon is trapped by a substitutional carbon, forming a relatively immobile carbon–carbon pair. The reduction in silicon interstitials, caused by the high carbon concentration (2×10^{19} cm^{-3}), results in a dramatic suppression of boron diffusion. Much lower carbon concentrations are expected to play an important role in dopant diffusion as well.

As shown in Figure 8.16, carbon and oxygen have diffusivities that fall between those of the rapid interstitial diffusers and the typical substitutional shallow dopants. Oxygen is only bound to two silicon neighbors, resulting in a much-reduced energy for oxygen interstitial formation. When one Si-O bond breaks, it can be reestablished with a different, neighboring silicon host atom. This kind of direct diffusion does not require native defects. Carbon, on the other hand, occupies substitutional sites, producing large strains because of its small size and strong, short bonds. The microscopic diffusion mechanism for carbon is not well understood but it is assumed that a minor fraction of carbon diffuses interstitially, via the kick-out or interstitialcy mechanism.

The possible interstitial diffusion of carbon in silicon is reminiscent of copper in germanium (Haller et al., 1981). This impurity occupies partially interstitial and partially substitutional sites. On the latter site, copper forms a triple acceptor with three well-defined hole binding energies (Section 2.2). The triple acceptor properties clearly indicate a substitutional site. However, the rapid diffusion behavior caused confusion in the early semiconductor days because it was not known that one kind of impurity could assume more than one kind of lattice site. Through high-temperature, electric field drift studies of radioactive ^{64}Cu, it became clear that copper travels as a positive ion (i.e., it must be a donor in its interstitial form). The energy level of this donor has not yet been determined. The diffusion of copper in germanium proceeds via the dissociative mechanism (Bracht et al., 1991) and is enhanced by the presence of dislocations.

Figure 8.19 shows experimentally established diffusion data for copper and other impurities in germanium. Diffusion in III-V compound semiconductors is summarized by Tuck (1988).

8.8 QUANTUM-WELL INTERMIXING

The diffusion of host and impurity atoms leads to interesting phenomena in quantum-well heterostructures. As shown in Figure 8.9, elevated temperatures cause sharp quantum-well profiles

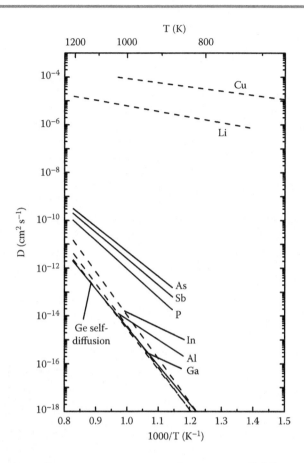

FIGURE 8.19 Diffusivities of impurities in germanium. Germanium self-diffusion is shown by the solid line. References: Cu (Seeger and Chik, 1968); Li (Fuller and Ditzenberger, 1953); As, Sb, and Bi (Dunlap, 1954); In and Al (Dorner et al., 1982a,b); Ga (Södervall et al., 1986); Ge self-diffusion (Vogel et al., 1983).

to broaden over time. In superlattices, broadening of the profiles due to interdiffusion is called *disordering*. Chang and Koma (1976) first observed this phenomenon in undoped MBE-grown $Al_xGa_{1-x}As$ heterostructures annealed under an arsenic overpressure. Using Equation 8.36, they found interdiffusion coefficients for various temperatures and compositions x. For an AlAs quantum well surrounded by GaAs, the interdiffusion coefficient at 992°C was $D = 2 \times 10^{-16}$ cm²/s. For an annealing time of 2 hr, the host Ga and Al atoms diffuse an average distance of ~10 nm.

It was later discovered that dopants could dramatically enhance the rate of interdiffusion. Zinc diffusion into AlAs/GaAs superlattices and $Al_xGa_{1-x}As$/GaAs heterostructures resulted in significant disordering at temperatures below 600°C (Laidig et al., 1981). The reason for this enhancement is that zinc diffusion is primarily governed by the kick-out mechanism, in which a gallium (or aluminum) interstitial kicks out a substitutional zinc acceptor (Equation 8.48). The interstitial zinc then assumes a new substitutional site by kicking out a host gallium atom. This process not only results in zinc diffusion; the group III interstitials involved in the kick-out process diffuse as well (Bracht et al., 2001). Zinc diffusion therefore increases the rate of gallium-aluminum interdiffusion.

A second effect is related to changes in the Fermi level. This effect was probed by comparing disordering in superlattices doped with silicon donors or beryllium acceptors (Kawabe et al., 1985). Because both of these dopants are slow diffusers at the annealing temperatures (750°C), they only

8.8 Quantum-Well Intermixing

influence interdiffusion through their effect on the Fermi level. It was found that silicon donors enhanced interdiffusion, whereas beryllium acceptors did not. The explanation for this doping effect was provided by the amphoteric defect model (Section 8.2). As shown in Figure 8.2, the formation enthalpy for gallium vacancies is low for *n*-type GaAs. This leads to a large concentration of gallium vacancies and enhanced interdiffusion. For *p*-type GaAs, on the other hand, the low concentration of negatively charged gallium vacancies results in suppressed interdiffusion.

Quantum-well intermixing leads to changes in the optical properties. When aluminum atoms diffuse into a GaAs quantum well, and gallium atoms diffuse out, the band gap of the quantum well increases. Controlled interdiffusion can therefore be used to tune the emission wavelength of optoelectronic devices (Hofstetter et al., 1998). By changing the band gap, interdiffusion also affects the refractive index, allowing one to produce well-defined waveguides for laser diodes.

Interdiffusion of indium and gallium in InGaN quantum wells was studied by McCluskey et al. (1998), for annealing temperatures of 1200 to 1400°C for 15 min. Large nitrogen pressures (1.5 GPa) had to be applied to prevent the loss of nitrogen from the samples. The heterostructure consisted of a 1 μm GaN:Mg layer, 20 1.6-nm thick $In_{0.18}Ga_{0.82}N$ quantum wells, and a 1 μm GaN:Si layer on a sapphire substrate. Annealing at 1200°C produced no detectable change in the heterostructure.

After annealing at 1300°C, there was noticeable quantum-well intermixing, as measured by x-ray diffraction (XRD). The main peaks in the XRD spectrum (Figure 8.20) are due to the GaN

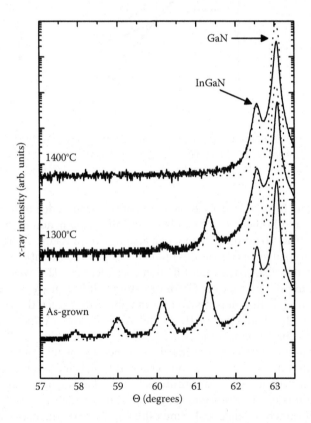

FIGURE 8.20 X-ray diffraction spectra of InGaN quantum wells annealed for 15 min at 1300 and 1400°C. Peaks corresponding to InGaN and GaN are indicated. The satellite peaks arise from the periodicity of the quantum-well superlattice. The loss of these peaks is a result of superlattice disordering. Simulated spectra are shown by the dotted lines. (After McCluskey, M.D. 1998. *Appl. Phys. Lett.* 73: 1281–1283.)

and InGaN lattice planes. The smaller satellite peaks arise from the periodicity of the multiple quantum wells (MQW). The satellite peaks diminish as a result of superlattice disordering. To simulate the XRD spectra, the quantum wells were modeled as Gaussians (Equation 8.33). After annealing at 1400°C, the quantum-well region was replaced by a uniformly disordered InGaN layer. Transmission electron microscope images of the heterostructure, before and after

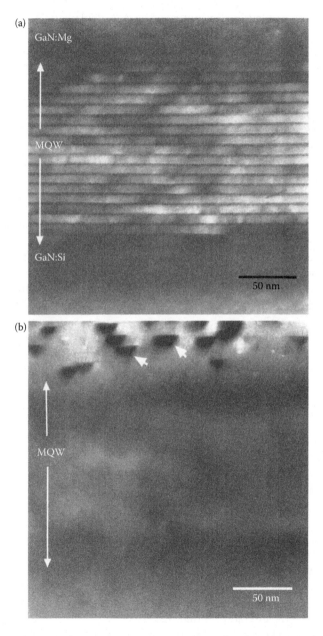

FIGURE 8.21 Transmission electron microscope image of InGaN/GaN multiple quantum well (MQW) structure, (a) before and (b) after annealing at 1400°C for 15 min. Due to In-Ga interdiffusion, the MQW structure is replaced by a uniform InGaN layer. (Pyramidal voids in the GaN:Mg layer are indicated by arrows.) (Reprinted with permission from McCluskey, M.D. et al. 1998. *Appl. Phys. Lett.* 73: 1281–1283. Copyright 1998, American Institute of Physics.)

8.8 Quantum-Well Intermixing

annealing, are shown in Figure 8.21. SIMS measurements showed that Mg acceptors diffused from the GaN:Mg layer into the quantum-well region. It is likely that the Mg diffusion enhanced intermixing.

PROBLEMS

8.1 Germanium has a lattice constant $a = 5.66$ Å. Calculate the density (cm^{-3}) of vacancies at the melting point (936°C), given a vacancy formation enthalpy of 2 eV.

8.2 A semiconductor has a band gap of 1.4 eV. When the Fermi level is at the valence-band maximum, the +1 charge state of a particular donor has an enthalpy 1.0 eV lower than the neutral charge state.
   a. Sketch the formation enthalpy versus Fermi energy for both charge states.
   b. What is the donor binding energy?

8.3 Silicon has a band gap of 1.1 eV. The substitutional Au impurity can exist in three charge states: +, 0, and −. When the Fermi level is at the valence-band maximum, the enthalpy of the + state is 0.35 eV lower than the 0 state, while the enthalpy of the − state is 0.63 eV higher than the 0 state.
   a. Sketch the formation enthalpy versus Fermi energy for all charge states.
   b. Where are the (0/+) and (0/−) levels relative to the valence-band maximum?

8.4 Germanium has a band gap of 0.7 eV. Substitutional Cu is a triple acceptor. The (0/−) and (−/2−) levels are 0.043 and 0.33 eV above the valence-band maximum, while the (2−/3−) level is 0.26 eV below the conduction-band minimum. Sketch the formation enthalpy versus Fermi energy for all charge states.

8.5 Consider a neutral oxygen vacancy in ZnO. Under Zn-rich conditions ($\mu_{Zn} = \mu_{Zn,bulk}$), the formation enthalpy is 3.7 eV. What is the formation enthalpy under O-rich conditions ($\mu_O = \mu_{O,gas}$)?

8.6 Suppose the concentration profile for a dopant is sinusoidal:

$$C = C_0(1 + \sin kx)$$

At an elevated temperature, the dopant atoms have a diffusivity D. Show that

$$C = C_0(1 + \sin kx \cdot e^{-\gamma t})$$

satisfies Fick's second law, and find γ.

8.7 Illustrate the diffusion of Zn in GaP, assuming the Frank–Turnbull mechanism.

8.8 The diffusion coefficient for hydrogen in silicon is given by Equation 8.59. A silicon wafer is 1 mm thick. At 1100°C, approximately how long does it take for hydrogen to diffuse from one side to the other?

8.9 Phosphorus in germanium diffuses by the vacancy mechanism. At 1210 K, the diffusion coefficient is $D = 1 \times 10^{-10}$ cm^2/s. At 860 K, $D = 1 \times 10^{-14}$ cm^2/s.
   a. Illustrate the diffusion mechanism.
   b. Calculate the migration enthalpy H^M.

8.10 Suppose a silicon wafer at 600 K contains 10^{10} cm^{-3} atomic hydrogen impurities. The Fermi level is 0.5 eV below the conduction-band minimum. Find the density (cm^{-3}) of H^0, H^-, and H^+.

REFERENCES

Alberi, K. and M.A. Scarpulla. 2016. Suppression of compensating native defect formation during semiconductor processing via excess carriers. *Sci. Rep.* 6: 27954 (10 pages).

Bracht, H. 2006. Diffusion mediated by doping and radiation-induced point defects. *Physica B* 376–377: 11–18.

Bracht, H. and E.E. Haller. 1998. Silicon self-diffusion in isotope heterostructures. *Phys. Rev. Lett.* 81: 393–396.

Bracht, H., N.A. Stolwijk, and H. Mehrer. 1991. Diffusion of copper, silver, and gold in germanium. *Phys. Rev. B* 43: 14465–14477.

Bracht, H., N.A. Stolwijk, and H. Mehrer. 1995. Properties of intrinsic point defects in silicon determined by zinc diffusion experiments under nonequilibrium conditions. *Phys. Rev. B* 52: 16542–16560.

Bracht, H., M.S. Norseng, E.E. Haller, and K. Eberl. 2001. Zinc diffusion enhanced Ga diffusion in GaAs isotope heterostructures. *Physica B* 308–310: 831–834.

Bracht, H., H.H. Silvestri, I.D. Sharp, and E.E. Haller. 2007. Self- and foreign-atom diffusion in semiconductor isotope heterostructures. II. Experimental results for silicon. *Phys. Rev. B* 75: 035211 (21 pages).

Chang, L. and A. Koma. 1976. Interdiffusion between GaAs and A1As. *Appl. Phys. Lett.* 29: 138–141.

Chroneos, A., H. Bracht, R.W. Grimes, and B.P. Uberuaga. 2008. Vacancy-mediated dopant diffusion activation enthalpies for germanium. *Appl. Phys. Lett.* 92: 172103 (3 pages).

Compaan, K. and Y. Haven. 1955. Correlation factors for diffusion in solids. *Trans. Faraday Soc.* 52: 786–801.

Crank, J. 1975. *The Mathematics of Diffusion*. Oxford, UK: Clarendon Press.

Dorner, P., W. Gust, A. Lodding, H. Odelius, B. Predel, and U. Roll. 1982a. SIMS-Untersuchungen zure Volumendiffusion von Al in Ge. *Acta Metall.* 30: 941–946.

Dorner, P., W. Gust, A. Lodding et al. 1982b. SIMS investigations on the diffusion of In in Ge single-crystals. *Zeitschrift fur Metallkunde* 73: 325–330.

Dunlap, W.C., Jr. 1954. Diffusion of impurities in germanium. *Phys. Rev.* 94: 1531–1540.

Eaglesham, D.J., P.A. Stolk, H.-J. Gossmann, and J.M. Poate. 1994. Implantation and transient B diffusion in Si: The source of the interstitials. *Appl. Phys. Lett.* 65: 2305–2307.

Fahey, P.M., P.B. Griffin, and J.D. Plummer. 1989. Point-defects and dopant diffusion in silicon. *Rev. Mod. Phys.* 61: 289–384.

Frank, F.C. and D. Turnbull. 1956. Mechanism of diffusion of copper in germanium. *Phys. Rev.* 104: 617–618.

Frank, W., U. Gösele, H. Mehrer, and A. Seeger. 1984. Diffusion in silicon and germanium. In *Diffusion in Crystalline Solids*, eds. G.E. Murch and A.S. Nowick, pp. 63–142. Orlando, FL: Academic Press.

Fuller, C.S. and J.A. Ditzenberger. 1953. Diffusion of lithium into germanium and silicon. *Phys. Rev.* 91: 193.

Ghez, R. 2001. *Diffusion Phenomena: Cases and Studies*. New York, NY: Kluwer Academic.

Goldstein, B. 1960. Diffusion in compound semiconductors. *Phys. Rev.* 121: 1305–1311.

Gösele, U. 2000. Surprising movements in solids. *Nature* 408: 38–39.

Gösele, U. and T.Y. Tan. 1985. The influence of point defects on diffusion and gettering in silicon. *Mater. Res. Soc. Symp. Proc.* 36: 105–116.

Gösele, U., W. Frank, and A. Seeger. 1980. Mechanisms and kinetics of the diffusion of gold in silicon. *Appl. Phys.* 23: 361–368.

Haller, E.E., W. Hansen, and F.S. Goulding. 1981. Physics of ultra-pure germanium. *Adv. Phys.* 30: 93–138.

Herring, C. and N.M. Johnson. 1991. Hydrogen migration and solubility in silicon. In *Semiconductors and Semimetals*, Vol. 34, eds. J.I. Pankove and N.M. Johnson, pp. 225–350. San Diego, CA: Academic Press.

Hofstetter, D., B. Maisenhölder, and H.P. Zappe. 1998. Quantum-well intermixing for fabrication of lasers and photonic integrated circuits. *IEEE J. Sel. Topics Quantum Electron.* 4: 794–802.

Janotti, A. and C.G. Van de Walle. 2007. Native point defects in ZnO. *Phys. Rev. B* 76: 165202 (22 pages).

Johnson, N.M. 1985. Mechanism for hydrogen compensation of shallow-acceptor impurities in single-crystal silicon. *Phys. Rev. B* 31: 5525–5528.

Johnson, N.M. and C.G. Van de Walle. 1999. Isolated monatomic hydrogen in silicon. In *Semiconductors and Semimetals*, Vol. 61, ed. N.H. Nickel, pp. 113–137. San Diego, CA: Academic Press.

Kawabe, M., N. Shimizu, F. Hasegawa, and Y. Nannichi. 1985. Effects of Be and Si on disordering of the AlAs/GaAs superlattice. *Appl. Phys. Lett.* 46: 849–850.

Kittel, C. 2005. *Introduction to Solid State Physics*, 8th ed. New York, NY: John Wiley & Sons.

220 References

Kizilyalli, I.C., T.L. Rich, F.A. Stevie, and C.S. Rafferty. 1996. Diffusion parameters of indium for silicon process modeling. *J. Appl. Phys.* 80: 4944–4947.

Laidig, W.D., N. Holonyak, Jr., M.D. Camras et al. 1981. Disorder of an AlAs-GaAs superlattice by impurity diffusion. *Appl. Phys. Lett.* 38: 776–778.

Li, S.X., K.M. Yu, J. Wu et al. 2005. Fermi-level stabilization energy in group-III nitrides. *Phys. Rev. B* 71: 161201 (4 pages).

McCluskey, M.D., L.T. Romano, B.S. Krusor, N.M. Johnson, T. Suski, and J. Jun. 1998. Interdiffusion of In and Ga in InGaN quantum wells. *Appl. Phys. Lett.* 73: 1281–1283.

Mizuo, S., T. Kusaka, A. Shintani, M. Nanba, and H. Higuchi. 1983. Effect of Si and SiO_2 thermal nitridation on impurity diffusion and oxidation induced stacking fault size in Si. *J. Appl. Phys.* 54: 3860–3866.

Palfrey, H.D., M. Brown, and A.F.W. Willoughby. 1981. Self-diffusion of gallium in gallium arsenide. *J. Electrochem. Soc.* 128: 2224–2228.

Pauling, L. 1988. *General Chemistry.* New York, NY: Dover.

Reif, F. 1965. *Fundamentals of Statistical and Thermal Physics.* New York, NY: McGraw-Hill.

Runyan, W.R. 1965. *Silicon Semiconductor Technology.* New York, UK: McGraw-Hill.

Seeger, A. and K.P. Chik. 1968. Diffusion mechanisms and point defects in silicon and germanium. *Phys. Stat. Solidi* 29: 455–542.

Shaw, D. 1973. *Atomic Diffusion in Semiconductors.* London, UK: Plenum Press.

Smith, R.C.T. 1953. Conduction of heat in the semi-infinite solid with a short table of an important integral. *Australian J. Phys.* 6: 127–130.

Södervall, U., H. Odelius, A. Lodding et al. 1986. Gallium tracer diffusion and its isotope effect in germanium. *Philisophical Mag. A* 54: 539–551.

Steen, C., A. Martinez-Limia, P. Pichler et al. 2008. Distribution and segregation of arsenic at the SiO_2/Si interface. *J. Appl. Phys.* 104: 023518 (11 pages).

Stolk, P.A., H.-J. Gossmann, D.J. Eaglesham, and J.M. Poate. 1996. The effect of carbon on diffusion in silicon. *Mat. Sci. Eng. B* 36: 275–281.

Tan, T.Y. and U. Gösele. 1985. Point defects, diffusion processes, and swirl defect formation in silicon. *Appl. Phys. A* 37: 1–17.

Tan, T.Y., H.M. You, S. Yu et al. 1992. Disordering in ^{69}GaAs/^{71}GaAs isotope superlattice structures. *J. Appl. Phys.* 72: 5206–5212.

Tuck, B. 1988. *Atomic Diffusion in III-V Semiconductors.* Bristol, UK: Adam Hilger.

Uematsu, M. 1997. Simulation of boron, phosphorus, and arsenic diffusion in silicon based on an integrated diffusion model, and the anomalous phosphorus diffusion mechanism. *J. Appl. Phys.* 82: 2228–2246.

Van Wieringen, A. and N. Warmholtz. 1956. On the permeation of hydrogen and helium in single crystal silicon and germanium at elevated temperatures. *Physica* 22: 849–865.

Vogel, G., G. Hettich, and H. Mehrer. 1983. Self-diffusion in intrinsic germanium and effects of doping on self-diffusion in germanium. *J. Phys. C* 16: 6197–6204.

Walukiewicz, W. 1989. Amphoteric native defects in semiconductors. *Appl. Phys. Lett.* 54: 2094–2096.

Walukiewicz, W. 2001. Intrinsic limitations to the doping of wide-gap semiconductors. *Physica B* 302–303: 123–134.

Wang, L., L. Hsu, E.E. Haller et al. 1996. Ga self-diffusion in GaAs isotope heterostructures. *Phys. Rev. Lett.* 76: 2342–2345.

Wang, L., J.A. Wolk, L. Hsu et al. 1997. Gallium self-diffusion in gallium phosphide. *Appl. Phys. Lett.* 70: 1831–1833.

Zhang, S.B. and J.E. Northrup. 1991. Chemical potential dependence on defect formation energies in GaAs: Application to Ga self-diffusion. *Phys. Rev. Lett.* 67: 2339–2342.

Electrical Measurements

Applications require control of semiconductor properties such as dopant and deep-level concentrations, compensation, point and line defects, and lifetime of minority carriers. These properties can be determined only if sufficiently sensitive and reliable analytical techniques are available. A good characterization method should allow the direct and independent measurement of a given property. In addition, one often wants to know exactly what is causing a semiconductor crystal to exhibit a certain property (e.g., which impurity—phosphorus, arsenic, or antimony—is making a silicon crystal n-type).

The range of characterization techniques and their level of sophistication are truly impressive. This sophistication has been possible in large part because of the development of more complex instrumentation—instruments that use, among other components, improved semiconductor materials. Characterization is therefore an essential part of the feedback loop that continually improves technology. In the following chapters, key experimental techniques are discussed. While the discussion is not all-inclusive, we have highlighted the most widely used methods for characterizing defects in semiconductors. The techniques are grouped into electrical measurements (Chapter 9), optical spectroscopy (Chapter 10), particle-beam methods (Chapter 11), and microscopy (Chapter 12). Additional information about semiconductor characterization may be found in books such as Kane and Larrabee (1970), Stavola (1998), and Schroder (2006).

In this chapter, we review the basics of electrical transport, followed by specific techniques for determining resistivity and free-carrier concentration. Capacitance-voltage and transient capacitance measurements are discussed. Deep-level transient spectroscopy is an especially popular technique that uses capacitive transients to characterize carrier traps. Minority carrier lifetime measurements, important for a range of device applications, are summarized. We conclude with a brief discussion of techniques that use the thermoelectric effect to characterize deep levels.

9.1 RESISTIVITY AND CONDUCTIVITY

Section 1.10 discussed the basics of carrier (electron or hole) transport and mobility. It is straightforward to relate mobility to electrical conductivity. The current that flows through a cross-sectional area A is

(9.1) $$I = \frac{Q}{dt}$$

where Q is the amount of charge (Coulombs) that flows through the area in a time dt (Figure 9.1). The charge Q is given by the charge density times the volume of the cylinder that passed through the area:

(9.2) $$Q = (ne)(Av_d dt)$$

where n is the free-carrier concentration, e is the charge ($\pm 1.6 \times 10^{-19}$ C), and v_d is the drift velocity. Plugging Equation 9.2 into Equation 9.1 yields

(9.3) $$I = nAv_d e$$

The *current density* is given by

(9.4) $$j = I/A = nv_d e$$

Because $\mathbf{v}_d = \mu \mathbf{E}$ (Equation 1.34), the current density can be expressed as

(9.5) $$\mathbf{j} = (ne\mu)\mathbf{E} \equiv \sigma \mathbf{E}$$

where σ is the electrical *conductivity*:

(9.6) $$\sigma = ne\mu$$

The *resistivity* ρ is defined as the inverse of the conductivity:

(9.7) $$\rho \equiv \frac{1}{\sigma} = \frac{1}{ne\mu}$$

From Ohm's law, $V = IR$, the *resistance* R of a sample is given by

(9.8) $$R = \rho l / A$$

where l is the length and A is the area through which the current flows.

By convention, resistance is measured in Ohms (Ω), resistivity has units of $\Omega \cdot$ cm, and conductivity has units of ($\Omega \cdot$ cm)$^{-1}$. Free-carrier concentration is expressed as cm^{-3}, and mobility has units of cm^2/Vs. Although electron mobilities are actually negative (Equation 1.34), the minus sign is often neglected when quoting values. For example, commercially available silicon and GaAs wafers have room-temperature electron mobilities of 1400 and 9000 cm^2/Vs, respectively. The electron mobility of ultrapure germanium at low temperatures is as high as 10^6 cm^2/Vs.

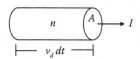

FIGURE 9.1 Cylinder defined by a free-carrier density n traveling through an area A.

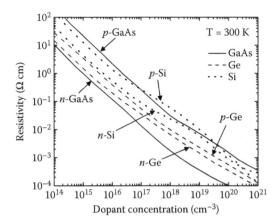

FIGURE 9.2 Resistivity versus dopant concentration for GaAs, germanium, and silicon samples at 300 K. (After Sze, S.M. and J.C. Irvin. 1968. *Solid State Electron.* 11: 599–602.)

Conductivity or resistivity measurements are used routinely to determine the net doping concentration $|N_d - N_a|$. The conductivity is given by

$$\sigma = ne\mu_n + pe\mu_p \tag{9.9}$$

where n and p are the free-electron/hole concentrations and μ_n and μ_p are electron/hole mobilities. For extrinsic semiconductors ($|N_d - N_a| > n_i$), one can neglect the minority carrier term to a good approximation. At a sufficiently high temperature, all the impurities are ionized. This is the saturation regime (Section 5.6), in which

$$n = \frac{\sigma}{e\mu_n} = N_d - N_a \quad (n\text{-type}) \tag{9.10}$$

or

$$p = \frac{\sigma}{e\mu_p} = N_a - N_d \quad (p\text{-type}) \tag{9.11}$$

An accurate knowledge of the mobilities $\mu_{n,p}$ is necessary to make the above equations useful for determining net dopant concentration. Mobility values can be found in tables and diagrams for crystals with a low degree of compensation (i.e., $N_d/N_a \ll 1$ for p-type and $N_a/N_d \ll 1$ for n-type crystals). Figures 9.2 and 9.3 display the resistivity and the mobility, respectively, for electrons and holes in germanium, silicon, and GaAs (Sze and Irvin, 1968).

9.2 METHODS OF MEASURING RESISTIVITY

The *four-point probe* in its collinear arrangement (Figure 9.4) is commonly used to determine the resistivity of semiconductor samples. The outer two probes are connected to a current source. The two inner probes are high-impedance voltage sensors. The sample thickness δ is assumed to be constant.

In order to determine the voltage V_{23} developed between the inner two electrodes, we first have to decide if the distance between the probes, s, is much smaller or much larger than the

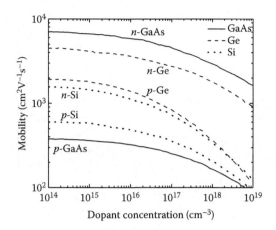

FIGURE 9.3 Mobility versus dopant concentration for GaAs, germanium, and silicon samples at 300 K. (After Sze, S.M. and J.C. Irvin. 1968. *Solid State Electron.* 11: 599–602.)

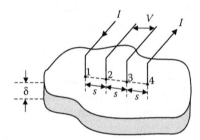

FIGURE 9.4 Four-point probe measurement on a sample of thickness δ. (After Wieder, H.H. 1979. *Nondestructive Evaluation of Semiconductor Materials and Devices*, pp. 68–105, ed. J.N. Zemel. New York, NY: Plenum Press.)

thickness of the semiconductor sample to be measured (Wieder, 1979). Let us first analyze the case for a *thick sample* such that $\delta \gg s$. The equipotential surfaces for a current entering at a point on the surface of a semi-infinite homogeneous material are hemispheres (Figure 9.5). The voltage drop dV across a hemispherical shell of radius r and thickness dr is

(9.12)
$$dV = IdR = I\rho \frac{dr}{2\pi r^2}$$

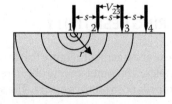

FIGURE 9.5 Four-point probe measurement on a thick (semi-infinite) sample. The solid lines are equipotential lines, for current flowing from point 1.

The voltage drop on the surface between points 2 and 3 is given by

$$(9.13) \qquad \int_s^{2s} dV = \left[-I\rho \frac{1}{2\pi r} \right]_s^{2s} = I\rho \frac{1}{4\pi s}$$

This is the voltage drop between 2 and 3 produced by current I flowing from point 1 into a semi-infinite space. An opposite current, of the same magnitude, flows from the semi-infinite space into point 4. This current generates the same voltage drop across 2 and 3 with the same polarity. V_{23}, the total voltage drop between 2 and 3, is twice the voltage drop given in Equation 9.13:

$$(9.14) \qquad V_{23} = I\rho \frac{1}{2\pi s}$$

From Equation 9.14, we can solve for the resistivity:

$$(9.15) \qquad \rho = 2\pi s \frac{V_{23}}{I}$$

The second case we consider is for a *thin sample, $s \gg \delta$*. The current I entering at point 1 produces circular equipotential lines. The voltage drop is given by

$$(9.16) \qquad dV = I\rho \frac{dr}{2\pi r \delta}$$

The voltage drop between points 2 and 3 is given by

$$(9.17) \qquad \int_s^{2s} I\rho \frac{dr}{2\pi r \delta} = \frac{I\rho}{2\pi\delta}\left[\ln(2s) - \ln(s)\right] = \frac{I\rho}{2\pi\delta}\ln(2)$$

As before, we superpose a current of the same magnitude that flows into point 4. This doubles the voltage drop:

$$(9.18) \qquad V_{23} = \frac{I\rho}{\pi\delta}\ln(2)$$

From Equation 9.18, we can solve for the resistivity:

$$(9.19) \qquad \rho = \frac{V_{23}}{I}\frac{\pi\delta}{\ln(2)}$$

where $\pi/\ln(2) = 4.5324$. Often one does not know the precise value of δ but is still interested in the *sheet resistance*:

$$(9.20) \qquad \frac{\rho}{\delta} = \frac{V_{23}}{I}\frac{\pi}{\ln(2)}$$

which has units of Ohms. However, to emphasize that it is sheet rather than bulk resistance, the units are usually called "Ohms per square" ($\Omega/\square$).

9.2 Methods of Measuring Resistivity

TABLE 9.1 Collinear Four-Point Probe Correction Factor F

δ/s	$F(\delta/s)$
0.4	0.9995
0.5	0.9974
0.5555	0.9948
0.6250	0.9898
0.7143	0.9798
0.8333	0.9600
1.0	0.9214
1.1111	0.8907
1.25	0.8490
1.4286	0.7938
1.666	0.7225
2.0	0.6336

Source: After Smits, F.M. 1958. *Bell Syst. Tech. J.* 37: 711–718.

We have discussed the two extremes, $\delta \gg s$ and $\delta \ll s$, but what about $\delta \approx s$? This situation is encountered in a research laboratory where semiconductor samples of a wide variety of sizes and shapes are studied. Following Equation 9.19, the resistivity is written as

$$(9.21) \qquad \rho = \frac{V_{23}}{I} \frac{\pi \delta}{\ln(2)} F(\delta/s)$$

where F is a correction factor (Smits, 1958). For $\delta/s = 0.4$, we are in the thin film regime. As shown in Table 9.1, the value of ρ deviates by 0.05% from the one obtained with the thin film equation (Equation 9.19). For $\delta/s = 2.0$, $F = 0.6336$, and we are close to the thick sample regime (Equation 9.15).

The *spreading resistance probe* enables measurements of spatially resolved resistivity profiles (Figure 9.6). Crowding of the current near a small-diameter contact results in $\sim$80% of the total potential drop occurring within a distance approximately five times the probe contact radius. The second contact may be a large-area Ohmic contact placed somewhere on the crystal sample (e.g., the back of a wafer). In that case, for a nonindenting contact and in the absence of any contact barrier, the resistance is given by

$$(9.22) \qquad R = \frac{\rho}{4a}$$

where a is the radius of the cylindrical contact.

More commonly, one uses a second identical fine probe in close proximity to the first one. If one wishes to measure, for example, the depth dependence of the resistivity in a diffusion or ion-implantation doped wafer, the wafer is beveled under a shallow angle (1 to 5°) to geometrically magnify the layer to be studied (Figure 9.7). The two probes are used to measure the voltage drop created by a complicated current flow pattern in the sample. With a detailed knowledge of the mobility as a function of dopant concentration, calibration measurements of known resistivity profiles, and significant modeling efforts, it has become possible to determine profiles as narrow as a few tens of nanometers.

The accuracy and repeatability of all characterization methods involving fine metal point contacts depend to a large degree on the quality of the area of the probe that is in direct contact

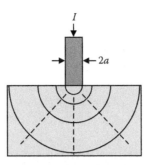

FIGURE 9.6 Spreading resistance probe of diameter 2a contacting a thick (semi-infinite) sample. The solid lines are equipotential lines, and the dashed lines indicate the electric-field directions. (After Schroder, D.K. 2006. *Semiconductor Material and Device Characterization*, 3rd ed. New York, NY: John Wiley & Sons.)

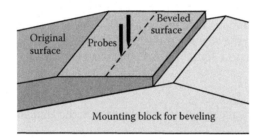

FIGURE 9.7 Spreading resistance measurement on a beveled wafer. The shallow angle of the beveled surface allows one to obtain resistivity versus depth information. (After Schroder, D.K. 2006. *Semiconductor Material and Device Characterization*, 3rd ed. New York, NY: John Wiley & Sons.)

with the semiconductor. In order to obtain satisfactory results, modern four-point and two-point spreading resistance probes make use of an automatic probe-lowering mechanism that places the probe tips without any lateral motion and with a well-calibrated force on the surface. The tips consist of very hard tungsten-osmium alloys. With an applied contact pressure exceeding 1000 kgf/mm^2, the probe tips deform elastically while the semiconductor gets slightly damaged. Such measurements are clearly invasive and may leave metal contamination on the surface. In order to avoid premature tip destruction, the lowering and raising of the contacts must proceed without any bias applied, because electrical bias may lead to spark erosion.

Contactless techniques have been developed that enable the measurement of conductivity without the need to apply electrical contacts to the sample. These methods are based on the free-carrier power absorption in a radio-frequency (RF) field (Miller et al., 1976). In one configuration, an inductor coil is placed near a semiconductor sample. The time-varying magnetic field generated by the coil induces eddy currents in the sample, which lead to resistive Joule heating. The power dissipated by the sample is proportional to V^2/ρ, where V is the amplitude of the RF voltage across the coil and ρ is the resistivity of the sample. A feedback circuit keeps V constant, while the current through the coil, I, is allowed to vary. The power is given by $P = IV$, so that

(9.23) $$I \sim V/\rho \sim V\sigma$$

Therefore, the feedback current I is proportional to the conductivity σ. Because conductivity is an additive property, one can measure the average conductivity in wafers with multiple doping layers.

There is significant engineering value in measuring the resistivity of a bulk semiconductor or, using a spreading method, in determining a resistivity-depth profile with high spatial resolution. However, we must recall that resistivity is proportional to the product of the concentration of the free carriers and their mobility. To decouple the two, we resort to Hall-effect measurements, discussed next.

9.3 HALL EFFECT

The Hall effect is used to determine the free-carrier type and concentration. It is the only method that yields both quantities directly and independently of other materials parameters. Figure 9.8 shows a standard Hall-effect configuration (Hall bar). A current is passed through a sample in the x direction and a magnetic field is applied along the z direction (Kittel, 2005). In the presence of an electric and magnetic field, the Lorentz force acting on a carrier is

$$\mathbf{F} = e(\mathbf{E} + \mathbf{v} \times \mathbf{B}) \tag{9.24}$$

The force due to the magnetic field tends to push carriers in the $-y$ direction, resulting in an accumulation of charge on the edges of the sample. This charge leads to an electric field, $\mathbf{E_H}$, that points along the y direction. At equilibrium, $\mathbf{E_H}$ balances the Lorentz force:

$$e(\mathbf{v} \times \mathbf{B}) = -e\mathbf{E_H} \tag{9.25}$$

For the simple geometrical configuration in Figure 9.8, the Hall effect is given by

$$E_H = B v_x \tag{9.26}$$

where v_x is the drift velocity along the x direction.

The *Hall coefficient* is defined

$$R_H = \frac{E_H}{j_x B} \tag{9.27}$$

where j_x is the current density along the x direction. In a typical experiment, j_x and B are set precisely by a current source and electromagnet, respectively. The Hall voltage, $V_H = E_H \delta$, is measured by a high-impedance voltmeter. From Equations 9.4, 9.26 and 9.27,

$$R_H = \frac{1}{ne} \tag{9.28}$$

This expression enables one to determine the carrier concentration n. Also, the sign of the Hall coefficient (positive or negative) reveals whether the carriers are predominantly holes or electrons. The mobility is obtained using Equation 9.5:

$$j_x = ne\mu E_x \tag{9.29}$$

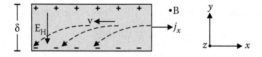

FIGURE 9.8 Standard Hall-effect configuration (Hall bar) for an *n*-type sample. The sample cross section is shown; the sample thickness is δ.

Hence, a Hall-effect measurement determines carrier concentration, type, and mobility.

The previous discussion assumed that only one type of carrier contributed to the conductivity. When both electrons *and* holes contribute to the conduction, one finds

$$(9.30) \qquad R_H = \frac{1}{|e|} \frac{p - b^2 n}{(p + bn)^2}$$

where $b = |\mu_n/\mu_p|$ is the ratio of the electron and hole mobilities. Additional complications arise from the statistical distribution of carrier velocities, which depend on temperature, electric field, the shape of the iso-energy surfaces, and scattering mechanisms (e.g., phonon scattering, ionized impurity scattering). An in-depth discussion of the Hall effect in semiconductors has been given by Beer (1990).

In many practical situations, a Hall-effect sample configuration called the *van der Pauw* geometry is used (van der Pauw, 1958). Van der Pauw applied a conformal transformation of an arbitrarily shaped sample of constant thickness δ, called a "lamella," onto a semi-infinite half-plane (Figure 9.9). The four contacts on the periphery of the lamella transform to four contacts on the boundary of the semi-infinite plane. Using this transformation, the resistivity can be expressed as (Appendix G)

$$(9.31) \qquad \rho = \frac{\pi \delta}{\ln(2)} \frac{R_{12,34} + R_{23,41}}{2} f(R_{12,34}/R_{23,41})$$

where $R_{12,34}$ is a resistance value obtained from the ratio of the voltage across contacts 3 and 4 to the current through contacts 1 and 2. $R_{23,41}$ is defined similarly. The factor f depends on the ratio of the resistance values (Figure 9.10). In most common cases, the samples are nearly circular or square shaped, and $f \approx 1$.

Hall-effect measurements can be performed with the van der Pauw geometry by passing the current from 1 to 3 and measuring the voltage between 2 and 4, or by sending the current from 2 to 4 and determining the voltage between 3 and 1. The applied magnetic field is perpendicular to the lamella. The Hall coefficient R_H is given by

$$(9.32) \qquad R_H = \delta \frac{\Delta R_{24,13}}{B}$$

where $\Delta R_{24,13}$ is change in resistance due to the B field. From Equation 9.28, the free-carrier concentration n (one carrier type) is given by

$$(9.33) \qquad n = \frac{1}{\delta} \frac{B}{\Delta R_{24,13} e}$$

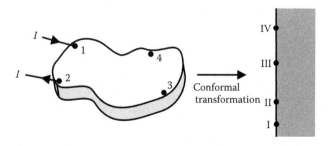

FIGURE 9.9 Van der Pauw geometry for resistivity and Hall-effect measurements. The conformal transformation mathematically maps the contacts onto the surface of a semi-infinite sample.

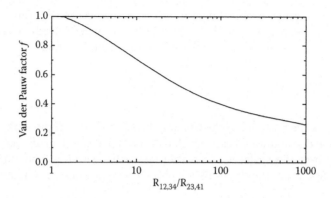

FIGURE 9.10 Van der Pauw factor f for calculating resistivity of a sample. (After Van der Pauw, L.J. 1958. *Philips Res. Rep.* 13: 1–9.)

With the knowledge of the resistivity and the carrier concentration, one can determine the mobility:

(9.34) $$\mu = \frac{1}{ne\rho} = \frac{2\ln(2)}{\pi B} \frac{\Delta R_{24,13}}{R_{12,34} + R_{23,41}} \frac{1}{f}$$

Van der Pauw and subsequent authors have evaluated the influence of contact size and contact position on the measurement results. These studies show that the van der Pauw geometry is very insensitive to deviations from the ideal case. Figure 9.11 shows that the error in sheet resistance

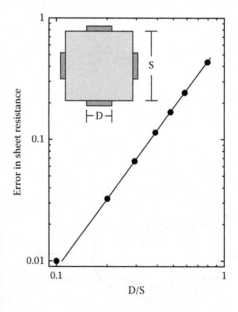

FIGURE 9.11 Error in sheet resistance measurements for contacts placed on the four edges of a square sample. The error is plotted as a function of the ratio of contact size to sample size. (After Bullis, W. Murray, ed. 1977. *NBS Special Publication* 400-29: 64–65.)

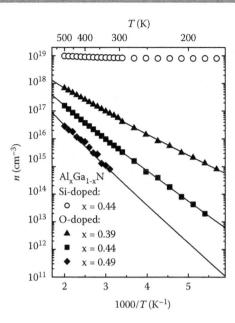

FIGURE 9.12 Free-electron concentration versus inverse temperature for silicon- and oxygen-doped AlGaN. The slopes of the solid lines yield the donor (*DX*) binding energies. (After McCluskey, M.D. et al. 1998. *Phys. Rev. Lett.* 80: 4008–4011.)

depends only to second order on the ratio of the contact length to the length of one side of a square-shaped sample (Bullis, 1977).

Variable-temperature Hall-effect measurements, with the van der Pauw geometry, provide a convenient way to obtain donor or acceptor binding energies and mobilities. The carrier concentrations in Figure 5.13 and mobilities in Figure 5.16 were obtained with a Hall-effect apparatus. Another example is the Arrhenius plot in Figure 9.12, which shows free-electron concentration versus inverse temperature for silicon- and oxygen-doped $Al_xGa_{1-x}N$ (McCluskey et al., 1998). The high concentration of shallow silicon donors results in metallic conduction, and the free-electron concentration is constant as a function of temperature. In the oxygen-doped material, the free electrons freeze out as the temperature is decreased. The donor binding energies were obtained from the slopes of the solid lines (Equation 5.62). It was found that the donor binding energies increased as a function of aluminum content (x), similar to *DX* centers in $Al_xGa_{1-x}As$.

It is important to emphasize that the results from Hall-effect measurements are not straightforward when a sample is inhomogeneous. Calculations have shown that samples with inhomogeneous doping concentrations can yield the wrong carrier type (Bierwagen et al., 2008). Low-mobility samples result in small Hall voltages that can fluctuate between negative and positive values. These experimental artifacts may account for some of the claims of *p*-type conductivity in ZnO (McCluskey and Jokela, 2009).

9.4 CAPACITANCE–VOLTAGE PROFILING

The methods discussed so far are based on charge transport (i.e., electrons or holes moving in a semiconductor). In an extrinsic semiconductor, the concentration of these free carriers depends intimately on the dopant concentrations. In the saturation region, where all the shallow donors and acceptors are fully ionized, the free-carrier concentration is equal to the net dopant concentration:

$$n = N_d^+ - N_a^- \quad (n\text{-type})$$
$$p = N_a^- - N_d^+ \quad (p\text{-type})$$

(9.35)

With this in mind, we consider measuring the net ionized dopant concentration instead of the free-carrier concentration. This can be best achieved by measuring the capacitance of a reverse biased p-n junction or a Schottky barrier junction (Schroder, 2006). The use of capacitance–voltage plots to obtain electrically active dopant concentrations is referred to as C-V profiling.

Let us return to the Schottky barrier on an n-type semiconductor (Figure 3.6). The space charge region, depleted of free carriers, is sandwiched between a conducting metal and the undepleted, conducting part of the semiconductor. This three-layer structure is a planar capacitor, with a charge $+Q$ in the semiconductor balanced by $-Q$ in the metal. Its capacitance is given by

$$C = \frac{dQ}{dV} = \varepsilon\varepsilon_0 \frac{A}{W}$$

(9.36)

where V is the reverse bias and A is the area defined by the metal contact (Equation 3.10).

Consider what happens when the depletion width increases by dW. The charge increases by an amount

$$dQ = |e| N(x) A dW$$

(9.37)

where $N(x)$ is the net donor concentration at $x = W$. From Equation 9.36, this expression for dQ yields a capacitance

$$C = |e| N(x) A \frac{dW}{dV}$$

(9.38)

Differentiating Equation 9.36 with respect to V leads to

$$\frac{dC}{dV} = -\varepsilon\varepsilon_0 A \frac{1}{W^2} \frac{dW}{dV} = -\frac{C^2}{\varepsilon\varepsilon_0 A} \frac{dW}{dV}$$

(9.39)

Using Equation 9.38 to substitute for dW/dV, we arrive at

$$\frac{dC}{dV} = -\frac{C^3}{\varepsilon\varepsilon_0 A^2 |e| N(x)}$$

(9.40)

From this expression, along with the depletion width (Equation 9.36), one can determine $N(x)$ from a C-V plot.

Using the identity

$$\frac{d(1/C^2)}{dV} = -\frac{2}{C^3} \frac{dC}{dV}$$

(9.41)

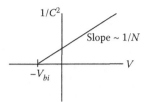

FIGURE 9.13 Plot of 1/C^2 versus applied voltage V for an abrupt, single-sided junction (e.g., Schottky diode or p^+-n diode). The slope is inversely proportional to the ionized net-dopant concentration. (After Ruge, I. 1975. *Halbleiter Technologie*, Springer-Verlag, Berlin.)

Equation 9.40 can be rewritten as

$$\frac{d(1/C^2)}{dV} = \frac{2}{\varepsilon\varepsilon_0 A^2 |e| N(x)} \tag{9.42}$$

This last equation shows that the slope of a 1/C^2 versus V plot is inversely proportional to the net-dopant concentration (only ionized centers). For a single-sided junction with constant N, one obtains a straight line. The intercept of the line with the abscissa gives the built-in potential (Figure 9.13). This occurs because, when $V = -V_{bi}$, the depletion width goes to zero (Equation 3.7).

The determination of $N(x)$ by C-V measurements is challenging because it requires extremely accurate data. This is due to the fact that differentiation has to be used to obtain the slope of a curve. Various schemes have been proposed to improve the accuracy. One such method has been published by Miller (1972), who modulated a reverse-biased diode with a small voltage ΔV. The resulting depletion layer modulation ΔW is obtained from the measured modulation in capacitance, using Equation 9.36. From Equation 9.38, the small modulated quantities are related by

$$\Delta V = W \Delta W \frac{|e| N(x)}{\varepsilon\varepsilon_0} \tag{9.43}$$

It can be seen that the magnitude of ΔW is inversely proportional to N. In this way, the *Miller profiling* technique obtains an accurate value of $1/N(x)$ (Miller et al., 1978).

A variation on C-V profiling is the electrochemical C-V profiling technique (Blood, 1986). A semiconductor is placed in contact with an electrolyte, and a DC reverse bias is applied. The electrolyte–semiconductor junction produces a depletion region, similar to a Schottky junction. As in the case of Miller profiling, N can be measured by modulating the voltage. Because the electrolyte etches the semiconductor surface over time, one can obtain N as a function of depth. Various electrolytes with well-defined etching rates are used for different semiconductors (Schroder, 2006). The primary disadvantage of this technique is that it is destructive, as it etches away a portion of the sample.

Summarizing capacitive measurements of the space charge region in a Schottky or p-n junction, we consider the advantages as well as disadvantages. First, most active devices involve junctions, making this type of measurement convenient. Second, it is easy with a modern capacitance bridge to measure low capacitances accurately, down to the femto-Farad (10^{-15} F) range. Third, the net-dopant concentration is measured directly—similar to the Hall effect, which measures the free-carrier concentration directly. Among the difficulties encountered are the effects of stray capacitances, temporal drifts in the capacitance due to surface states, and deep levels in the bulk of the semiconductor.

In general, measurements should be taken in a temperature range where all the dopant atoms are ionized and the semiconductor is extrinsic ($N > n_i$). The former condition is not fulfilled for p-type GaN:Mg because the shallow acceptors are quite deep, with an acceptor binding energy of 180 meV (Götz and Johnson, 1999). At room temperature, less than 1% of the acceptors are ionized.

9.5 CARRIER EMISSION AND CAPTURE

The previous sections in this chapter considered equilibrium conditions for which the time derivatives of the measured quantities were zero. Additional information about defects in semiconductors can be gained by measuring time-dependent phenomena. One such example of this is optical generation and recombination of excess carriers (Section 7.5).

As shown in Figure 9.14, a deep level may do the following:

- Emit an electron into the conduction band (e_n)
- Capture an electron from the conduction band (c_n)
- Capture a hole from the valence band (c_p)
- Emit a hole into the valence band (e_p)

Consider N_T deep donors per cm^3 with an energy level E_T below the conduction-band minimum. The concentrations of empty and occupied deep levels are N_T^+ and N_T^0, respectively. Empty donor levels can capture electrons while occupied donor levels can emit electrons. The rate of change in free-electron concentration is

$$(9.44) \qquad dn/dt = e_n N_T^0 - c_n n N_T^+$$

where n is the free-electron concentration, e_n is the *emission coefficient*, and c_n is the *capture coefficient* (Sah et al., 1970). The capture coefficient is

$$(9.45) \qquad c_n = \sigma_n v_{th}$$

where σ_n is the electron capture cross section (Section 7.5) and v_{th} is the thermal velocity of free electrons (Equation 5.63):

$$(9.46) \qquad v_{th} = \sqrt{\frac{3k_B T}{m^*}}$$

The emission coefficient is related to the capture coefficient (Problem 9.7) and is given by

$$(9.47) \qquad e_n = \sigma_n v_{th} N_C e^{-E_T/k_B T}$$

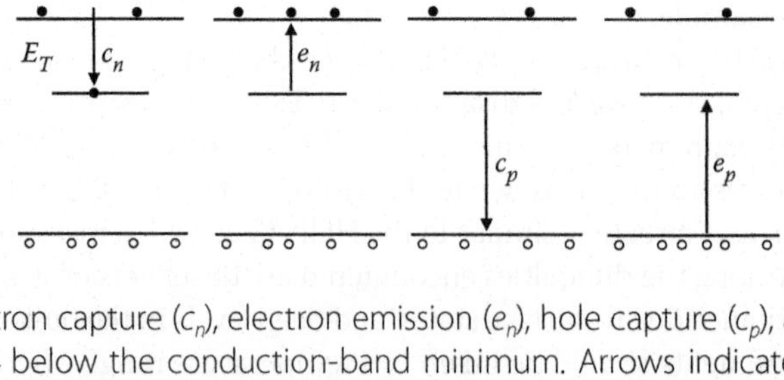

FIGURE 9.14 Electron capture (c_n), electron emission (e_n), hole capture (c_p), and hole emission (e_p) from a deep level E_T below the conduction-band minimum. Arrows indicate the direction of the electron transitions. For example, the transition of an electron from the valence band to the defect is equivalent to the emission of a hole (e_p) from the defect to the valence band.

Notice how a deep level (large E_T) has a low emission coefficient. The same results hold for deep acceptors, with n replaced by p, N_C replaced by N_V, and with E_T measured relative to the valence-band maximum. Typical cross sections are listed in Table 7.3.

Suppose we have an n-type region with neutral deep donors. A reverse bias is applied rapidly, sweeping away (depleting) the free electrons so that n drops to zero. The deep donors in the depletion region will begin to emit electrons and become positive. The rate at which this happens is governed by the emission coefficient:

$$dN_T^0/dt = -e_n N_T^0 \tag{9.48}$$

Equation 9.48 is solved by a decaying exponential,

$$N_T^0(t) = N_T^0(0)e^{-t/\tau_e} \tag{9.49}$$

where $1/\tau_e = e_n$. This process is important for characterizing deep defects, discussed in the following section.

In the case of DX centers and other large-relaxation defects, there is an energy barrier E_c in the configuration-coordinate diagram (Figure 7.18). The barrier must be surmounted in order for an electron to be captured, resulting in a capture cross section that is strongly temperature dependent. From Equation 7.69, the capture cross section is given by

$$\sigma_{DX} = \sigma_\infty e^{-E_c/k_B T} \tag{9.50}$$

When this value is substituted for σ_n in Equation 9.47, the exponents add such that the activation energy for emission is given by

$$E_e = E_T + E_c \tag{9.51}$$

In practice, E_T can be obtained from variable-temperature Hall-effect measurements (Figure 9.12). In the next section, we discuss how one measures the thermal emission rate of electrons or holes from deep traps, by monitoring the capacitance of a reverse-biased diode. From an Arrhenius plot of emission rate versus inverse temperature, one can obtain E_e. Equation 9.51 then gives us E_c. For many deep levels, the capture barrier E_c can also be measured independently. For DX centers, however, the low capture rates make such a determination challenging (Mooney, 1990).

9.6 DEEP-LEVEL TRANSIENT SPECTROSCOPY

Deep-level transient spectroscopy (DLTS) enables the observation of the charging and discharging of deep levels in a semiconductor through the measurement of capacitance transients. Three properties of deep levels can be determined: concentration, position in the band gap, and capture cross section for free carriers. With some additional sophistication, the thermal activation energy of the capture cross section can be measured as well.

To explore the principles of DLTS, let us discuss an n^+-p junction. Assume that the temperature is high such that all shallow dopants are ionized. We start with a junction that is reverse biased during the time interval (1) (Figure 9.15). At time (2), the bias is rapidly reduced (pulsed) to a lower value, and the depletion layer width shrinks significantly. This shrinkage is accomplished quickly, by a flood of free holes that neutralize the negatively charged acceptors. While the shallow acceptors remain ionized, the deep acceptors capture holes and become neutral. At time (3),

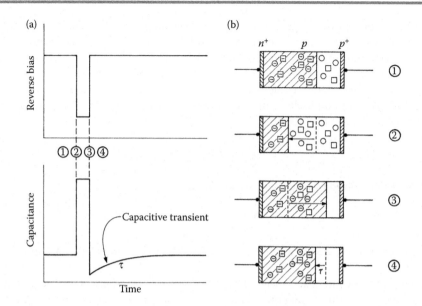

FIGURE 9.15 (a) Change in capacitance in a deep-level transient spectroscopy (DLTS) measurement, as the reverse bias is pulsed. (b) The depletion width during each time interval. Circles and squares represent shallow and deep traps, respectively. (Reprinted from Haller, E.E. et al., *IEEE Trans. Nucl. Sci.* NS-26, 265–270. © 1979 IEEE.)

the bias reduction pulse is removed, and the original bias is reapplied. Because the deep acceptors are now neutral, the depletion region is wider than during time interval (1). During time interval (4), we observe the recovery of the *p-n* junction as the deep acceptors release their holes.

The change of the charge state of the deep levels is observed by measuring the capacitance. Figure 9.15 schematically displays what happens to the space charge during and after the bias reduction pulse. The deep traps that become neutral during the bias reduction pulse no longer contribute any space charge. Because of the lower total space charge, the depletion layer widens beyond the steady-state width immediately after the bias pulse is removed at time (3). Every deep acceptor trap that thermally releases its hole becomes negatively charged again, leading to a progressive reduction of the depletion layer width. When all the deep acceptors have released their holes, we have returned to the pre-pulse condition.

Parameters that characterize a deep level can be determined from a measurement of the capacitance as a function of time (Figure 9.15). The amplitude of the capacitive transient ΔC starting at time (3) yields the concentration of the acceptor traps:

$$N_T^0 \approx 2(N_a - N_d)\frac{\Delta C}{C} \quad (9.52)$$

where $N_a - N_d$ is the shallow net-acceptor concentration in the *p*-region, ΔC is the amplitude of the capacitance transient, and C is the steady-state capacitance (Lang, 1974). If the time between (2) and (3) is sufficiently long, then all the acceptor traps capture a hole and $N_T^0 = N_T$. Equation 9.52 is valid for $N_T \ll N_a - N_d$, which is the case for most semiconductors that have not been deliberately doped with deep levels. Samples with high concentrations of defects can be illuminated, in order to photo-ionize the deep defects and provide a high concentration of carriers (Mooney, 1983).

The position of the acceptor trap level above the valence band, E_T, can be determined from the emission rate as a function of temperature (Equation 9.47). D.V. Lang at Bell Labs proposed

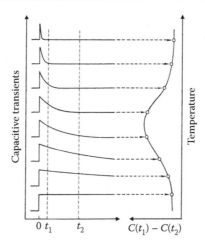

FIGURE 9.16 Implementation of a rate window by means of a double boxcar integrator. The output corresponds to the difference of the signals at times t_1 and t_2. (After Miller, G.L., D.V. Lang, and L.C. Kimerling. 1977. *Ann. Rev. Mater. Sci.* 7: 377–448.)

an electronic scheme to select capacitive transients with a given time constant (Lang, 1974). Lang used a double boxcar integrator, an instrument that measures repetitive signals, C_1 and C_2, at two times, t_1 and t_2, and outputs the difference of these two signals. The difference signal is averaged over many signal periods, leading to an excellent signal-to-noise ratio and high sensitivity. Figure 9.16 graphically conveys the formation of a peak at a specific temperature, for traps (all of the same type) emitting their charges exponentially with time. At the bottom of the figure, the temperature is very low and emission proceeds slowly, leading to long time constants; C_1–C_2 is small. At the upper edge of the figure, the temperature is high enough for the emission event to be completed before C_1 is recorded; again, C_1–C_2 is close to zero. Somewhere between these two extremes, there will be a maximum in the difference C_1–C_2. This maximum occurs for

$$\tau_e = \frac{t_2 - t_1}{\ln(t_2 / t_1)}. \tag{9.53}$$

When a diode with several kinds of deep traps is cooled, one obtains signals only at the temperatures for traps with the emission rate corresponding to the preset time constant τ_e of the DLTS apparatus. Each kind of trap leads to a peak at a specific temperature in a DLTS spectrum, making DLTS a useful technique for identifying deep levels in a sample. Figure 9.17 shows a DLTS spectrum for ultrapure germanium diodes that were intentionally doped with copper. This work was the very first DLTS study on ultrapure germanium (Haller et al., 1979). The peaks in the spectrum were attributed to deep levels introduced by copper, copper-hydrogen, and oxygen impurities.

Along with the double boxcar integrator, other time-constant selection methods have been developed. One approach for selecting a particular time constant uses a lock-in amplifier with a square wave (Auret, 1986). Another method uses a Laplace transform to analyze the capacitive transient. Mathematically, an exponential decay is transformed into a delta function by an inverse Laplace transformation. Performing an inverse Laplace transform on a DLTS signal can provide extremely high resolution and enable one to discriminate between traps with similar time constants. However, exponentials are not orthogonal functions like trigonometric functions, and this inverse transformation does not cope well with noise (Nolte and Haller, 1987). A review of Laplace-transform DLTS is given by Dobaczewski et al. (2004).

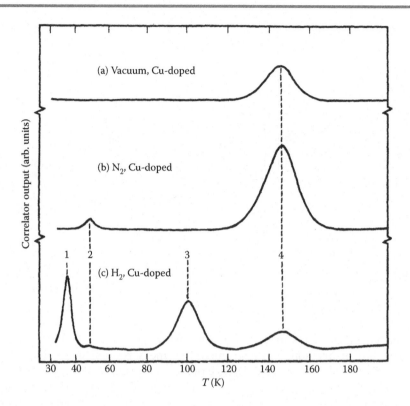

FIGURE 9.17 Deep-level transient spectroscopy (DLTS) spectrum of deep-level acceptors in copper-doped germanium. Peaks (1) and (3) are due to copper–hydrogen complexes, (2) is related to oxygen impurities, and (4) is from doubly ionized copper acceptors. (Reprinted from Haller, E.E. et al., *IEEE Trans. Nucl. Sci.* NS-26, 265–270. © 1979 IEEE.)

The capture cross section can be determined from the dependence of the capacitive transient amplitude ΔC on the length of the bias reduction pulse (time 2 to time 3). It takes several time constants τ_c to fill most of the traps during the bias reduction pulse. By going to ever-shorter pulse lengths, fewer deep traps get filled during the bias reduction period, and the capacitive transient ΔC becomes smaller. By measuring the neutral acceptor concentration (Equation 9.52) as a function of pulse time, one can determine the carrier capture rate c_p. Equation 9.45 is then used to obtain the capture cross section.

9.7 MINORITY CARRIERS AND DEEP-LEVEL TRANSIENT SPECTROSCOPY

The discussion so far has focused on traps that capture and release majority carriers (i.e., holes in a *p*-type semiconductor or electrons in *n*-type material). However, the capture of *minority* carriers is an important process in many devices. For example, consider a photovoltaic solar cell (Section 3.5). A photon that is absorbed in the depletion region of the *p-n* junction will create an electron and a hole. The built-in electric field will push the electron toward the *n* side and the hole toward the *p* side. When the electron and hole are created in the *n*-type layer, the hole is the minority carrier. If the minority hole lifetime is too short, then it will recombine with an electron before it can reach the *p* side, reducing the efficiency of the solar cell. Minority electron lifetimes affect the performance in a similar fashion.

Along with solar cell efficiency, minority carrier lifetimes affect the gain of transistors, the leakage current in *p-n* junctions, and the time a MOS or CCD can store charge (Schroder, 1997). As discussed in Section 7.5, deep-level defects play a dominant role in determining minority carrier lifetimes. Given the importance of these deep traps, the ability to characterize them accurately is essential for a range of electronic devices.

Can DLTS be used to characterize traps that are of minority type (e.g., a deep donor in a *p*-type semiconductor)? The answer is "yes" if a way is found to provide a supply of minority carriers. This can be achieved in at least two ways. First, the bias reduction pulse can be made larger than the steady state bias, leading to forward current that injects minority carriers. Second, above-bandgap light can be used to illuminate the junction, creating electrons and holes. Minority traps then capture the appropriate free carrier, turning neutral.

Whereas the net space charge is *reduced* when majority traps become neutral, the net charge *increases* when minority traps capture a charge and become neutral. This is because ionized minority deep centers are charged opposite to ionized majority traps. With their opposite charge, they decrease the net space charge. This compensating space charge disappears when the minority traps become neutral, leading to an increased majority space charge. The consequence of this space charge increase is a decrease in depletion layer width and an increase in capacitance. We see that the capacitance transient ΔC of minority traps has the opposite sign of the majority traps. Figure 9.18 shows a DLTS spectrum of ultrapure germanium (Haller et al., 1979). Peak 11 is due to minority traps. In this case, light was shone on the diode, generating free holes and free electrons.

Great precaution is taken in every step of semiconductor device fabrication to avoid contamination of the semiconductors with deep-level impurities or defects. DLTS is one of the tools of choice to characterize the traps. It is a parametric technique, and its sensitivity tracks the net shallow-level space charge concentration. A well-designed and well-constructed DLTS system can detect 1 deep-level impurity in 10^5 shallow levels, and the best systems are ~10 times more

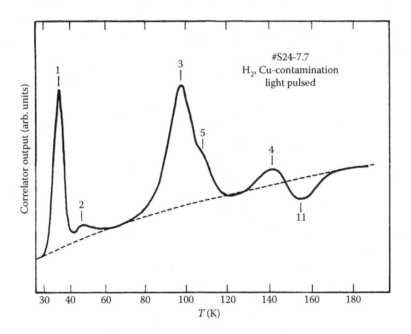

FIGURE 9.18 Deep-level transient spectroscopy (DLTS) spectrum of deep-level complexes in ultrapure germanium. Peak 11 corresponds to a minority trap. (Reprinted from Haller, E.E. et al., *IEEE Trans. Nucl. Sci.* NS-26, 265–270. © 1979 IEEE.)

sensitive. Such a system should be able to detect 10^9 cm^{-3} deep levels in a semiconductor doped to 10^{15} cm^{-3}. However, DLTS requires a Schottky or *p-n* junction, which may not be feasible for some high-impedance samples.

9.8 MINORITY CARRIER LIFETIME

In Section 7.5, we discussed free-carrier generation and recombination processes. The dominant physical parameter describing the return of a semiconductor to its equilibrium condition is the minority carrier lifetime. Using deep-level defects and impurities, this lifetime can be changed and controlled over many orders of magnitude. Short lifetimes are important for fast switches. Long lifetimes lead to efficient *p-n* junction rectifiers and solar cells, as well as high-gain *n-p-n* and *p-n-p* transistors.

The *Haynes–Shockley experiment* (Haynes and Shockley, 1951) measures the transport of minority carriers. Figure 9.19 shows a version of the basic experimental setup. A relatively small bias is applied across a long bar of an *n*-type semiconductor, generating a constant electric field along the bar. A contact on a *p*-type layer is at $x = 0$. When a forward-bias pulse is applied, holes are injected into the semiconductor. The holes drift in the electric field toward the right and are collected at $x = d$. The contact at $x = d$ is on a *p*-type layer that is under reverse bias in order to efficiently collect the holes.

The drift mobility μ is obtained from the drift time t_0, which is the mean time it takes the holes to move from $x = 0$ to $x = d$:

$$(9.54) \qquad v_d = d/t_0 = \mu E$$

where E is the electric field along the bar (McKelvey, 1966). Because the minority carrier lifetime is finite, not all of the holes make it to the contact at $x = d$. Figure 9.19 illustrates the propagation of the hole pulse, which is assumed to be a delta-function at injection. The hole packet drifts and broadens with time, described by a Gaussian shape:

$$(9.55) \qquad \Delta p(x,t) = \frac{\Delta p_0}{\sqrt{4\pi Dt}} e^{-t/\tau_p} e^{-(x-v_d t)^2/4Dt}$$

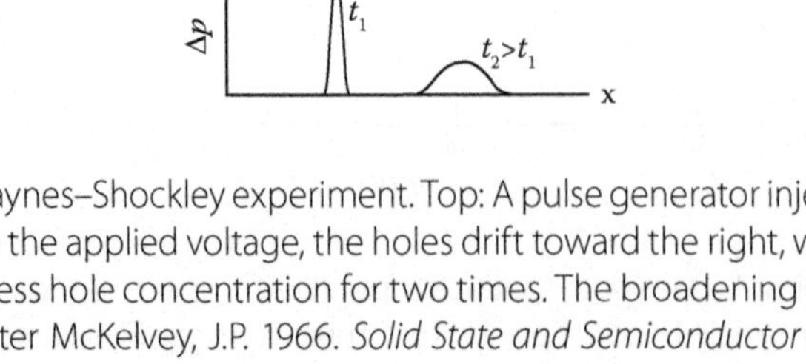

FIGURE 9.19 The Haynes–Shockley experiment. Top: A pulse generator injects holes into an *n*-type semiconductor. Due to the applied voltage, the holes drift toward the right, where they are collected. Bottom: Plot of the excess hole concentration for two times. The broadening of the distribution is due to carrier diffusion. (After McKelvey, J.P. 1966. *Solid State and Semiconductor Physics, Chapter 10.* New York, NY: Harper and Row.)

where Δp is the excess hole concentration as a function of time and space, Δp_0 is proportional to the total number of holes injected at $t = 0$, D is the hole diffusion coefficient, and τ_p is the minority hole lifetime.

The holes arriving at the depletion layer at $x = d$ are collected rapidly, producing a voltage drop across the load resistor. This voltage drop is displayed on the y-axis of an oscilloscope. The peak of the voltage occurs at the drift time t_0. The hole diffusion coefficient D can be obtained from the width of the pulse. Alternatively, one can use the *Einstein relation*, which relates drift mobility and diffusivity:

$$(9.56) \qquad D = \mu k_B T / e$$

From the exponentially decaying amplitude in Equation 9.55, τ_p can be determined by measuring the peak of Δp as the distance d is changed.

Photoconductivity decay measurements, requiring optical excitation, are often used to determine the minority carrier lifetime. Optical excitation combines several advantages over other methods. It covers a very wide range of time scales. It is noninvasive and the intensity can be controlled over a very wide range. A semiconductor bar is illuminated by short optical pulses ($h\nu > E_g$) produced by the light source and a motor-driven chopper wheel. The increase in conductivity is measured by applying a constant current source across the semiconductor bar. The change in the bias voltage across the bar, ΔV, decays exponentially as the excess carriers disappear through recombination, to the point where the steady state bias value is reached. This exponential decay is given by

$$(9.57) \qquad \Delta V = \Delta V_0 e^{-t/\tau}$$

where ΔV_0 is the change in bias at $t = 0$ and τ is the carrier lifetime.

These two examples suffice to illustrate how the decay of excess carrier concentrations can be measured. A word of caution, however, should be added. The lifetimes that one measures typically depend on many important details of the excitation and recombination processes. Many different physical processes, each with a characteristic time constant, may proceed over some time interval in parallel.

In summary, it should be stated that the minority carrier lifetime of a semiconductor is often a good measure of its quality. For silicon, one finds that the lower the deep-level concentration caused by impurities and defects is, the longer is the minority carrier lifetime (Queisser and Haller, 1998). The best floating-zone silicon single crystals exhibit lifetimes exceeding 100 ms.

9.9 THERMOELECTRIC EFFECT

The application of a temperature gradient induces a small voltage in a semiconductor. This is called the thermoelectric effect. The *thermoprobe*, or "hot probe," is a simple tool that uses the thermoelectric effect to determine the type of conduction (Ruge, 1975). One measures a current in the external circuit generated by charge carrier flow in the semiconductor from the area under the hot tip to the area under the cold tip. External currents of opposite sign are generated for n- and p-type crystals. In the case of germanium, the method is useful at room temperature with specimens up to a resistivity of $\sim$30 to 40 Ω cm ($\geq 4 \times 10^{13}$ carriers cm^{-3}). For higher-purity crystals, the intrinsic carrier concentration dominates, leading to an indication of n-type conductivity. The reason for this is the carrier mobility, which is higher for electrons than for holes. The type of conduction of silicon crystals can be determined over a resistivity range from 1mΩ cm to $\sim$100 Ω cm.

It is possible to obtain quantitative information about the thermoelectric effect by carefully regulating the temperature gradient. A schematic illustration of a measurement system is shown

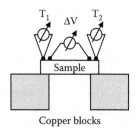

FIGURE 9.20 A thermopower (thermoelectric effect) measurement. (After Ager III, J.W. et al. 2008. *Phys. Stat. Sol. B*, 245, 873–877.)

in Figure 9.20 (Ager et al., 2008). A sample is suspended between two copper blocks, which are held at specific temperatures. The open-circuit voltage between two contacts, ΔV, is measured by a high-impedance multimeter, and the temperature at each contact is measured by a thermocouple. The temperature difference ΔT is varied while keeping the average temperature constant.

The *Seebeck coefficient* quantifies the thermoelectric effect. It is defined as

$$S = \frac{\Delta V}{\Delta T} \qquad (9.58)$$

In the case of a *p*-type semiconductor, a negative temperature gradient ($\Delta T < 0$) causes the holes to move toward the right, leaving behind a negative charge on the left. This results in a negative voltage ΔV (i.e., if a resistive load were connected to the contacts, then current would flow toward the left). Holes therefore give rise to a *positive* Seebeck coefficient *S*. Electrons, on the other hand, result in a *negative* Seebeck coefficient. Through modeling of the free carriers, one can estimate the carrier concentration from the value of the Seebeck coefficient (Seeger, 1982). This technique was used to measure hole concentrations in InN and InGaN (Miller et al., 2010).

The thermoelectric effect can also be used to study deep levels. In *thermoelectric effect spectroscopy* (TEES), a sample at low temperature is illuminated with light, causing deep traps to capture electrons or holes (Šantić and Desnica, 1990). After the traps are filled, the sample is warmed up at a well-defined rate, while a small temperature gradient is applied across the sample. Similar to DLTS, specific traps release their carriers at specific temperatures. When the carriers are released, a current is measured, the sign of which reveals whether the carriers are electrons or holes. In a related technique, *thermally stimulated current* (TSC), a bias instead of a temperature gradient is applied (Milnes, 1973). TSC produces currents when carriers are released but does not distinguish between electrons and holes. While these thermal techniques are not as quantitative as DLTS, they use relatively simple equipment and do not require a Schottky or *p-n* junction.

PROBLEMS

9.1 An *n*-type semiconductor has a free-electron density 10^{16} cm^{-3} and mobility 200 cm^2/Vs. An electric field 20 V/cm is applied. Calculate the following:
   a. Drift velocity
   b. Current density
   c. Conductivity
   d. Resistivity

9.2 A scientist performs a 4-point probe measurement on a 1 μm thin film. The probes are spaced at intervals of 1 mm. For an applied current of 5 mA, the inner 2 probes indicate a voltage drop of 1.5 V. What is the resistivity?

Chapter 9—Electrical Measurements 243

9.3 Sketch a Hall-effect experiment for the Hall bar (not van der Pauw) geometry. Show the current direction and Hall field, assuming a p-type sample.

9.4 A Hall-effect measurement is performed on a Hall bar, which has dimensions $0.5 \times 0.5 \times 5$ mm. A current of 1.0 mA is sent through the sample (along the long axis). The measurement indicates that the sample is p-type with a hole density of 6.25×10^{15} cm^{-3} and mobility 100 cm^2/Vs. Calculate:
   a. Current density
   b. Conductivity
   c. Resistivity
   d. Applied voltage
   e. Drift velocity

9.5 A resistivity measurement is performed on a 0.5 mm thick sample in the van der Pauw geometry. When 5 mA flows from contact 1 to 2, the voltage drop between contacts 4 and 3 is 0.40 V. When 5 mA flows from 2 to 3, the voltage drop between 1 to 4 is 0.05 V. Calculate the resistivity.

9.6 A C-V profile is taken on an n-type silicon sample ($\varepsilon = 11.7$) with a metal contact of area 1 mm^2. The following linear relation is obtained:

$$1/C^2(\mathrm{F}^{-2}) = (2.4 \times 10^{20})V + 1.7 \times 10^{20}$$

Find the net donor concentration (cm^{-3}) and Schottky barrier height (V).

9.7 Consider a semiconductor with N_T deep donors and N_a compensating acceptors. Assume there are no shallow donors.
   a. For steady state ($dn/dt = 0$), show that

$$e_n(N_T - N_a - n) = c_n n(n + N_a)$$

   b. From part (a) and Equation 5.56, derive the emission coefficient (Equation 9.47).

9.8 Table 7.3 lists capture cross sections for substitutional impurities in silicon. Consider Au in its neutral charge state at room temperature. (Approximate the silicon effective electron and hole masses as $m^* = m$.)
   a. Calculate the electron and hole capture coefficients.
   b. Calculate the electron and hole emission coefficients.

9.9 A DLTS apparatus uses a double boxcar integrator with a preset time constant of $\tau_e = 10$ μs. A deep acceptor is known to have hole capture coefficient $c_p = 10^{-7}$ cm^3/s, which is independent of temperature. The deep acceptor gives rise to a peak in the DLTS signal at 150 K. Calculate the acceptor binding energy. (Approximate the effective hole mass as $m^* = m$.)

9.10 The positive lead of a voltmeter is attached to contact #1 of a semiconductor sample, while the ground lead is attached to contact #2. Contact #1 is heated to 40°C while contact #2 is 20°C. The voltmeter reads −0.5 mV.
   a. Calculate the Seebeck coefficient.
   b. Are the dominant charge carriers electrons or holes?

REFERENCES

Ager III, J.W., N. Miller, R.E. Jones et al. 2008. Mg-doped InN and InGaN—Photoluminescence, capacitance-voltage and thermopower measurements. *Phys. Stat. Sol. B* 245: 873–877.

Auret, F.D. 1986. Considerations for capacitance DLTS measurements using a lock-in amplifier. *Rev. Sci. Instrum.* 57: 1597–1613.

244 References

Beer, A.C. 1990. Hall effect and the beauty and challenges of science. In *The Hall Effect and Its Applications*, eds. C.L. Chien and C.R. Westgate, pp. 299–338. New York, NY: Plenum Press.

Bierwagen, O., T. Ive, C.G. Van de Walle, and J.S. Speck. 2008. Causes of incorrect carrier-type identification in van der Pauw–Hall measurements. *Appl. Phys. Lett.* 93: 242108 (3 pages).

Blood, P. 1986. Capacitance-voltage profiling and the characterization of III-V semiconductors using electrolyte barriers. *Semicond. Sci. Technol.* 1: 7–27.

Bullis, W. Murray, ed. 1977. Semiconductor measurement technology. *NBS Special Publication* 400-29: 64–65.

Dobaczewski, L., A.R. Peaker, and K. Bonde Nielsen. 2004. Laplace-transform deep-level spectroscopy: The technique and its applications to the study of point defects in semiconductors. *J. Appl. Phys.* 96: 4689–4728.

Götz, W. and N.M. Johnson. 1999. Characterization of dopants and deep level defects in gallium nitride. In *Semiconductors and Semimetals* Vol. 57, eds. J. Pankove and T. Moustakas, pp. 185–207. New York, NY: Academic Press.

Haller, E.E., P.P. Li, G.S. Hubbard, and W.L. Hansen. 1979. Deep level transient spectroscopy of high-purity germanium diodes/detectors. *IEEE Trans. Nucl. Sci.* NS-26: 265–270.

Haynes, J.R. and W. Shockley. 1951. The mobility and life of injected holes and electrons in germanium. *Phys. Rev.* 81: 835–843.

Kane, P.F. and G.B. Larrabee. 1970. *Characterization of Semiconductor Materials*. New York, NY: McGraw-Hill.

Kittel, C. 2005. *Introduction to Solid State Physics*, 8th ed. New York, NY: John Wiley & Sons.

Lang, D.V. 1974. Deep-level transient spectroscopy: A new method to characterize traps in semiconductors. *J. Appl. Phys.* 45: 3023–3032.

McCluskey, M.D. and S.J. Jokela. 2009. Defects in ZnO. *J. Appl. Phys.* 106: 071101 (13 pages).

McCluskey, M.D., N.M. Johnson, C.G. Van de Walle, D.P. Bour, M. Kneissl, and W. Walukiewicz. 1998. Metastability of oxygen donors in AlGaN. *Phys. Rev. Lett.* 80: 4008–4011.

McKelvey, J.P. 1966. *Solid State and Semiconductor Physics, Chapter 10*. New York, NY: Harper and Row.

Miller, G.L. 1972. A feedback method for investigating carrier distributions in semiconductors. *IEEE Trans. Elec. Dev.* ED-19: 1103–1108.

Miller, G.L., D.A.H. Robinson, and J.D. Wiley. 1976. Contactless measurement of semiconductor conductivity by radio frequency–free carrier power absorption. *Rev. Sci. Instrum.* 47: 799–805.

Miller, G.L., D.V. Lang, and L.C. Kimerling. 1977. Capacitance transient spectroscopy. *Ann. Rev. Mater. Sci.* 7: 377–448.

Miller, G.L., D.A.H. Robinson, and S.D. Ferris. 1978. In *Semiconductor Characterization Techniques, Electrochemical Society Proceedings*, Vol. 78–3, eds. P.A. Barnes and G.A. Rozgonyi, p. 22. Pennington, NJ: Electrochemical Society.

Miller, N., J.W. Ager III, H.M. Smith III et al. 2010. Hole transport and photoluminescence in Mg-doped InN. *J. Appl. Phys.* 107: 113712 (8 pages).

Milnes, A.G. 1973. *Deep Impurities in Semiconductors, Chapter 9*. New York, NY: John Wiley & Sons.

Mooney, P.M. 1983. Photo-deep level transient spectroscopy: A technique to study deep levels in heavily compensated semiconductors. *J. Appl. Phys.* 54: 208–213.

Mooney, P.M. 1990. Deep donor levels (*DX* centers) in III-V semiconductors. *J. Appl. Phys.* 67: R1–R26.

Nolte, D.D. and E.E. Haller. 1987. Optimization of the energy resolution of deep level transient spectroscopy. *J. Appl. Phys.* 62: 900–906.

Queisser, H.-J. and E.E. Haller. 1998. Defects in semiconductors: Some fatal, some vital. *Science* 281: 945–950.

Ruge, I. 1975. *Halbleiter Technologie*. Berlin: Springer-Verlag.

Sah, C.T., L. Forbes, L.L. Rosier, and A.F. Tasch, Jr. 1970. Thermal and optical emission and capture rates and cross sections of electrons and holes at imperfect centers in semiconductors from photo and dark junction current and capacitance experiments. *Solid State Electron.* 13: 759–788.

Šantić, B. and U.V. Desnica. 1990. Thermoelectric effect spectroscopy of deep levels—application to semi-insulating GaAs. *Appl. Phys. Lett.* 56: 2636–2638.

Schroder, D.K. 1997. Carrier lifetimes in silicon. *IEEE Trans. Electron. Devices* 44: 160–170.

Schroder, D.K. 2006. *Semiconductor Material and Device Characterization*, 3rd ed. New York, NY: John Wiley & Sons.

Seeger, K. 1982. *Semiconductor Physics: An Introduction*. Berlin: Springer-Verlag.

Smits, F.M. 1958. Measurement of sheet resisitivities with the four-point probe. *Bell Syst. Tech. J.* 37: 711–718.

Stavola, M., ed. 1998. *Semiconductors and Semimetals*. Vols. 51A and 51B. San Diego, CA: Academic Press.

Sze, S.M. and J.C. Irvin. 1968. Resistivity, mobility and impurity levels in GaAs, Ge, and Si at 300°K. *Solid State Electron*. 11: 599–602.

Van der Pauw, L.J. 1958. A method of measuring specific resistivity and Hall effect of discs of arbitrary shape. *Philips Res. Rep*. 13: 1–9.

Wieder, H.H. 1979. Four-terminal nondestructive electrical and galvanomagnetic measurements. In *Nondestructive Evaluation of Semiconductor Materials and Devices*, pp. 68–105, ed. J.N. Zemel. New York, NY: Plenum Press.

Optical Spectroscopy

Optical studies belong to one of the richest areas of solid-state and semiconductor physics. Characterization techniques utilize a wide range of the electromagnetic spectrum, from radio waves to the ultraviolet. Methods involving photons with higher energies, x-rays and gamma rays, are discussed in Chapter 11. Energy absorption and emission processes may involve photons alone, but in many cases, electrons or ions play a role in the interactions. Photoelectron spectroscopy, for example, uses photons to emit electrons out of a semiconductor surface while cathodoluminescence is the process of light emitted from a semiconductor surface upon bombardment with energetic electrons.

There are many reasons why photons are exceptionally well suited for semiconductor defect characterization (Perkowitz, 1993). Photons essentially provide "pure" energy and can be produced monochromatically at high intensities (e.g., lasers, electron synchrotrons) over a wide energy range. Photons can be manipulated easily and sent over large distances. At energies exceeding ~1 eV, single photons can be counted. Photons travel fast and offer high time resolution, providing insight into phenomena occurring on femtosecond (10^{-15} s) time scales. Additional sensitivity can be achieved by modulating a physical parameter, such as stress or magnetic field, and using a lock-in amplifier to measure the corresponding modulation in optical properties.

A discussion of all possible techniques involving photons would fill many volumes. In this chapter, techniques central to semiconductor characterization are summarized. First, we discuss the fundamentals of absorption and emission spectroscopy, including photoluminescence and Raman scattering. Fourier transform infrared spectroscopy, useful for studying vibrational and electronic properties of defects with high spectral resolution, is discussed. Methods for time-resolved measurements and the application of stress are outlined. The chapter concludes with a summary of magnetic resonance techniques, which utilize microwaves and radio waves in combination with a static magnetic field.

10.1 ABSORPTION

Perhaps the simplest use of photons for the characterization of a semiconductor is linear absorption spectroscopy. A monochromatic photon beam arrives at a semiconductor sample with intensity I_0. After passing through the sample, the intensity is I. From the Beer–Lambert Law (Equation 1.25), the optical *transmission* is given by

$$(10.1) \qquad T \equiv I/I_0 = e^{-\alpha x}$$

where x is the sample thickness and α is the linear absorption coefficient. The absorbance is defined as

$$(10.2) \qquad \text{Absorbance} = \log_{10}(I_0/I) = -\log_{10}(T)$$

Equation 10.1 neglects reflection from the sample surfaces. Taking into account multiple reflections from identical surfaces in a vacuum, at normal incidence,

$$(10.3) \qquad T = \frac{(1-R)^2 e^{-\alpha x}}{1 + R^2 e^{-2\alpha x} - 2R \cdot e^{-\alpha x} \cos(2knx)}$$

where the reflectance R is given by Equation 7.8. The cosine term accounts for interference between the multiple reflections. This results in *Fabry–Perot* interference fringes that have maxima when $2x$ is a multiple m of the wavelength in the material:

$$(10.4) \qquad 2x = m\lambda/n$$

In a transmission spectrum, the spacing between adjacent maxima,

$$(10.5) \qquad \Delta(1/\lambda) = 1/(2nx)$$

allows one to determine the thickness of the sample if the refractive index n is known. Fabry–Perot fringes can be seen in Figure 7.9. The amplitude of Fabry–Perot fringes is reduced by absorption, scattering, or surfaces that are rough or nonparallel. Because of these effects, in many situations, multiple reflections are negligible and the denominator in Equation 10.3 can be replaced by 1.

A typical optical transmission system consists of a broadband light source, monochromator, and detector. One example of a light source is a quartz tungsten halogen lamp, which emits light from 300 nm to 2 μm. The tungsten filament is contained in a silica bulb filled with inert gas, along with a small amount of bromine or iodine to suppress the deposition of tungsten on the bulb. Another type of light source is the arc lamp, which consists of a tungsten cathode and anode, separated by ~1 mm, in a silica envelope filled with an inert gas. To turn the arc lamp on, a high voltage is applied to the cathode–anode pair, igniting a spark and gas discharge. A constant current is then maintained by the power supply. Arc lamps may contain deuterium (~180 to 400 nm) or xenon (~200 nm to 2.5 μm). Mercury arc lamps, which contain mercury and a rare gas, emit lines in the ultraviolet (UV)/visible region and a continuum for wavelengths longer than 2.5 μm. For infrared (IR) measurements, discussed in Section 10.4, a heated piece of silicon carbide called a "globar" is used for the mid-IR range (2 to 50 μm), while a mercury lamp is preferred for the far-IR (50 to 2000 μm).

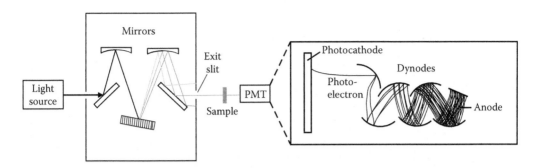

FIGURE 10.1 Monochromator and photomultiplier tube (PMT).

The *monochromator* uses a grating to disperse the light spectrally (Figure 10.1). The spacing between the lines on the grating determines the dispersion. For a line spacing a and wavelength λ:

(10.6) $$a \sin\theta = m\lambda$$

where θ is the angle of reflection, and m is an integer that denotes the order of the reflection (Hecht, 2002). The grating angle is rotated to scan through a range of wavelengths, using the $m = 1$ reflection. Higher-order reflections ($m > 1$) can be rejected with filters. Gratings are optimized for a *blaze wavelength*, for which the reflected intensity is a maximum.

The spectral width of the beam exiting the monochromator is determined by a and the width of the exit slit. A small value of a (many lines per mm) results in high spectral dispersion, so that the light traveling through the exit slit has a narrow spectral width. Highly monochromatic light allows one to take spectra with high spectral *resolution*. A narrow slit results in high resolution, because it rejects light that does not have the selected wavelength. The downside is that, in order to improve resolution, the photon flux must be reduced. Another factor that affects resolution is the distance that the light travels. A 1 m path length, for example, will disperse light more than a 0.25 m path length, resulting in finer resolution for a given slit width.

After passing through the sample, the intensity of the light I is measured by a detector. One example of a sensitive detector is a *photomultiplier tube* (PMT) (Figure 10.1). In a PMT, light impinges on a photocathode such as GaAs, generating free electrons. A high voltage (up to 1 to 2 kV) causes these photoelectrons to accelerate toward the first anode, or "dynode," liberating secondary electrons. These electrons in turn bombard the next dynode, liberating more electrons, resulting in an electron cascade. The electrons are collected at the last anode, resulting in a fast, large electrical pulse. A discriminator rejects pulses below a set threshold, allowing for single-photon counting. The reference intensity I_0 can be recorded in a separate measurement. Alternatively, the light can be split into two beams, only one of which passes through the sample. The sample and reference beams are then measured simultaneously.

Figure 1.20 shows the extremely strong dependence of α on photon energy for silicon, germanium, and GaAs. The linear absorption coefficient α changes by over six orders of magnitude near the fundamental absorption edge, caused by the excitation of electrons across the band gap. The change in α is especially abrupt for the direct-band-gap semiconductor GaAs. The rapid ascent of α to values exceeding $\sim 10^4$ cm^{-1} for germanium and silicon is preceded by an energy region of lower α values. As discussed in Section 1.9, this is the indirect-band-gap range where, in addition to the photon, phonons are required for the conservation of crystal momentum (Figure 1.19). It is fortuitous that α reaches values as high as 10^6 cm^{-1}, because it allows absorption (and reflection) measurements of very thin semiconductor films. This is especially advantageous for semiconductor alloys, such as InGaN, that can only be grown in thin-film form.

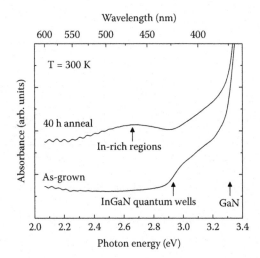

FIGURE 10.2 Optical absorption from InGaN quantum wells and GaN layers. The broad absorption peak at ~2.65 eV was attributed to indium-rich InGaN precipitates. (After McCluskey, M.D. et al. 1998. *Appl. Phys. Lett.* 72: 1730–1732.)

UV/visible spectroscopy has been used to observe the optical properties of InGaN quantum wells (McCluskey et al., 1998). The heterostructure consisted of a 2 nm InGaN/4 nm GaN superlattice sandwiched between a 0.2 μm GaN:Mg layer and a 4 μm GaN:Si layer, grown on a sapphire substrate. Figure 10.2 shows the absorption onset from the InGaN quantum wells as well as the thicker GaN layers. Fabry–Perot interference fringes arise from multiple reflections in the heterostructure. The small amplitude of the fringes is likely due to interface and surface roughness. After heating the sample at 950°C for 40 hr, a broad absorption peak appears at ~2.65 eV. This peak was attributed to indium-rich InGaN precipitates that formed in the quantum-well region. To verify that InGaN precipitates had actually formed, transmission electron microscopy (Chapter 12) was performed. The microscopy images showed clear evidence of nanoscale InGaN precipitates with a high indium content.

10.2 EMISSION

As discussed in Section 7.5, the recombination of electrons and holes can lead to the emission of photons with characteristic wavelengths. The emission spectra provide valuable information about the band structure and defect levels in a semiconductor. In a photoluminescence (PL) experiment, a monochromatic light source is focused onto a sample (Figure 10.3). To generate electron–hole pairs, the photon energy is greater than the band gap. In many cases, defect emission is also caused by sub-band-gap light. The excitation light can be produced using a lamp and monochromator, or a continuous-wave (cw) laser. The light emitted by the sample is collected and focused into a monochromator, which disperses the light spectrally. The monochromator plus detector is sometimes referred to as a spectrometer.

The emission spectrum can be obtained with a scanning monochromator and a PMT. Alternatively, *multichannel detection* can be employed. One popular configuration uses a monochromator with a CCD array at the exit (Figure 10.3). Photons of different wavelengths strike different channels of the array, causing charge to build up. After a period of time, the charges are read out, resulting in a plot of photons per channel. By calibrating the channels, using a known light source, one obtains a plot of photons versus wavelength.

Chapter 10—Optical Spectroscopy   251

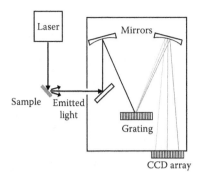

FIGURE 10.3 PL experiment. Light emitted by the sample is dispersed by a grating monochromator and detected by a charge-coupled device (CCD) array.

If the light intensity is low, then an image intensifier can be placed in front of the CCD array. One such intensifier uses a *microchannel plate*, which is an array of glass tubes that operate similarly to PMTs. Each tube is a few microns in diameter and is coated with a metal. In an image intensifier, a photon strikes a photocathode, and an electron is accelerated down a tube by an applied bias. This electron knocks out secondary electrons from the metal coating. These electrons strike a phosphor at the end of the tube, generating photons. In this way, one photon is converted into many photons, resulting in increased sensitivity. These photons are then detected by the CCD array.

Exciton binding energies are typically low such that cryogenic temperatures are required to obtain useful exciton emission spectra. Examples of low-temperature PL spectra of GaN, taken with a 325 nm He-Cd laser and 1 m monochromator, are shown in Figure 7.13 (Shan et al., 1995). As the temperature is raised, the donor-bound exciton lines decrease relative to the free exciton lines. Figure 7.16 shows a PL spectrum of GaP, where the rich fine structure results from donor–acceptor pair transitions for specific donor–acceptor distances. In that study, the exciting source was a high-pressure mercury lamp with filters that passed light wavelengths shorter than 480 nm (Thomas et al., 1964). The PL spectrum of isoelectronic pairs in GaP is shown in Figure 7.17. The PL emission from iron in InP is shown by the dashed line in Figure 7.20.

Photoluminescence excitation (PLE) spectroscopy examines the intensity of an emission peak as a function of *excitation* wavelength. PLE gives us the absorption profile for a particular emission process. The combination of PL and PLE was used to establish that nitrogen is a deep acceptor in ZnO (Tarun et al., 2011). The experiments were performed on *n*-type samples doped with nitrogen acceptors (N_O^-). Monochromatic light excites the N_O^- electron into the conduction band. When the electron is recaptured by a neutral nitrogen acceptor, a photon is emitted. The calculated configuration-coordinate diagram (Lyons et al., 2009) estimated absorption and emission energies of 2.4 and 1.7 eV, respectively (Figure 10.4). The energy difference of 0.7 eV is due to large lattice relaxation of the deep acceptor.

Figure 10.4 shows a room-temperature PL spectrum at an excitation wavelength of 490 nm (2.53 eV). A broad "red" emission band was detected at ~730 nm (1.70 eV), in agreement with the deep-acceptor model. The PLE spectrum was obtained by monitoring the red emission band as a function of excitation wavelength. An excitation onset of ~2.2 eV was observed, again consistent with the configuration coordinate diagram. The agreement between experiment and theory showed that, contrary to numerous previous reports, nitrogen is a deep acceptor and therefore cannot produce *p*-type ZnO.

Besides PL, other emission spectroscopies utilize different methods for generating electron–hole pairs (Pankove, 1971). *Electroluminescence* (EL), relevant to laser diodes and LEDs, uses electrical injection of electrons and holes. EL can result from Schottky junctions (albeit inefficiently), *p-n* junctions, and heterostructures (Schubert, 2006). Generally, the electron-hole densities

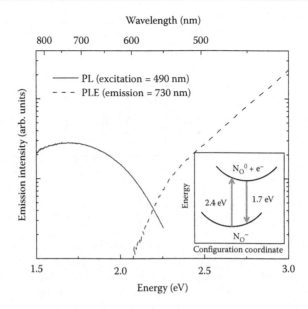

FIGURE 10.4 Photoluminescence (PL) and photoluminescence excitation (PLE) spectra from nitrogen acceptors in ZnO. The dashed line is the PL intensity (at 730 nm) as a function of excitation wavelength. (After Tarun, M.C., M. Zafar Iqbal, and M.D. McCluskey, *AIP Advances* 1, 022105, 2011.) Inset: Configuration-coordinate diagram calculated by Lyons et al. (2009).

produced by EL are more difficult to control than in a PL experiment. High electron-hole densities result in quasi-Fermi levels in the bands (Figure 7.11) and light emission primarily originates from band-to-band recombination. Therefore, EL is not as sensitive to defect levels as PL.

In *cathodoluminescence* (CL), the electron–hole pairs are generated by a "primary" electron beam with electron energies up to ∼200 keV. Each primary electron produces hundreds or thousands of electron–hole pairs. CL is especially useful for wide-band-gap semiconductors, where short-wavelength laser sources and optics are expensive or inconvenient. It also offers good spatial resolution. CL microscopy experiments typically are performed in a scanning electron microscope, discussed in Chapter 12.

Thermoluminescence can be thought of as delayed PL. First, a sample at low temperature is exposed to light, producing electrons and holes that fill traps. As the temperature is increased, specific traps release their carriers at specific temperatures. In a *p*-type semiconductor, for example, an electron could transition from a deep donor to a shallow acceptor, resulting in donor–acceptor pair emission. For a constant warming rate dT/dt, the PL intensity versus time gives peaks that correspond to certain traps, similar to the TEES and TSC techniques (Section 9.9).

10.3 RAMAN SPECTROSCOPY

Raman scattering (Section 6.5) provides a means for investigating the vibrational properties of defects and the semiconductor host. In a Stokes process, an incoming photon gives some of its energy to a vibrational excitation. The energy of that excitation is inferred from the difference between the incident and scattered photon energy. Experimentally, a Raman experiment is similar to a PL experiment (Figure 10.3). The main difference is that because vibrational energies are relatively low, the Raman peaks lie quite close to the excitation wavelength. An experimental

Chapter 10—Optical Spectroscopy **253**

challenge is the low intensity of the Raman signal, which can be four to six orders of magnitude weaker than the elastically scattered light (Yu and Cardona, 1996).

To allow for the detection of weak signals, it is essential to reject stray light, defined as light that does not have the selected wavelength but still travels to the detector. A *notch filter* is used to block the laser wavelength but transmit wavelengths that are a few nm away. Commercially available holographic notch filters have a strong dip in transmission at the laser wavelength (e.g., 532 nm) and a spectral width of $\sim$10 nm. Double or triple monochromators are used to reject stray light emitted or scattered by the sample. As discussed in Section 10.2, sensitive detectors and image intensifiers are used to enable single-photon counting.

By using polarizers, one can determine the symmetry of a Raman-active vibrational mode. From Equation 6.55, the Stokes process produces an oscillating dipole:

$$\mathbf{p} \sim \alpha_1 \cdot \hat{\mathbf{e}}_L \cos[(\omega_L - \omega_0)t] \tag{10.7}$$

where α_1 is a tensor that has the symmetry of the vibrational mode, $\hat{\mathbf{e}}_L$ is the polarization of the laser light, ω_L is the laser frequency, and ω_0 is the vibrational frequency. A polarizer $\hat{\mathbf{e}}_S$ is placed in front of the detector. If $\hat{\mathbf{e}}_S$ is orthogonal to $\mathbf{p}$, then the detected intensity I_S will be zero. If $\hat{\mathbf{e}}_S$ is parallel to $\mathbf{p}$, then I_S will be a maximum. In general, the time-averaged intensity is given by

$$I_S \sim \left| \hat{\mathbf{e}}_S \cdot \mathbf{R} \cdot \hat{\mathbf{e}}_L \right|^2 \tag{10.8}$$

where $\mathbf{R}$ is proportional to α_1 and is referred to as the *Raman tensor*. By measuring the intensity of a Raman line for various polarizations, it is possible to determine the symmetry of the Raman tensor.

An example of Raman scattering from a localized defect is the nitrogen–hydrogen complex in zinc selenide (Wolk et al., 1993). In this experiment, Raman spectra were obtained in a pseudo-backscattering geometry, where an argon-ion laser beam (514.5 nm) was incident nearly normal to the (100) surface. The backscattered light was dispersed by a single monochromator and detected by a PMT. The incident and backscattered light propagated approximately along the [100] and [$\bar{1}$00] directions, respectively.

A comprehensive list of Raman tensors can be found in Cardona (1982). In the present example, the point-group symmetry was C_{3v} and the stretch mode (A_1) was measured. The Raman tensor for this mode is given by

$$\mathbf{R} = \begin{bmatrix} a & & \\ & a & \\ & & b \end{bmatrix} \tag{10.9}$$

where a and b are empirical constants. The basis vectors of the Raman tensor are given by the symmetry of the defect. Let us consider a N-H complex aligned along the $z = [111]$ direction. The unit basis vectors are (Section 2.6)

$$\hat{\mathbf{x}}' = \frac{1}{\sqrt{2}} \begin{bmatrix} 1 \\ -1 \\ 0 \end{bmatrix} \quad \hat{\mathbf{y}}' = \frac{1}{\sqrt{6}} \begin{bmatrix} 1 \\ 1 \\ -2 \end{bmatrix} \quad \hat{\mathbf{z}}' = \frac{1}{\sqrt{3}} \begin{bmatrix} 1 \\ 1 \\ 1 \end{bmatrix} \tag{10.10}$$

Consider the conditions where the laser light and scattered light are both polarized along [001]. This polarization must be expressed in terms of the basis vectors (Equation 10.10):

$$(10.11) \qquad \hat{e}_S = \hat{e}_L = \begin{bmatrix} [001] \cdot \hat{x}' \\ [001] \cdot \hat{y}' \\ [001] \cdot \hat{z}' \end{bmatrix} = \begin{bmatrix} 0 \\ -2/\sqrt{6} \\ 1/\sqrt{3} \end{bmatrix}$$

Now, we insert Equations 10.11 and 10.9 into Equation 10.8 to obtain the Raman scattering efficiency:

$$(10.12) \qquad I_S \sim \begin{bmatrix} 0 & -2/\sqrt{6} & 1/\sqrt{3} \end{bmatrix} \begin{bmatrix} a & & \\ & a & \\ & & b \end{bmatrix} \begin{bmatrix} 0 \\ -2/\sqrt{6} \\ 1/\sqrt{3} \end{bmatrix} = \frac{2}{3}a + \frac{1}{3}b$$

Next, consider the case where the laser light is polarized along [011] and the scattered light is polarized along $[01\bar{1}]$. Similar to Equation 10.11, the polarization unit vectors are given by

$$(10.13) \qquad \hat{e}_S = \begin{bmatrix} -1/2 \\ 3/\sqrt{12} \\ 0 \end{bmatrix} \qquad \hat{e}_L = \begin{bmatrix} -1/2 \\ -1/\sqrt{12} \\ 2/\sqrt{6} \end{bmatrix}$$

Inserting Equations 10.13 and 10.9 into Equation 10.8 yields

$$(10.14) \qquad I_S \sim \begin{bmatrix} -1/2 & 3/\sqrt{12} & 0 \end{bmatrix} \begin{bmatrix} a & & \\ & a & \\ & & b \end{bmatrix} \begin{bmatrix} -1/2 \\ -1/\sqrt{12} \\ 2/\sqrt{6} \end{bmatrix} = 0$$

For this cross-polarized geometry, the Raman intensity is zero. Figure 10.5 shows the Raman intensities for these and other polarizations. The experimental results established that the symmetry of the N-H complex is C_{3v}.

The derivation of Raman scattering given in Section 6.5 was based on classical electrodynamics. The intensity of Raman scattering can be enhanced when the incident light excites a quantum-mechanical transition to a higher electronic state. This effect is called *resonant Raman* scattering. Resonant Raman occurs, for example, when above-band-gap photons excite an electron and hole, which lose energy via the emission of one or more phonons. When the electron and hole recombine, the emitted Stokes photon has an energy that is lower by $n\hbar\omega$, where n is the number of phonons emitted and ω is the optical phonon frequency.

An example of resonant Raman scattering is shown in Figure 7.4. In that case, the resonant Raman signal for LO phonons in GaN:O was suppressed by a high free-electron concentration. A resonant Raman spectrum for ZnO:Cu is shown in Figure 10.6, for a laser wavelength of 325 nm (3.8 eV). The deep acceptors formed by Cu provided recombination pathways for the electrons and holes, leading to a green PL emission at the expense of band-edge emission. The suppression of band-edge emission improved the sensitivity of the Raman spectroscopy and allowed the observation of LO phonons up to $n = 9$.

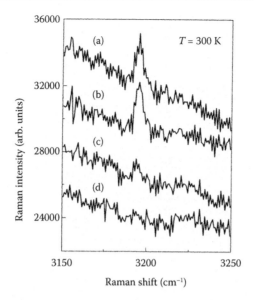

FIGURE 10.5 Raman scattering spectra for N-H complexes in ZnSe. (a) Laser and scattered light polarized along [001]. (b) Laser and scattered light polarized along [011]. (c) Laser light [001], scattered light [010]. (d) Laser light [011], scattered light [01$\bar{1}$]. (Reprinted with permission from Wolk, J.A. et al. 1993. Local vibrational mode spectroscopy of nitrogen–hydrogen complex in ZnSe. *Appl. Phys. Lett.* 63: 2756–2758. Copyright 1993, American Institute of Physics.)

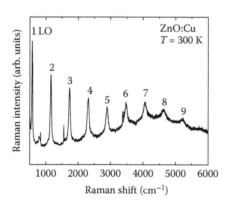

FIGURE 10.6 Resonant Raman spectrum of ZnO:Cu, showing up to 9 LO phonons. (Data courtesy of L. Bergman and J. Huso.)

10.4 FOURIER TRANSFORM INFRARED SPECTROSCOPY

Fourier transform infrared (FTIR) spectroscopy has three special advantages: high spectral resolution, good signal-to-noise ratios, and the ability to measure a broad region of the spectrum quickly (Bell, 1972; Griffiths and de Haseth, 1986). Central to an FTIR spectrometer is a Michelson interferometer (Figure 10.7). A parallel beam of collimated light from a broadband source is directed at a semitransparent beamsplitter. One of the two beams reflects off a movable mirror while the other beam reflects off a fixed mirror. The two beams recombine at the beamsplitter, travel through the sample, and finally impinge upon a detector. The detector signal

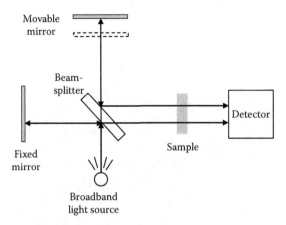

FIGURE 10.7 Fourier transform infrared (FTIR) spectrometer.

is proportional to the intensity of the interfered beam and the plot of intensity versus optical path difference is the *interferogram*. In practice, to maximize the signal-to-noise ratio, several hundred to several thousand interferograms are taken and averaged.

To see how the interferogram is related to the IR spectrum, first consider a monochromatic light source. The two beams have an optical path difference δ. When they recombine, the electric field is the sum of the two,

$$E = E_0 e^{i(kz-\omega t)} + E_0 e^{i(kz-\omega t + k\delta)} \tag{10.15}$$

and the time-averaged intensity is given by

$$I \sim \frac{1}{2}|E|^2 = E_0^2 [1 + \cos(k\delta)] \tag{10.16}$$

Now, consider a broadband source of IR light. The intensity received by the detector is the integral over all the wavelengths,

$$I \sim \int_0^\infty E_0(k)^2 \, dk + \int_0^\infty E_0(k)^2 \cos(k\delta) \, dk \tag{10.17}$$

The second term, which varies with δ, is referred to as the interferogram. We write it as

$$I(\delta) = \int_0^\infty I_0(k) \cos(k\delta) \, dk \tag{10.18}$$

Equation 10.18 is simply the Fourier transform of the spectrum $I_0(k)$. To obtain the spectrum, we perform an inverse Fourier transform:

$$I_0(k) \sim \int_0^\infty I(\delta) \cos(k\delta) \, d\delta \tag{10.19}$$

Chapter 10—Optical Spectroscopy   **257**

In practice, a computer uses a fast Fourier transform (FFT) algorithm (Cooley and Tukey, 1965) to evaluate the integral (10.19).

The finite maximum optical path difference, L, introduces instrumental broadening. Again, consider a monochromatic source with $k_0 = 2\pi/\lambda_0$. The interferogram from this source is given by Equation 10.16:

$$(10.20) \qquad I(\delta) = I_0 \cos(k_0\delta)$$

The inverse Fourier transform of this interferogram is

$$(10.21) \qquad I_0(k) \sim \int_0^L I_0 \cos(k_0\delta)\cos(k\delta)d\delta \sim \frac{\sin\left[(k-k_0)L\right]}{(k-k_0)L} + \frac{\sin\left[(k+k_0)L\right]}{(k+k_0)L}$$

For a typical mid-IR spectrum, $L \sim 1$ cm and $k_0 \sim 1000$ rad/cm. Therefore, $k_0 L \gg 1$ and the second term can be neglected. The computed spectrum is then given by

$$(10.22) \qquad I_0(k) \sim \mathrm{sinc}\left[(k-k_0)L\right]$$

where sinc $z = \sin(z)/z$. The sinc function has a maximum at k_0 with sidelobes. The full width of the central peak is $\Delta k = 2\pi/L$. Because $k = 2\pi/\lambda$, the resolution in wavenumbers (cm^{-1}) is

$$(10.23) \qquad \Delta(1/\lambda) \sim 1/L$$

Thus, the resolution improves as the scanning length increases. One can multiply the interferogram by an *apodization* function to suppress the sidelobes. The most common apodization is a linear function that goes to zero at $\delta = L$.

There are two major advantages to FTIR spectroscopy over scanning grating spectrometers, especially in the far-IR. The *Jacquinot advantage* (Girard and Jacquinot, 1967) is gained because the optical throughput of an interferometer is, to first order, independent of the resolution. The resolution is simply proportional to the number of times a certain frequency is "probed" by the moving mirror. In contrast, a grating spectrometer has a slit which has to be closed down as the required resolution is increased. The *Fellgett advantage* (Fellgett, 1958) is related to the signal-to-noise ratio at the detector. For IR measurements, the detector is not generally in the photon-counting regime. It is therefore optimal to take the entire spectrum "at once" rather than scan through the wavelengths.

FTIR spectroscopy has had a major impact in identifying impurities and their vibrational properties. Examples of FTIR vibrational absorption spectra are shown for ZnO:H (Figure 6.9 and 6.22), AlSb:Se,H (Figure 6.13), Si:O (Figure 6.15), Ge:O (Figure 6.17), GaAs:C (Figure 6.18), and GaAs:Si,H (Figure 6.20). Electronic absorption spectra shown for the phosphorus donor in silicon (Figure 5.7) and acceptors in ZnTe (Figure 5.8) were also obtained with FTIR spectroscopy. Along with absorption, FTIR spectrometers can also be used to observe PL emission. The PL spectrum in Figure 7.14 was obtained with a high-resolution FTIR spectrometer, with no apodization. The sidelobes around the extremely sharp P bound exciton peak are due to the finite mirror travel (Equation 10.22) (Karaiskaj et al., 2001).

The combination of high-resolution FTIR spectroscopy with isotopic enrichment has led to extraordinarily precise measurements (Haller, 1995; Cardona and Thewalt, 2005). An example of FTIR PL from copper-related defects in isotopically pure silicon-28 is shown in Figure 10.8 (Thewalt et al., 2007). The PL emission arises from the recombination of an exciton bound to the copper complex. Silicon samples doped with copper-63 or copper-65 show sharp lines at 8183.1

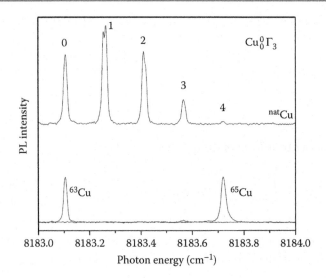

FIGURE 10.8 Photoluminescence (PL) spectra from silicon-28 doped with natural copper (^{nat}Cu) and isotopically pure copper (^{63}Cu and ^{65}Cu). (Reprinted from *Physica B*, 401–2, Thewalt, M.L.W. et al. Can highly enriched ^{28}Si reveal new things about old defects?, 587–592. Copyright 2007, with permission from Elsevier.)

or 8183.7 cm^{-1}, respectively. Silicon doped with natural copper, which contains a mixture of copper-63 and copper-65, shows *five* lines. These lines correspond to a complex with four copper atoms that contains zero, one, two, three, or four copper-65 atoms. The change in isotope mass results in small changes in photon energy that are resolved by these experiments. The lines cannot be resolved in natural silicon, due to isotopic broadening (Steger et al., 2011).

10.5 PHOTOCONDUCTIVITY

As discussed in Section 9.8, photoconductivity measurements can be used to determine minority carrier lifetime. For intrinsic photoconductors, photoconductivity results from the excitation of electrons across the band gap. The increased number of electrons and holes increases the conductivity of the sample. When an external bias is applied, a photocurrent is generated. The excitation of carriers from impurity levels to the conduction or valence band will also result in photoconductivity. The phenomenon of photoconductivity has led to the development of highly sensitive photon detectors, transducers that convert a photon signal into an electrical signal.

Semiconductors doped with specific impurities can be optimized for photon energy ranges over a wide spectral region. Spectacular scientific results have been published with far-IR photoconductors that were mounted and cooled to 2 K inside the Infrared Astronomical Satellite (Neugebauer et al., 1984). Various stages of star formation, intergalactic dust, and over 20,000 new galaxies were discovered. The Spitzer telescope, launched in 2003, has four optical instruments observing the universe in the 1 to 160 μm wavelength range. The Multiband Imaging Photometer (MIPS) provides long wavelength capability for the mission, in imaging bands at 24, 70, and 160 μm and measurements of spectral energy distributions between 52 and 100 μm at a spectral resolution of ∼7% (Rieke et al., 2003). The two longer-wavelength arrays use Ge:Ga detectors, resulting in good photometry with rms relative errors of less than 10%. The improvements of the IR images of the Messier 81 (M81) galaxy over the visible light image demonstrate the power of far-IR astronomy (see www.spitzer.caltech.edu).

MIPS has one 30 × 30 array of Ge:Ga photoconductors and two 25-element uniaxially stressed Ge:Ga photoconductor arrays. Ga forms shallow acceptors in Ge with a hole binding energy of 11 meV, corresponding to a wavelength of ~120 μm. When uniaxially stressed, the binding energy of the Ga acceptors drops to 5.5 meV (170 μm). The Ge:Ga photoconductors for MIPS were fabricated in our laboratories (EEH). A critical component was the Ohmic low-noise contacts made with boron ion implantation followed by thermal annealing and metallization, with a thin adhesion layer of Pd and a thick layer of Au.

Shallow dopants produce characteristic photoconductivity spectra that serve as unique "fingerprints" for impurity identification. As discussed in Chapter 5, shallow donors and acceptors are well described by the hydrogenic model. As hydrogen atom analogs, the electrons or holes have a ground state and bound excited states. Photons can induce transitions between the s-like ground state and p-like bound excited states. For silicon, germanium, and GaAs, the ground-state binding energy for electrons or holes lies in the range from a few meV to a few tens of meV. Far-infrared FTIR spectroscopy is ideally suited for the study and characterization of shallow level impurities.

In order to induce IR transitions in donors or acceptors, the dopants must have their carrier occupy the ground state. This means the semiconductor has to be deep in the freeze-out regime, which for shallow dopants is near liquid-helium temperatures. Compensating donors or acceptors remain ionized at all temperatures and are not accessible with standard far-IR spectroscopy. This limitation can be overcome by shining low-intensity band edge light on the sample. Many of the generated free electrons and holes get trapped by acceptors and donors, rendering them neutral.

When Ohmic contacts are attached to the sample, one can measure the conductivity as a function of IR wavelength. In this way, the optical generation of free carriers from neutral donors or acceptors can be probed directly. In a cooled semiconductor that contains very few carriers in the absence of a photon flux, we can create free-carrier changes of many orders of magnitude. For low shallow-level concentrations, the sensitivity of photoconductivity measurements is typically much better than absorption.

Figure 10.9 illustrates how photoconductivity experiments can probe shallow levels. At low sample temperatures (4.2 K), the photoconductive continuum is observed, due to the excitation of a bound carrier from the impurity into the conduction or valence band. As the temperature is raised, we begin to detect lines below the onset of the photoconductive continuum (Figure 10.10). These peaks are due to a *two-step process*: (1) excitation from the ground state of an impurity into one of the bound excited states, followed by (2) the absorption of a phonon that transports the bound carrier into the band, thereby creating a free charge carrier. These steps form the basis for *photothermal ionization spectroscopy* (PTIS).

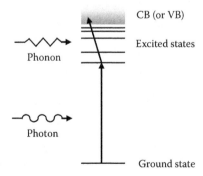

FIGURE 10.9 Two-step photothermal ionization spectroscopy (PTIS) process. First, an infrared (IR) photon promotes the electron or hole to an excited state. Second, a phonon excites the carrier into the conduction or valence band.

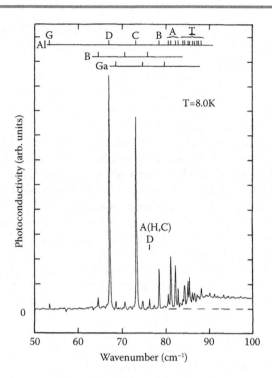

FIGURE 10.10 Photothermal ionization spectroscopy (PTIS) spectrum of residual acceptors in ultrapure germanium. Excited-state lines are indicated for Al, B, Ga, and A(H,C). Excitation into the valence-band continuum results in absorption above the dashed line. (Reprinted from Haller, E.E., W.L. Hansen, and F.S. Goulding. 1981. *Adv. Phys.* 30: 93–138. With permission from Taylor & Francis.)

The PTIS technique has been used extensively to characterize high-purity semiconductors and has led to the discovery of numerous novel acceptor and donor levels (Haller et al., 1981). Spectra for residual acceptors in ultrapure germanium are shown in Figure 10.10. The lines are sufficiently sharp to allow a clear separation between aluminum, boron, and gallium excited-state transitions. The major lines belong to the aluminum acceptors, which are present at a concentration of $<10^{10}$ cm^{-3}. An additional line denoted A(H,C) originates from a shallow acceptor carbon–hydrogen complex. The activation by hydrogen of neutral dopants, such as substitutional silicon or carbon in germanium, was one of the interesting discoveries made possible by the development of ultrapure germanium.

10.6 TIME-RESOLVED TECHNIQUES

The availability of lasers with short pulse durations and high intensities has enabled a host of time-resolved experiments that probe the femtosecond-to-nanosecond time regimes (Nurmikko, 1992). One widely used source is the Nd-doped yttrium aluminum garnet (Nd:YAG) laser, with an emission wavelength of 1064 nm, pulse durations below 100 ps, and repetition rates up to 100 MHz. Second harmonic generation (frequency doubling) produces a wavelength of 532 nm. Ultrafast Ti:sapphire amplifiers are used to emit 800 nm laser pulses with durations below 100 fs.

The intensity of short laser pulses can be detected with photodiodes and oscilloscopes. For time-resolved PL emission, fast PMTs are preferred because they are sensitive enough for single photon counting. In *time-correlated single photon counting* experiments (O'Connor and Phillips,

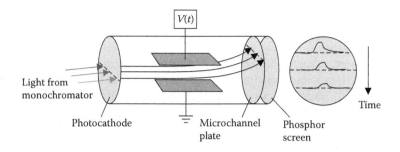

FIGURE 10.11 Streak camera. Electrons from the photocathode are bent by a sweep voltage $V(t)$. The result, illustrated on the phosphor screen, is a set of time-resolved spectra.

1984), a laser or LED pulse excites the semiconductor, and the emission is detected by a monochromator and PMT. Fast electronics compare the time elapsed between the laser pulse and the detected PMT signal. As the pulses are repeated, a histogram of photons versus time is built up. In the case of single exponential decay, a plot of log(photons) versus time is a straight line.

The ultimate time-resolved PL experiment is performed with a *streak camera*, a device that records the spectrum (intensity vs. wavelength) as a function of time. To accomplish this, a streak tube is placed at the monochromator exit. The spectrally dispersed photons strike a photocathode, producing electrons (Figure 10.11). These electrons are multiplied by a microchannel plate (Section 10.2) and impinge on a phosphor screen. The time resolution is accomplished by applying a sweep voltage in the vertical direction. This voltage deflects the electrons vertically as a function of time. The final image on the phosphor screen has wavelength on the horizontal axis and time on the vertical axis. A sensitive CCD camera takes a picture of the image.

Time-resolved PL of GaN was performed by Shan et al. (1995). The excitation source was a pulsed dye laser pumped by a Nd:YAG laser, which produced pulse durations of 5 ps at a wavelength of 600 nm. A nonlinear crystal frequency "doubled" the pulse to 300 nm. The PL emission spectra were dispersed spectrally by a monochromator and temporally by a streak camera with 2 ps resolution. The intensities of the free and donor-bound excitons peaks (Figure 7.13) were measured as a function of time. The results indicated effective lifetimes for free and bound excitons of ~35 and 55 ps, respectively. Nonradiative recombination at defect sites was believed to play an important role in the exciton decay. A review of time-resolved PL processes in III-V semiconductors is given by Ahrenkiel (1993).

Another time-resolved optical technique is the *pump-probe* experiment. The output of an ultrafast laser is split into a strong "pump" beam and a weak "probe" beam (Figure 10.12). The probe pulse is delayed by sending it on a longer path than the pump pulse. The two beams spatially

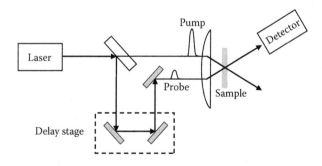

FIGURE 10.12 Pump-probe experiment. The path length difference between the pump and probe beams results in a delay. The intensity of the probe pulse is measured as a function of delay time.

overlap on the sample. The intensity of the transmitted probe pulse is recorded by a detector as a function of the delay time. An optical chopper and a lock-in amplifier may be used to improve the signal-to-noise ratio.

IR pump-probe experiments have proven fruitful for measuring the lifetimes of local vibrational modes (LVMs) in semiconductors (Lüpke et al., 2003). The IR laser pulses for these studies are provided either by an optical parametric oscillator (Tang and Cheng, 1995) or free-electron laser (Colson et al., 2002). The wavelength is tuned to match a particular LVM. The pump pulse excites a fraction of the defects into their vibrationally excited state. Due to anharmonicity (Section 6.2), defects in the excited state cannot absorb a second pulse of the same wavelength. Therefore, the effect of the pump is to increase the probe signal. Over time, the defects relax to their ground state and the probe signal returns to normal. The time dependence for this "transient bleaching" provides the ground-state recovery lifetime.

The transient bleaching signal was measured for the 2072 cm^{-1} mode of a divacancy–hydrogen complex in proton-implanted silicon (Lüpke et al., 2002). The complex consists of two neighboring silicon vacancies and two hydrogen atoms that passivate dangling bonds. The long ground-state recovery lifetime of 295 ps is related to the weak coupling of the hydrogen atoms to the lattice. Bond-centered hydrogen, in contrast, is crowded by the neighboring silicon atoms and has a lifetime of only 8 ps (Budde et al., 2000).

10.7 APPLIED STRESS

The application of mechanical stress to a semiconductor provides information about defect symmetry and interactions with the host lattice. This section summarizes methods to produce three types of deformation: uniaxial stress, hydrostatic pressure, and uniaxial strain.

Uniaxial stress is generated when a sample is compressed along the longitudinal direction and allowed to relax along the transverse directions (Ramdas and Rodriguez, 1992). In a typical uniaxial-stress rig, a gas-driven piston exerts a force on a hollow steel push rod. The push rod provides the force that squeezes the sample. In practice, paper is sometimes inserted between the metal and sample, to improve the homogeneity of the stress. The yield strength of semiconductors is low such that the maximum uniaxial stress is limited to 1 to 2 GPa.

Under the application of stress, acceptor–hydrogen complexes may orient along a preferred direction. A well-characterized example is the boron–hydrogen complex in silicon (Stavola et al., 1988). Under ambient conditions, the complexes are oriented randomly along the <111> directions. When stress is applied along the [110] direction, the two <111> orientations at the bottom of the ball-and-stick diagram (Figure 10.13) are inequivalent to the two <111> orientations at the top. At temperatures above 60 K, the hydrogen atom is mobile enough to jump from one <111> bond-centered site to another. The preferential orientation can be quantified by measuring the *dichroism*

$$(10.24) \qquad D = \frac{\alpha_\perp - \alpha_{//}}{\alpha_\perp + \alpha_{//}}$$

where $\alpha_\perp$ and $\alpha_{//}$ are the absorption coefficients of the IR bond-stretching peaks for light polarized perpendicular and parallel to the applied stress. When the stress is released, D decays as the hydrogen atoms become randomly oriented again. The temperature dependence of this decay for boron–hydrogen complexes indicates that the reorientation kinetics are non-Arrhenius (Cheng and Stavola, 1994).

Uniaxial stress can also be used to probe the electronic structure of impurities. The photoconductivity spectrum of Ge:Cu under [100] stress, at a temperature of 12 K, is shown in Figure 10.14 (Dubon et al., 1994). The Cu absorption onset shows a shift to lower energy up to a stress of 4 kbar

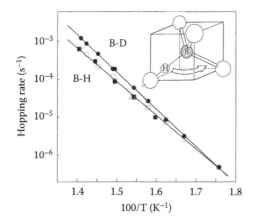

FIGURE 10.13 Reorientation kinetics for hydrogen in Si:B-H. When stress is released, the dichroism (Equation 10.24) decays as the mobile hydrogen atoms become randomly oriented. This decay rate is plotted as a function of inverse temperature. (Reprinted with permission from Cheng, Y.M. and M. Stavola. 1994. Non-Arrhenius reorientation kinetics for the B-H complex in Si: Evidence for thermally assisted tunneling. *Phys. Rev. Lett.* 73: 3419–3422. Copyright 1994 by the American Physical Society.)

(0.4 GPa). For larger stresses, the absorption onset remains constant. The explanation for this novel behavior is shown in the energy-level diagram in Figure 10.14. In the ground state, all three holes in the neutral copper triple acceptor are in $1s$-like hydrogenic states. Under [100] stress, the valence band splits into two doubly degenerate bands. Because the holes bound to Cu are in hydrogenic states, the levels follow the shifts in the respective valence bands. For stresses above 0.4 GPa, it is energetically favorable for the third hole to occupy a $2s$ state, and the Cu acceptor transforms into a Li-like $1s^2 2s^1$ atomic configuration.

The preferred apparatus for generating large *hydrostatic* pressures is the diamond-anvil cell (DAC) (Jayaraman, 1983). In a DAC, the flat parallel faces of two diamonds press on a metal gasket. The sample is placed in a hole in the gasket, along with a pressure transmitting fluid such as liquid nitrogen or 4:1 methanol-ethanol mixture. The transparency and strength of diamonds

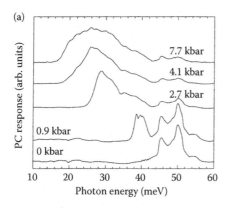 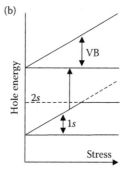

FIGURE 10.14 (a) Photoconductivity (PC) spectrum of copper in germanium at a temperature of 12 K, under [100] uniaxial stress. (b) The splitting of the valence band (VB) and copper acceptor levels. The transition corresponding to the PC onset, which shifts to lower energy under stress, is indicated by the vertical arrow. (After Dubon, O.D. et al. 1994. *Phys. Rev. Lett.* 72: 2231–2234.)

allow researchers to perform optical studies over a wide spectral range at pressures exceeding 100 GPa. Examples of high-pressure optical spectroscopy include studies of *DX* centers in GaAs (Section 7.8) and GaN (Section 7.2).

Hydrostatic pressure was used to investigate the interaction between localized and extended vibrational modes in solids. Interstitial oxygen (O_i) in silicon (Section 6.7) is a model system for studying such interactions. Using hydrostatic pressure, Hsu et al. (2003) brought the asymmetric stretch mode of $^{18}O_i$ into resonance with the second harmonic of the $^{18}O_i$ resonant mode. IR spectroscopy was used to observe an anticrossing between these two vibrational modes at pressures near 4 GPa (Figure 10.15). This effect is an example of a localon, a novel vibrational excitation that has localized and extended components (Section 6.6). For pressures above 4 GPa, a dramatic increase in the linewidth was observed as the LVM entered the two-phonon continuum.

Although hydrostatic pressure can generate large pressures, it does not alter the symmetry of the crystal and therefore is limited in probing the structure of defects. Symmetry breaking at high stresses can be achieved through *shock compression* (Peng et al., 2005). In these single-event experiments, a projectile accelerated by a gas gun impacts the target, producing a shock wave. Behind the shock wave, the target material is in a state of high stress, the magnitude of which depends on the mechanical properties of the impactor and the target crystal, and the velocity of the projectile. The projectile velocity is measured to an accuracy of 0.1% by a velocity gauge in the gun barrel, and the stress value is calculated to an accuracy of a few percent from the well-established shock velocities of the impactor materials (e.g., sapphire, quartz, or poly(methyl methacrylate) [PMMA]).

Shock compression produces uniaxial strain, where the atoms are compressed along the direction of the shock-wave propagation. The atoms in the interior of the sample are unable to move due to inertial confinement. Only the atoms on the edge are able to relax. An edge wave

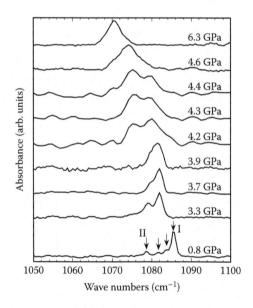

FIGURE 10.15 Infrared (IR) spectra of Si:^{18}O as a function of hydrostatic pressure at ~10 K. IR transitions are indicated by the arrows. At 4.2 GPa, an avoided crossing is observed. At higher pressures, the peak broadens as the frequency enters the two-phonon continuum. (Reprinted with permission from Hsu, L., M.D. McCluskey, and J.L. Lindstrom. 2003. Resonant interaction between localized and extended vibrational modes in Si:^{18}O under pressure. *Phys. Rev. Lett.* 90: 095505. Copyright 2003 by the American Physical Society.)

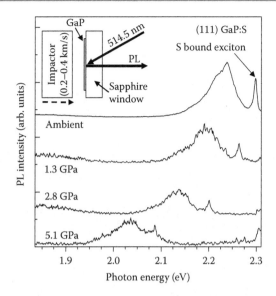

FIGURE 10.16 Photoluminescence (PL) spectra of GaP:S at ~80 K, under ambient conditions and shock compressed to several longitudinal stresses. The shift of the bound exciton peak provides a measure of the band-gap shift. (After Grivickas, P., M.D. McCluskey, and Y.M. Gupta. 2008. *Appl. Phys. Lett.* 92: 142104.)

propagates toward the center of the sample at approximately the speed of sound. All measurements are obtained in the time window between the arrival of the shock wave and the encroachment of the edge waves into the probed area. Hence, the measurements probe the sample while it is under *uniaxial strain*.

Grivickas et al. (2008) investigated the effect of uniaxial strain on donor-bound excitons in GaP:S, using shock compression and time-resolved PL measurements. The excitation source was a dye laser with a wavelength of 514.5 nm and the PL emission was detected by a streak camera system with a time resolution of 20 ns. Under ambient conditions, the PL spectrum consisted of a sharp peak at 2.301 eV due to excitons bound to neutral sulfur donors (Figure 10.16). A broad peak at 2.24 eV is due to the transition of a hole from the valence band to a neutral donor. Under uniaxial compression, both of these peaks shift to lower energy. Because the sulfur-bound exciton follows the band gap closely, the PL shift was nearly equal to the band-gap shift. Shock compression was also used to probe excitons bound to isoelectronic nitrogen (Grivickas et al., 2009).

10.8 ELECTRON PARAMAGNETIC RESONANCE

The previous sections discussed spectroscopy from the UV to IR. The remaining sections in this chapter discuss magnetic resonance techniques, which utilize microwaves and radio waves in combination with a static magnetic field. In general, magnetic resonance provides a useful way to characterize electrically active defects in semiconductors. A comprehensive treatment of the subject is given by Spaeth et al. (1992). Electron paramagnetic resonance (EPR) is a technique that detects spins. Single donors, for example, have an unpaired spin in their neutral charge state, making them EPR active. When they are ionized, however, they do not have an EPR signal.

The basic mechanism for EPR can be seen for a single electron in an external magnetic field that points in the *z* direction (Wertz and Bolton, 1986). An electron spin produces a magnetic moment, the *z* component of which is given by

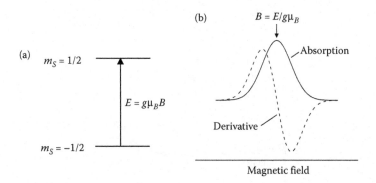

FIGURE 10.17 (a) Electron paramagnetic resonance (EPR) transition for a spin-1/2 electron. (b) Absorption peak and derivative peak obtained in an EPR experiment.

(10.25) $$\mu_z = -g\mu_B m_S$$

where the Landé g-factor is close to 2 (for a free electron, $g = 2.00232$), μ_B is the Bohr magneton (9.274 × 10^{-24} Joule/Tesla or 14 GHz/Tesla), and $m_S = \pm\frac{1}{2}$. Zeeman splitting results from the energy of interaction between the electronic spin and the magnetic field:

(10.26) $$E_e = -\boldsymbol{\mu} \cdot \mathbf{B} = g\mu_B B m_S$$

The magnetic field lifts the spin degeneracy, causing the spin-up state to lie higher in energy than the spin-down state (Figure 10.17). This energy difference is given by

(10.27) $$\Delta E_e = g\mu_B B$$

A photon of energy ΔE_e can promote an electron from a spin-down to a spin-up state. For a B-field of 0.34 Tesla and $g = 2$, Equation 10.27 yields a photon frequency of 9.5 GHz, which is in the microwave region of the electromagnetic spectrum. In practice, the microwave frequency is held fixed while the magnetic field is varied. Typical frequencies lie in the X band (8 to 12 GHz) and W band (75 to 110 GHz).

In a solid, the g-factor for an electron spin differs from the free-electron value. This difference arises from orbital angular momentum induced on the neighboring atoms by the spin (Watkins, 1998). For a localized defect like a vacancy, the surrounding atoms have filled orbital shells, so the magnitude of this orbital angular momentum is small. Such systems have $g \approx 2$. Delocalized carriers, weakly bound to shallow effective mass (EM) donors or acceptors, may have strong orbital angular momentum components, resulting in values that may depart significantly from the free-electron value. EM donors in InP, for example, have $g = 1.20$ (Kennedy and Glaser, 1998). Some transition-metal impurities with unfilled d shells have g values that are greater than 2. Cobalt (Co^{2+}) ions in II-VI semiconductors, for example, have $g \approx 2.3$ (Beaulac et al., 2010).

In a typical EPR apparatus, a microwave bridge contains the microwave source and detector (Figure 10.18). The intensity of the microwave source is controlled with an attenuator. The microwaves then travel into the cavity, which is placed between the poles of an electromagnet. When the cavity is empty, its quality factor Q is high and the reflected microwave power is low. When a sample that absorbs microwaves is placed in the cavity, the loss in Q causes microwaves to be reflected. These reflected microwaves are detected by a Schottky diode. The detected EPR signal is found to be proportional to the sample absorption. Typically, the magnetic field is modulated at low frequencies. A lock-in detector produces an output that is proportional to the corresponding

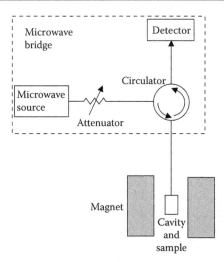

FIGURE 10.18 An electron paramagnetic resonance (EPR) apparatus. The circulator delivers microwaves to the sample and directs the reflected microwaves to the detector.

modulation in the EPR signal. This results in a line shape that is the first derivative of a peak (Figure 10.17).

In addition to the applied magnetic field, the electron spin also interacts with magnetic fields produced by nearby nuclei. This interaction results in a small *hyperfine* splitting:

(10.28) $$\Delta E_{HF} = A m_S m_I$$

where A is the hyperfine coupling constant and m_I is the nuclear spin along the z axis. EPR transitions obey the selection rules

(10.29) $$\Delta m_S = \pm 1/2, \quad \Delta m_I = 0$$

that is, the nuclear spin does not change. The hyperfine interaction therefore results in $2I + 1$ EPR resonances, one for each value of m_I (Bourgoin and Lannoo, 1983). For a nuclear spin of $I = \frac{1}{2}$ ($m_I = \pm\frac{1}{2}$), EPR resonance signals occur at energies

(10.30) $$\Delta E = g \mu_B B \pm A/2$$

The hyperfine interaction provides useful information about the chemical identity of a defect. Nitrogen-14, for example, has a nuclear spin of $I = 1$ and is nearly 100% abundant. In ZnO, the neutral ^{14}N acceptor gives rise to three EPR lines, corresponding to $m_I = -1$, 0, and 1 (Carlos et al., 2001).

The discussion so far has been for isotropic centers, for which g and A are scalar quantities. In general, g and A are tensors, with symmetries that correspond to the point-group symmetry of the defect. Consider the example of the silicon vacancy (Watkins, 1963), which has D_{2d} symmetry and is oriented along a $z = <100>$ direction (Figure 5.6). The positively charged vacancy (V^{1+}) has one unpaired electron in the B_2 level. Suppose that a magnetic field is applied along the [100] direction. V^{1+} centers that are aligned along [100] exhibit an EPR resonance with a Landé factor of $g_{//} = 2.0087$. Centers that are oriented along [010] or [001] directions, on the other hand, are not parallel to the magnetic field and show a different EPR resonance, $g_\perp = 1.9989$. Because the centers are randomly oriented, the intensities of the EPR lines at $g_{//}$ and $g_\perp$ have a ratio of 1:2. If the magnetic field is applied along the [111] direction, then all of the orientations are equivalent, and there is only one EPR resonance.

EPR measurements were essential in determining the structural and electronic properties of vacancies in silicon. Using this technique, Watkins and Troxell (1980) showed that V^{1+} is a metastable state (i.e., the V^0 or V^{2+} state is energetically preferred). As discussed in Section 2.3, such a defect is called a negative-U center. From DLTS measurements, the (0/2+) level lies 0.13 eV above the valence-band maximum. In p-type silicon, V^{2+} is the equilibrium charge state, because it is energetically favorable to give two electrons to shallow acceptors. When the sample is illuminated with sub-band-gap light, an electron is promoted from the valence band to V^{2+}, transforming it to V^{1+}. At liquid-helium temperatures, the V^{1+} state persists and can be measured by EPR. When the sample is warmed up, the V^{1+} EPR signature disappears as the centers transform back into EPR-inactive V^{2+}.

Another interesting application of EPR was in determining the structure of Li-O donors in ultrapure germanium doped with lithium by diffusion (Haller and Falicov, 1979). At low temperatures (2 K), the germanium single crystals acted as resonant cavities with Q-factors greater than 5×10^5. The g-factors obtained from the EPR spectra are plotted in Figure 10.19. By rotating the sample about a <110> axis, the **B**-field direction passes through the major crystallographic directions. The solid lines are fits for Li-O complexes aligned along <111> directions. For a defect complex aligned along a particular <111> axis, the g-value is given by (Spaeth et al., 1992)

(10.31) $$g = \sqrt{g_{//}^2 \cos^2\varphi + g_\perp^2 \sin^2\varphi}$$

where φ is the angle between the complex and the **B** vector. Different g-values arise from complexes aligned along different <111> directions.

In most cases, a defect has a single, isolated electron spin that gives rise to an EPR resonance. However, there are instances where two or more paramagnetic electrons are coupled, resulting in a total spin $S > ½$. Two electrons in a triplet state, for example, have $S = 1$ and $m_S = -1, 0$, and 1. Consider a defect with axial symmetry along the z axis. This symmetry lifts the spin degeneracy: the $m_S = 0$ state has a different energy than the $m_S = \pm 1$ states. In an applied magnetic field, EPR transitions occur from $m_S = -1 \rightarrow 0$ and $0 \rightarrow 1$. Because of the defect symmetry, these two

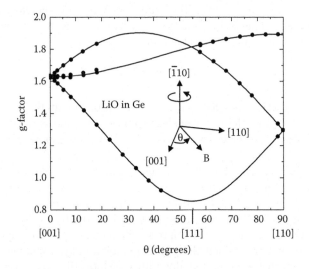

FIGURE 10.19 Measured g-factors of the Li-O donor in germanium. (After Haller, E.E. and L.M. Falicov. 1979. *Proc. 14th Intl. Conf. Phys. Semicond., Inst. Phys. Conf. Ser. No. 43*, ed. B.H.L. Wilson, pp. 1039–1042.)

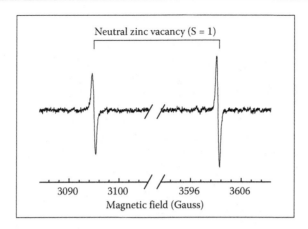

FIGURE 10.20 Electron paramagnetic resonance (EPR) spectrum of neutral zinc vacancies in ZnO. (After Evans, S.M. et al. 2008. *J. Appl. Phys.* 103: 043710. Courtesy of L. Halliburton.)

transitions have slightly different energies. This is the case for the neutral zinc vacancy in ZnO illuminated by UV light (Evans et al., 2008). As shown in Figure 10.20, a magnetic field applied along the *c* axis results in two EPR lines.

10.9 OPTICALLY DETECTED MAGNETIC RESONANCE

The merger between PL and EPR is referred to as optically detected magnetic resonance (ODMR). This technique relies on the fact that many optical transitions are spin-dependent. As in EPR, microwaves are sent into the sample cavity, which sits between the poles of an electromagnet. A laser is directed onto the sample, and the emitted light at a given wavelength is detected. When the magnetic field is tuned to EPR transitions, the intensities of PL transitions increase or decrease. These changes result in peaks or dips in an intensity-versus-*B* plot. The microwave power can be modulated at a low frequency (e.g., 77 Hz). The resultant modulation in PL intensity is then measured with a lock-in amplifier.

In a PL emission process, an electron falls to a lower-energy state, emitting a photon. Consider a donor–acceptor pair transition (Section 7.6). Before the transition, the neutral donor–acceptor pair ($D^0 A^0$) may have a total electron spin of $m_S = -1$, 0, or 1. For a weakly coupled spin system, the four possible spin states are designated:

(10.32)
$$|m_D, m_A\rangle = \begin{cases} |\tfrac{1}{2}, \tfrac{1}{2}\rangle & (m_S = 1) \\ |-\tfrac{1}{2}, \tfrac{1}{2}\rangle & (m_S = 0) \\ |\tfrac{1}{2}, -\tfrac{1}{2}\rangle & (m_S = 0) \\ |-\tfrac{1}{2}, -\tfrac{1}{2}\rangle & (m_S = -1) \end{cases}$$

where m_D and m_A are the electron spins for the neutral donor and acceptor, respectively. The Zeeman energy levels are approximately equal to the sum of the electron energies:

(10.33)
$$E = g_D \mu_B B m_D + g_A \mu_B B m_A$$

These energy levels are shown in Figure 10.21.

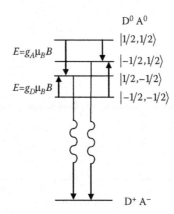

FIGURE 10.21 Optically detected magnetic resonance (ODMR) process for a donor–acceptor pair transition. Radiative transitions to the ground state (D⁺ A⁻) are indicated. Electron paramagnetic resonance (EPR) transitions, which result in enhanced emission, are shown by the arrows. (After Spaeth, J.-M., J.R. Niklas, and R.H. Bartram. 1992. *Structural Analysis of Point Defects in Solids*. Berlin: Springer-Verlag.)

After the transition, the donor and acceptor are ionized (D⁺ A⁻), and spin is zero. The selection rule for photon emission is that the spin cannot change; therefore, the initial spin state must have $m_S = 0$. The $m_S = 0$ states are optically active, whereas the $m_S = \pm 1$ levels are "dark" states. This spin-dependent emission forms the basis for ODMR. In most cases, the PL emission rate is higher than the spin-lattice relaxation rate, so that the spins do not have time to thermally relax to their ground state. The PL emission results in an overpopulation of the dark states. When the B-field is tuned to a resonance, then the spins are excited from a dark state to an optically active state, resulting in an increase in PL intensity (Figure 10.21). The energies of these ODMR peaks correspond to the Landé factors g_D and g_A.

An interesting example of ODMR is represented by bulk, p-type InP:Zn, compensated by shallow donors and deep phosphorus antisite (P_{In}) donors (Kennedy and Glaser, 1998). The unoccupied antisite is P_{In}^{2+}, the singly occupied antisite is P_{In}^{1+}, and the (+/2+) donor level lies 1.1 eV above the valence band maximum (Figure 10.22). The PL spectrum shows a shallow donor to Zn acceptor transition at 1.37 eV and a P_{In}^{1+} to Zn transition at 0.85 eV. In one experiment, the PL intensity at 0.85 eV was monitored as a function of magnetic field, at a microwave frequency of 24 GHz. Two ODMR peaks are observed near 0.85 Tesla, corresponding to a P_{In}^{1+} donor electron with $g = 2.004$. The splitting is due to the hyperfine interaction between the electron and the 100% abundant, spin-½ $^{31}P_{In}$ nucleus. The Zn acceptor ODMR signal was not observed, due to broadening from random strains in the crystal.

In a second experiment with InP:Zn, the PL intensity at 1.37 eV was monitored as a function of magnetic field, at a microwave frequency of 35 GHz. Here, the dominant spin-dependent process was the capture of a shallow donor electron by an unoccupied P_{In}^{2+} donor. When that capture occurs, the PL intensity at 1.37 eV is diminished, resulting in *negative* peaks in the ODMR spectrum. These negative peaks occur when the magnetic field is tuned to resonances corresponding to $g = 2.004$ (the P_{In}^{1+} donor) or $g = 1.20$ (the EM donor).

10.10 ELECTRON NUCLEAR DOUBLE RESONANCE

In addition to EPR transitions, electromagnetic radiation can also induce nuclear magnetic resonance (NMR) transitions. A nuclear spin produces a magnetic moment, the z component of which is given by

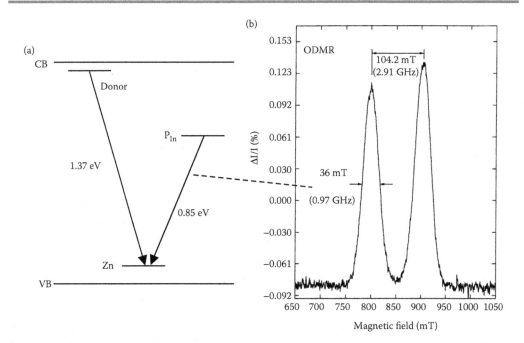

FIGURE 10.22 (a) Energy-level diagram of donor–acceptor pair transitions in InP. (b) ODMR spectrum for the 0.85 eV transition. (Reprinted from Kennedy, T.A. and E.R. Glaser. 1998. *Semiconductors and Semimetals*, Vol. 51A, ed. M. Stavola, pp. 93–136. San Diego, CA: Academic Press. Copyright 1998, with permission from Elsevier.)

(10.34) $$\mu_z = g_N \mu_N m_I$$

where μ_N is the nuclear magneton (5.051 × 10^{-27} Joule/Tesla or 7.62 MHz/Tesla) and m_I is the nuclear spin along the z axis. Owing to the heavier mass of the proton as compared to the electron, the nuclear magneton is 1836 times smaller than the Bohr magneton. In an external magnetic field, the nuclear Zeeman shift is given by

(10.35) $$\Delta E_N = -g_N \mu_N B m_I$$

The small value of the nuclear magneton means that NMR transitions lie in the radio frequency (RF) range of the spectrum.

Nuclei with nonzero spin have different values of g_N that have been cataloged (Spaeth et al., 1992). Nuclear spins therefore provide a unique chemical "fingerprint" that is useful for identifying impurities. A typical NMR apparatus, however, does not have the sensitivity to characterize trace impurities in semiconductors. Instead, researchers make use of electron nuclear double resonance (ENDOR). An ENDOR experiment consists of an EPR spectrometer, with the addition of an RF coil. When the RF frequency matches specific nuclear-spin transitions, the EPR signal is enhanced. This enhancement versus RF frequency is the ENDOR spectrum.

To see how ENDOR works, consider a typical EPR experiment (Section 10.8). The measurement relies on the excitation of an electron spin by microwave radiation. As the microwave power is increased, the EPR signal grows. For high powers, though, the levels become saturated. The populations of the two-level system become close to 50%/50%. The EPR signal strength is proportional to the population difference. In the saturation regime, the electron spins can no longer absorb the microwaves efficiently.

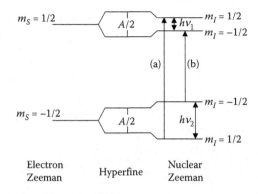

FIGURE 10.23 Electron Zeeman, hyperfine, and nuclear Zeeman shifts for a spin-1/2 nucleus and $A > 0$. Allowed electron paramagnetic resonance (EPR) transitions are shown by the arrows (a) and (b).

Now, we include the effect of the nuclear spin. Figure 10.23 shows the electron Zeeman splitting, hyperfine splitting, and nuclear Zeeman splitting for a spin $I = ½$ nucleus. A prominent example of such a system is phosphorus in silicon (Feher, 1956). The selection rule for an EPR transition requires that $\Delta m_I = 0$. Consider a saturated transition for $m_I = ½$, shown by the arrow in Figure 10.23a. When the RF photon energy equals the nuclear energy splitting $h\nu_1$, a transition is induced from $m_I = ½$ to $-½$. By exciting the nuclear spin transition, we have depopulated the excited state. This reduction in saturation leads to an enhanced EPR signal and a peak in the ENDOR spectrum. An ENDOR peak also appears when the RF photon energy equals $h\nu_2$. The sum of $h\nu_2$ and $h\nu_1$ yields the hyperfine parameter A. By varying the magnetic field, one can obtain g_N from Equation 10.35.

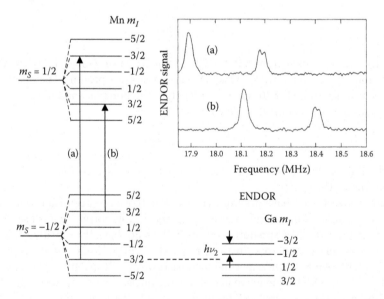

FIGURE 10.24 Left: Hyperfine splitting for Mn^{2+}. Two allowed electron paramagnetic resonance (EPR) transitions are shown by the arrows (a) and (b). Bottom right: Splitting of one of the levels, due to the hyperfine and nuclear Zeeman interactions from a neighboring ^{71}Ga nucleus. Upper right: Electron nuclear double resonance (ENDOR) spectra corresponding to the EPR transitions (a) and (b). (Reprinted with permission from Van Gisbergen, S.J.C.H.M. et al. 1994. Ligand ENDOR on substitutional manganese in GaAs. *Phys. Rev.* B 49: 10999–11004. Copyright 1994 by the American Physical Society.)

Chapter 10—Optical Spectroscopy   **273**

The richness and utility of magnetic resonance techniques are illustrated by the example of substitutional manganese (Mn_{Ga}) in GaAs (van Gisbergen et al., 1994). The Mn^{2+} charge state has five $3d$ electrons, with a total spin $S = 5/2$. Here, we consider the transition from $m_S = -\frac{1}{2}$ to $\frac{1}{2}$. The hyperfine interaction with the ^{55}Mn nucleus ($I = 5/2$, 100% abundant) results in six well-resolved EPR lines (Figure 10.24). One can perform ENDOR experiments by setting the magnetic field to correspond to any one of these six resonances.

ENDOR was used to probe the interaction with the ^{71}Ga neighbors ($I = 3/2$, 39.6% abundant). The hyperfine interaction due to a ^{71}Ga nucleus is shown in Figure 10.24. The RF frequency was tuned to excite the lower ($m_S = -\frac{1}{2}$) transitions. To first order, these transitions all have an energy $h\nu_2$. This ENDOR peak is split by various perturbations, including the crystal field of the Mn impurity, configurations of neighboring ^{71}Ga atoms, and nuclear quadrupole interactions. Some of these peaks are shown in the figure. The two spectra were taken at different magnetic fields, corresponding to the Mn $m_I = 3/2$ and $-3/2$ transitions. The shift in the peaks versus B yielded a g_N value characteristic of ^{71}Ga ($g_N\mu_N = 12.984$ MHz/Tesla). This work provided definitive evidence that manganese is a substitutional impurity in GaAs.

PROBLEMS

10.1 A semiconductor thin film has a band gap in the UV. In the visible/near-IR range, the refractive index is 2.5. The transmitted light spectrum shows interference fringes with maxima at 800, 533, and 400 nm. What is the film thickness?

10.2 Light with wavelengths near 500 nm reflects off a grating with 600 lines/mm. The dispersed light travels 0.25 m and exits a 100-μm wide slit.
   a. What is the angle of reflection?
   b. Estimate the resolution (nm).

10.3 A defect has a PL band centered at 2.1 eV with the zero-phonon line at 2.4 eV. The PL is due to a transition of an electron from the conduction-band minimum to the defect level. A PLE spectrum shows an excitation onset at 2.8 eV. The band gap is 3.5 eV.
   a. Where is the defect level relative to the valence-band maximum?
   b. What is the Stokes shift?
   c. What is the Franck–Condon shift for the excited state?
   d. What is the Franck–Condon shift for the ground state?

10.4 In a Raman scattering experiment, light of wavelength 532 nm is incident on a GaN crystal. The Raman-scattered light has a wavelength of 512 nm.
   a. Was a phonon created or destroyed in the scattering process?
   b. Find the frequency (cm^{-1}) of the phonon that was created or destroyed.
   c. Sketch the Raman spectrum. Set the laser line frequency to equal 0 cm^{-1}.

10.5 For a N-H complex aligned along the [111] direction in ZnSe, calculate the Raman intensity for laser light polarized along [001] and scattered light polarized along [010].

10.6 An IR spectrum has an intensity that is constant for frequencies less than 1000 cm^{-1} and zero greater than 1000 cm^{-1}. Calculate an expression for the interferogram obtained by an FTIR spectrometer. Plot this interferogram.

10.7 An interferogram is given by

$$I(\delta) = A\cos(k_0\delta)e^{-b\delta}$$

   a. Sketch the interferogram.
   b. Calculate the spectrum $I_0(k)$ in the region $k \approx k_0$, assuming infinite mirror travel (Equation 10.19). Hint: Expand the cosine terms using Euler's identity. After evaluating the integral, discard terms with ($k + k_0$) in the denominator.

274 References

10.8 An EPR apparatus uses a microwave frequency of 9 GHz. An isotropic defect has $g = 2.015$ and $A = 1.4$ GHz. The electron and nuclear spins are both ½. At what values of the magnetic field will the EPR resonances occur?

10.9 The ODMR spectrum in Figure 10.22 was obtained with a microwave frequency of 24 GHz. Estimate g from this spectrum.

10.10 The ENDOR spectra in Figure 10.24 were taken at (a) $B = 816.1$ mT and (b) 833.47 mT. Estimate $g_N\mu_N$ (MHz/T).

REFERENCES

Ahrenkiel, R.K. 1993. Minority-carrier lifetime in III-V semiconductors. In *Semiconductors and Semimetals*, Vol. 39, eds. R.K. Ahrenkiel and M.S. Lundstrom, pp. 40–150. San Diego, CA: Academic Press.

Beaulac, R., S.T. Ochsenbein, and D.R. Gamelin. 2010. Colloidal transition-metal-doped quantum dots. In *Nanocrystal Quantum Dots*, 2nd ed, ed. V.I. Klimov, Chapter 11. Boca Raton, FL: CRC Press.

Bell, R.J. 1972. *Introductory Fourier Transform Spectroscopy*. New York, NY: Academic Press.

Bourgoin, J. and M. Lannoo. 1983. *Point Defects in Semiconductors II*. Berlin: Springer-Verlag.

Budde, M., G. Lüpke, C. Parks Cheney, N.H. Tolk, and L.C. Feldman. 2000. Vibrational lifetime of bond-center hydrogen in crystalline silicon. *Phys. Rev. Lett.* 85: 1452–1455.

Cardona, M. 1982. Resonance phenomena. In *Light Scattering in Solids II*, eds. M. Cardona and G. Guntherodt, 19–98. Berlin: Springer.

Cardona, M. and M.L.W. Thewalt. 2005. Isotope effects on the optical spectra of semiconductors. *Rev. Mod. Phys.* 77: 1173–1224.

Carlos, W.E., E.R. Glaser, and D.C. Look. 2001. Magnetic resonance studies of ZnO. *Physica B* 308-10: 976–979.

Cheng, Y.M. and M. Stavola. 1994. Non-Arrhenius reorientation kinetics for the B-H complex in Si: Evidence for thermally assisted tunneling. *Phys. Rev. Lett.* 73: 3419–3422.

Colson, W.B., E.D. Johnson, M.J. Kelley, and H.A. Schwettman. 2002. Putting free-electron lasers to work. *Physics Today* 55(1): 35–41.

Cooley, J.W. and J.W. Tukey. 1965. An algorithm for the machine calculation of complex Fourier series. *Math. Computat.* 19: 297–301.

Dubon, O.D., J.W. Beeman, L.M. Falicov, H.D. Fuchs, E.E. Haller, and C. Wang. 1994. Copper acceptors in uniaxially stressed germanium: $(1s)^3$ to $(1s)^2(2s)^1$ ground-state transformation. *Phys. Rev. Lett.* 72: 2231–2234.

Evans, S.M., N.C. Giles, L.E. Halliburton, and L.A. Kappers. 2008. Further characterization of oxygen vacancies and zinc vacancies in electron-irradiated ZnO. *J. Appl. Phys.* 103: 043710 (7 pages).

Feher, G. 1956. Observation of nuclear magnetic resonances via the electron spin resonance line. *Phys. Rev.* 103: 834–835.

Fellgett, P. 1958. I.—Les principes généraux des méthodes nouvelles en spectroscopie interférentielle— A propos de la théorie du spectromètre interférential multiplex. *J. Phys. Radium* 19: 187–191.

Girard, A. and P. Jacquinot. 1967. Principles of instrumental methods in spectroscopy. In *Advanced Optical Techniques*, ed. A.C.S. Van Hell, pp. 71–121. Amsterdam: North-Holland.

Griffiths, P.R. and J.A. de Haseth. 1986. *Fourier Transform Infrared Spectrometry*. New York, NY: John Wiley & Sons.

Grivickas, P., M.D. McCluskey, and Y.M. Gupta. 2008. Band-gap luminescence of GaP:S shock compressed to 5GPa. *Appl. Phys. Lett.* 92: 142104. (3 pages).

Grivickas, P., M.D. McCluskey, Y. Zhang, J. F. Geisz, and Y.M. Gupta. 2009. Bound exciton luminescence in shock compressed GaP:S and GaP:N. *J. Appl. Phys.* 106: 023710 (7 pages).

Haller, E.E. 1995. Isotopically engineered semiconductors. *J. Appl. Phys.* 77: 2857–2878.

Haller, E.E. and L. M. Falicov. 1979. High-resolution EPR and piezospectroscopy studies of the lithium-oxygen donor in germanium. *Proc. 14th Intl. Conf. Phys. Semicond., Inst. Phys. Conf. Ser. No. 43*, ed. B.H.L. Wilson, pp. 1039–1042.

Haller, E.E., W.L. Hansen, and F.S. Goulding. 1981. Physics of ultra-pure germanium. *Adv. Phys.* 30: 93–138.

Hecht, E. 2002. *Optics*, 4th ed. Reading, MA: Addison-Wesley.

Hsu, L., M.D. McCluskey, and J.L. Lindstrom. 2003. Resonant interaction between localized and extended vibrational modes in Si:^{18}O under pressure. *Phys. Rev. Lett.* 90: 095505 (4 pages).

Jayaraman, A. 1983. Diamond-anvil cell and high-pressure physical investigations. *Rev. Mod. Phys.* 55: 65–107.

Karaiskaj, D., M.L.W. Thewalt, T. Ruf et al. 2001. Photoluminescence of isotopically purified silicon: How sharp are the bound exciton transitions? *Phys. Rev. Lett.* 86: 6010–6013.

Kennedy, T.A. and E.R. Glaser. 1998. Magnetic resonance of epitaxial layers detected by photoluminescence. In *Semiconductors and Semimetals*, Vol. 51A, ed. M. Stavola, pp. 93–136. San Diego, CA: Academic Press.

Lüpke, G., X. Zhang, B. Sun, A. Fraser, N.H. Tolk, and L.C. Feldman. 2002. Structure-dependent vibrational lifetimes of hydrogen in silicon. *Phys. Rev. Lett.* 88: 135501 (4 pages).

Lüpke, G., N.H. Tolk, and L.C. Feldman. 2003. Vibrational lifetimes of hydrogen in silicon. *J. Appl. Phys.* 93: 2317–2336.

Lyons, J.L., A. Janotti, and C.G. Van de Walle. 2009. Why nitrogen cannot lead to p-type conductivity in ZnO. *Appl. Phys. Lett.* 95: 252105 (3 pages).

McCluskey, M.D., L.T. Romano, B.S. Krusor, D.P. Bour, N.M. Johnson, and S. Brennan. 1998. Phase separation in InGaN/GaN multiple quantum wells. *Appl. Phys. Lett.* 72: 1730–1732.

Neugebauer, G., H.J. Habing, R. Vanduinen et al. 1984. The infrared astronomy satellite (IRAS) mission. *Astrophys. J.* 278: L1–L6.

Nurmikko, A.V. 1992. Transient spectroscopy by ultrashort laser pulse techniques. In *Semiconductors and Semimetals*, Vol. 36, eds. D.G. Seiler and C.L. Littler, pp. 85–135. San Diego, CA: Academic Press.

O'Connor, D.V. and D. Phillips. 1984. *Time-Correlated Single Photon Counting*. London, UK: Academic Press.

Pankove, J.I. 1971. *Optical Processes in Semiconductors*. New York: Dover.

Peng, H.Y., M.D. McCluskey, Y.M. Gupta, M. Kneissl, and N.M. Johnson. 2005. Shock-induced band-gap shift in GaN: Anisotropy of the deformation potentials. *Phys. Rev. B* 71: 115207 (5 pages).

Perkowitz, S. 1993. *Optical Characterization of Semiconductors*. New York, NY: Academic Press.

Ramdas, A.K. and S. Rodriguez. 1992. Piezospectroscopy of semiconductors. In *Semiconductors and Semimetals*, Vol. 36, eds. D.G. Seiler and C.L. Littler, pp. 137–220. San Diego, CA: Academic Press.

Rieke, G., E.T. Young, A. Alonso-Herrero et al. 2003. Far infrared photoconductors for space-borne astronomy: A review based on the MIPS 70 micron array. *Proceedings of the Far-IR, Sub-MM and MM Detector Technology Workshop*, eds. J. Wolf, J. Farhoomand, and C.R. McCreight, NASA/CP-211408, paper 2-01, pp. 47–53.

Schubert, E.F. 2006. *Light-Emitting Diodes*. Cambridge, UK: Cambridge University Press.

Shan, W., X.C. Xie, J.J. Song, and B. Goldenberg. 1995. Time-resolved exciton luminescence in GaN grown by metalorganic chemical vapor deposition. *Appl. Phys. Lett.* 67: 2512–2514.

Spaeth, J.-M., J.R. Niklas, and R.H. Bartram. 1992. *Structural Analysis of Point Defects in Solids*. Berlin: Springer-Verlag.

Stavola, M., K. Bergman, S.J. Pearton, and J. Lopata. 1988. Hydrogen motion in defect complexes: Reorientation kinetics of the B-H complex in silicon. *Phys. Rev. Lett.* 61: 2786–2789.

Steger, M., A. Yang, T. Sekiguchi et al. 2011. Photoluminescence of deep defects involving transition metals in Si: New insights from highly enriched ^{28}Si. *J. Appl. Phys.* 110: 081301 (25 pages).

Tang, C.L. and L.K. Cheng. 1995. *Fundamentals of Optical Parametric Processes and Oscillators*. Amsterdam: Harwood Academic.

Tarun, M.C., M. Zafar Iqbal, and M.D. McCluskey. 2011. Nitrogen is a deep acceptor in ZnO. *AIP Advances* 1: 022105 (7 pages).

Thewalt, M.L.W., M. Steger, A. Yang et al. 2007. Can highly enriched ^{28}Si reveal new things about old defects? *Physica B* 401-2: 587–592.

Thomas, D.G., M. Gershenzon, and F.A. Trumbore. 1964. Pair spectra and "edge" emission in gallium phosphide. *Phys. Rev.* 133: A269–A279.

Van Gisbergen, S.J.C.H.M., A.A. Ezhevskii, N.T. Son, T. Gregorkiewicz, and C.A.J. Ammerlaan. 1994. Ligand ENDOR on substitutional manganese in GaAs. *Phys. Rev. B* 49: 10999–11004.

Watkins, G.D. 1963. An EPR study of the lattice vacancy in silicon. *J. Phys. Soc. Jpn.* 18(Suppl. 2): 22–27.

Watkins, G.D. 1998. EPR and ENDOR studies of defects in semiconductors. In *Semiconductors and Semimetals*, Vol. 51A, ed. M. Stavola, pp. 1–43. San Diego, CA: Academic Press.

Watkins, G.D. and J.R. Troxell. 1980. Negative-U properties for point defects in silicon. *Phys. Rev. Lett.* 44: 593–596.

Wertz, J.E. and J.R. Bolton. 1986. *Electron Spin Resonance*. New York, NY: Chapman and Hall.

Wolk, J.A., J.W. Ager III, K.J. Duxstad et al. 1993. Local vibrational mode spectroscopy of nitrogen-hydrogen complex in ZnSe. *Appl. Phys. Lett.* 63: 2756–2758.

Yu, P.Y. and M. Cardona. 1996. *Fundamentals of Semiconductors*. Berlin: Springer-Verlag.

Particle-Beam Methods

Advances in nuclear and particle physics have provided a range of novel techniques for semiconductor characterization and doping (Feldman and Mayer, 1986). Particles used to probe solids include protons, helium nuclei, heavy ions, high-energy photons (x-rays and gamma rays), electrons, positrons, muons, and neutrons. In this chapter, particle-beam methods that are relevant to impurities and defects in semiconductors are discussed. We begin with techniques that involve ion beams; namely, Rutherford backscattering, ion implantation, and secondary ion mass spectrometry. Techniques that utilize x-ray emission and absorption are summarized. Basic interactions of electrons with matter are discussed; electron microscopy is covered in Chapter 12. The chapter concludes with techniques that involve exotic particles and decay: positron annihilation, muon spin resonance, perturbed angular correlation, and nuclear reactions.

11.1 RUTHERFORD BACKSCATTERING SPECTROMETRY

Ernest Rutherford discovered the atomic nucleus when he observed that most energetic helium ions (alpha particles) would pass through a very thin gold foil unaffected, except for a few that suffered large-angle scattering. The observed backscattering indicated that most of the atomic matter is concentrated in the nucleus. Rutherford backscattering spectrometry (RBS), therefore, was born around the turn of the 20th century. It took a surprisingly long time for RBS to be introduced to electronic materials characterization (Chu et al., 1978). One of the pioneers in this field, J.W. Mayer, saw RBS as a great opportunity to use the large number of early 1960s 2 to 4 MeV accelerators that had been developed for nuclear physics experiments. RBS has yielded crucial information on the composition of thin films (up to ~1 μm), interface reactions, and crystallinity. RBS is ideally suited for many thin-film or near-surface phenomena.

The RBS process can be modeled as an elastic collision, like billiard balls (Figure 11.1). The *kinematic factor* K_M gives the kinetic energy of an elastically scattered particle of mass m:

11.1 Rutherford Backscattering Spectrometry

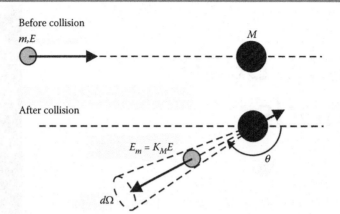

FIGURE 11.1 Backscattering collision between a helium particle of mass m and kinetic energy E with a target atom of mass M.

$$(11.1) \quad K_M = E_m/E = \left(\frac{\sqrt{M^2 - m^2 \sin^2\theta} + m\cos\theta}{M+m} \right)^2$$

where E_m is the kinetic energy, E is the energy before the collision, M is the mass of the target nucleus, and θ is the scattering angle. The largest energy transfer occurs for $\theta = 180°$, perfect backscattering:

$$(11.2) \quad K_M = \left(\frac{M-m}{M+m} \right)^2$$

For RBS experiments, θ is usually chosen to be very close to 180°. From Equation 11.2, the highest scattered energy occurs for an infinite target mass M.

To obtain scattering rate information, we consider the forces acting between the colliding particles. For $M \gg m$, the Rutherford scattering *differential cross section* is given by (Eisberg and Resnick, 1985)

$$(11.3) \quad \frac{d\sigma}{d\Omega} = \left(\frac{1}{4\pi\varepsilon_0} \right)^2 \left(\frac{zZe^2}{4E} \right)^2 \frac{1}{\sin^4(\theta/2)}$$

where ze and Ze are the charges of the helium ($z = 2$) and target nucleus, respectively. The differential cross section $d\sigma/d\Omega$ is proportional to the number of helium nuclei that are scattered into a solid angle $d\Omega$ (Figure 11.1). Equation 11.3 is proportional to the square of the product of nuclear charges divided by the energy:

$$(11.4) \quad \frac{d\sigma}{d\Omega} \sim \frac{(zZ)^2}{E^2}$$

Another effect to be considered is the energy loss of the projectile during passage through a medium (Nicolet, 1979). Large energy losses from large-angle scattering, the basis of RBS, are rare compared to the numerous small energy losses that do not influence the trajectory appreciably.

Chapter 11—Particle-Beam Methods   **279**

Small energy losses occur when the helium particle collides with electrons in the material. For a layer of infinitesimal thickness Δx, the average amount of energy lost by a particle is ΔE. The *stopping cross section* is defined

$$(11.5) \qquad \varepsilon = \frac{1}{N}\frac{\Delta E}{\Delta x}$$

where N is the concentration of atoms in the layer material. Typical values for ε are in the range 20–150 eV/(10^{15} atoms/cm^2). Stopping cross sections are additive. Consider a molecule consisting of m A-atoms and n B-atoms. The stopping cross section is given by

$$(11.6) \qquad \varepsilon = m\varepsilon^A + n\varepsilon^B$$

where ε^A and ε^B are the stopping cross sections for atoms A and B, respectively. This additive property is known as Bragg's rule (Bragg and Kleeman, 1905). The energy loss is then given from Equation 11.5:

$$(11.7) \qquad \Delta E = \varepsilon N \Delta x$$

The stopping of energetic helium nuclei is composed of a random sequence of independent small energy losses. Niels Bohr modeled this process by assuming that a helium nucleus collides with stationary free electrons (Bohr, 1915). The average energy loss ΔE over a length Δx is given by (Nicolet, 1979)

$$(11.8) \qquad \Delta E = \left[\frac{1}{64\pi\varepsilon_0^2} Z(ze^2)^2 \frac{m}{m_e E} \right] N \Delta x$$

where m/m_e is the ratio of the helium to free-electron mass. Because this is a statistical process, the energy loss has a random distribution. The energy *straggling* can be estimated by

$$(11.9) \qquad \langle (\delta \Delta E)^2 \rangle = \left[\frac{1}{4\pi\varepsilon_0^2} Z(ze^2)^2 \right] N \Delta x$$

where $\langle (\delta \Delta E)^2 \rangle$ is the variance of ΔE. This variance leads to a broadening in ΔE, typically a few percent, that limits the energy resolution of an RBS measurement.

With knowledge of the kinematic factor K_M, the scattering cross section $d\sigma/d\Omega$, and the stopping cross section ε, one can analyze RBS spectra. Multichannel pulse height analyzers (PHAs), a thin window silicon surface barrier detector, and electronic pulse amplifiers are used to record the number and energies of backscattered particles. In a typical PHA, the energies are displayed on the x axis and the number of counts per channel (energy slice) is displayed on the y axis. Figure 11.2 shows three RBS measurements of metal films on silicon for a helium energy of $E = 2$ MeV. The lightest element of the three, titanium, results in the most energy loss (Equation 11.2). The height of each spectrum is proportional to the backscattering cross section; from Equation 11.4, high-Z elements result in strong RBS signals. The width of each spectrum (ΔE) is proportional to the film thickness and the stopping cross section (Equation 11.7).

The ratio of concentrations of elements in a thin compound film can be obtained with RBS. From Equation 11.4, the backscatter yield H is proportional to

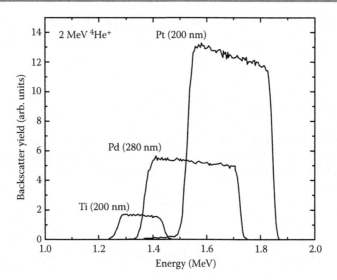

FIGURE 11.2 Rutherford backscattering spectrometry (RBS) measurements of three different metal films. The signals from the silicon substrates are not shown. (After Nicolet, M.-A. 1979. *Nondestructive Evaluation of Semiconductor Materials and Devices*, pp. 581–630. New York, NY: Plenum Press.)

$$H \sim N \frac{d\sigma}{d\Omega} \sim NZ^2 \tag{11.10}$$

For a binary compound, the composition can be estimated as

$$\frac{N_A}{N_B} = \frac{H_A/Z_A^2}{H_B/Z_B^2} \tag{11.11}$$

Figure 11.3 shows the various stages of platinum-silicide formation (Hiraki et al., 1971). The unannealed sample consists of a layer of Pt on a Si substrate. After the 400°C annealing step, there is a backscattering signal at the Si-edge, indicating that the Pt layer has transformed into a Pt-Si compound. Using Equation 11.11 to determine the composition, we find

$$\frac{N_{Pt}}{N_{Si}} = \frac{7300/78^2}{130/14^2} \approx 2 \quad (Pt_2Si) \tag{11.12}$$

After the 450°C annealing step, a similar analysis shows that the compound is PtSi.

In the preceding discussion, we implicitly assumed noncrystalline materials. In the majority of semiconductor applications, however, one uses single-crystal bulk and thin films. As discussed in Section 4.9, channeling occurs when a collimated ion beam hits a crystal along one of the major axes. The ion range for a given energy increases; for RBS measurements, the backscattering yield decreases. The number of backscattered helium ions can be 100 times less for the <110> directions than for a random orientation. This *channeling RBS* technique is used to study the regrowth of amorphous layers (rendered amorphous by ion implantation), to determine the radiation damage in thin layers, and to find the position of impurities by triangulation (channeling in several different channels).

An example of channeling RBS is shown in Figure 11.4 for GaAs coimplanted with various ions (Moll et al., 1992). A GaAs sample with a random orientation (a) shows a higher backscattering rate than those oriented along a <111> direction. Unimplanted samples (e) or samples implanted with

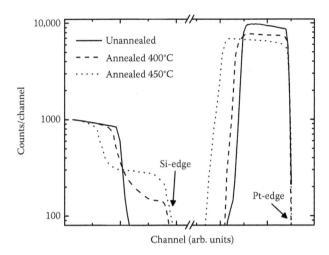

FIGURE 11.3 Rutherford backscattering spectrometry (RBS) spectra of platinum-silicide formation. Following the annealing step at 400°C, the Pt thin film has turned into Pt$_2$Si. The 450°C anneal results in the formation of PtSi. (After Hiraki, A., M.-A. Nicolet, and J.W. Mayer. 1971. *Appl. Phys. Lett.* 18: 178–181.)

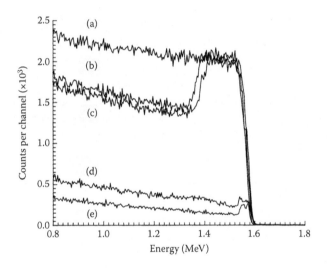

FIGURE 11.4 Rutherford backscattering spectrometry (RBS) spectra for GaAs samples. Spectrum (a) is for a random orientation. <111> aligned spectra are shown for (b) C + Ga implant, (c) C + Al implant, (d) C + B implant, and (e) unimplanted. The amorphous layers in (b) and (c) are 140 and 120 nm thick, respectively. (After Moll, A.J. et al. 1992. *Appl. Phys. Lett.* 60: 2383–2385.)

light ions (d) are single crystal, resulting in a significant decrease in backscatter due to channeling. Samples implanted with heavy ions [(b) and (c)] have an amorphous layer. The loss of crystalline channels in the amorphous layer results in an increase in the backscattering rate.

11.2 ION RANGE

The previous discussion for RBS considered MeV ion energies. For such large energies, the model for the large-angle collision between the ion and target can neglect the electrons. This is not the typical situation we encounter in ion implantation of semiconductors (Ziegler, 1984). At energies of one to a few hundred keV, the effects of the electrons bound to the ion and the target atom have to be taken into account. The degree to which the electrons screen the nuclear Coulomb potential is a complicated matter that depends on many physical parameters.

Building on the work of Bohr (1915) and others, Lindhard, Scharff, and Schiøtt (LSS) developed a comprehensive theoretical framework for ions moving in an amorphous solid (Lindhard et al., 1963). First, they defined dimensionless energy (ε) and length (ρ) parameters:

(11.13)
$$\varepsilon = E\left(\frac{4\pi\varepsilon_0 a}{e^2}\right)\left[\frac{M_2}{Z_1 Z_2 (M_1 + M_2)}\right]$$

and

(11.14)
$$\rho = R\pi a^2 N\left[\frac{M_1 M_2}{(M_1 + M_2)^2}\right]$$

with

(11.15)
$$a = \frac{0.8853 a_0}{(Z_1^{2/3} + Z_2^{2/3})^{1/2}}$$

where E is the ion energy; R is the ion range; a_0 is the Bohr radius (0.53 Å); $Z_{1,2}$ and $M_{1,2}$ are the atomic numbers and atomic masses of the projectile (1) and the target atoms (2), respectively. Using these parameters, LSS derived a universal relationship for the *nuclear stopping power* $(d\varepsilon/d\rho)_n$, shown in Figure 11.5 as a function of $\varepsilon^{1/2}$. The peak energy for nuclear stopping is denoted ε_1.

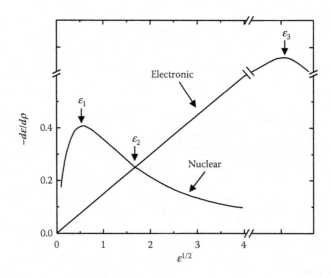

FIGURE 11.5 Nuclear and electronic stopping-power curves from LSS theory, expressed in terms of the reduced variables ρ and ε. The electronic stopping is plotted for $k = 0.15$. (After Lindhard, J., M. Scharff, and H.E. Schiøtt. 1963. *Mat. Fys. Medd. Dan. Vid. Selsk.* 33(14): 1–42.)

Collisions between energetic ions and electrons are very different from nuclear collisions because of the small mass of the electrons. Collisions with electrons lead to small-angle scattering. For keV ions, LSS theory shows that the *electronic stopping power* $(d\varepsilon/d\rho)_e$ can be described by

$$(11.16) \qquad -\left(\frac{d\varepsilon}{d\rho}\right)_e = k\varepsilon^{1/2}$$

with

$$(11.17) \qquad k = \frac{0.0793\zeta_e Z_1^{1/2} Z_2^{1/2}(M_1 + M_2)^{3/2}}{(Z_1^{2/3} + Z_2^{2/3})^{3/4} M_1^{3/2} M_2^{1/2}}$$

and

$$(11.18) \qquad \zeta_e \approx Z_1^{1/6}$$

The electronic stopping power cannot be expressed by a universal curve, as for nuclear stopping, but by a family of curves that depend on k, which typically lies between 0.1 and 0.2. The peak in electronic stopping power is denoted ε_3 and the crossover between electronic and nuclear stopping powers is ε_2 (Figure 11.5). The characteristic energies are given in Table 11.1 for various elements in silicon, germanium, and tin.

In summary, for low energies nuclear interactions dominate, while at high energies electronic energy loss is more important. It follows that the path of an energetic ion ($\varepsilon > \varepsilon_2$) into a solid is rather straight at first, due to low-angle electronic scattering. When the ion has slowed down sufficiently ($\varepsilon < \varepsilon_2$), nuclear or large-angle scattering will become dominant, leading to a random walk (Figure 11.6).

In order to obtain the total ion range, one has to integrate over the nuclear and electronic losses. Table 11.2 contains numerical values for the path lengths of various ions implanted into silicon and germanium. Inspection of a few relevant cases (e.g., boron and phosphorus in silicon) illustrates that depths of a few hundred to a few thousand Angstroms can be reached with ion energies of tens to hundreds of keV.

There is one more refinement that is required before we can predict the average depth reached by an ion implanted into a target. The total ion path length R traveled by an ion in a solid is longer than the depth reached perpendicular to the surface, often called the projected range R_p. This discrepancy arises because the ion changes direction many times before coming to rest, especially during the nuclear collision phase. In Table 11.3, we see that a low-energy lithium ion

---

TABLE 11.1 Characteristic Energies (keV) for Ions Implanted into Silicon, Germanium (Similar to GaAs), and Tin (Similar to CdTe)

Ion	ε_1			ε_2			ε_3
	in Si	in Ge	in Sn	in Si	in Ge	in Sn	
B	3	7	12	17	13	10	3×10^3
P	17	29	45	140	140	130	3×10^4
As	73	103	140	800	800	800	2×10^5
Sb	180	230	290	2000	2000	2000	6×10^5
Bi	530	600	700	6000	6000	6000	2×10^6

11.2 Ion Range

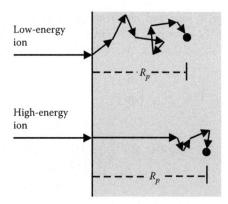

FIGURE 11.6 Trajectories for low- and high-energy ions implanted into a material. (After Carter, G. and W.A. Grant. 1976. *Ion Implantation of Semiconductors*. New York, NY: John Wiley & Sons.)

TABLE 11.2 Total Path Length R (μm) Calculated from Lindhard, Scharff, and Schiøtt (LSS) Theory

	^{7}Li	^{11}B	^{31}P	^{75}As	^{115}In
In Silicon:					
1 keV	0.020	0.011	0.004	0.002	0.002
10 keV	0.131	0.069	0.023	0.012	0.010
100 keV	0.826	0.527	0.184	0.067	0.052
1 MeV	3.41	2.55	1.43	0.631	0.415
In Germanium:					
1 keV	0.030	0.016	0.005	0.002	0.002
10 keV	0.153	0.083	0.023	0.010	0.008
100 keV	0.810	0.501	0.141	0.056	0.040
1 MeV	3.27	2.32	1.03	0.457	0.288

TABLE 11.3 Projected Range Corrections (R_p/R) Calculated from Lindhard, Scharff, and Schiøtt (LSS) Theory

	^{7}Li	^{11}B	^{31}P	^{75}As	^{121}Sb
In Silicon:					
20 keV	0.54	0.57	0.72	0.83	0.88
40 keV	0.62	0.64	0.75	0.84	0.88
100 keV	0.72	0.73	0.79	0.86	0.89
500 keV	0.86	0.86	0.86	0.89	0.91
In Germanium:					
20 keV	0.33	0.34	0.50	0.67	0.76
40 keV	0.40	0.40	0.52	0.69	0.76
100 keV	0.53	0.50	0.58	0.72	0.78
500 keV	0.74	0.71	0.71	0.77	0.81

implanted into germanium has a projected range $\sim$30% of the total distance traveled (R). In contrast, 500-keV antimony ions implanted into silicon reach a depth of 91% of R.

As discussed in Section 11.1, ion scattering processes are statistical in nature, and therefore lead to different values of R for each ion. To first order, LSS theory predicts a Gaussian distribution centered at R_p with a width ΔR_p. The higher the ion energy, the smaller is the range straggling. For example, boron implanted into silicon has $\Delta R_p/R_p \approx 0.5$ for 10–20 keV ions and $\Delta R_p/R_p \approx 0.2$ for 200 keV ions (Mayer et al., 1970). A practical consequence of this is the need for several implantation runs with a number of ion energies and doses if one wants to produce a constant dopant concentration profile over a large depth. Numerical computation programs have been developed for ion implantation (Ziegler, 1984). These programs are convenient for the prediction of multiple implant profiles over a wide range of energies, targets, and ions.

The LSS theory assumes an amorphous target. In crystalline targets, channeling causes large deviations from the LSS theory, especially for small ion energies (Feldman et al., 1982). The widest channels in the diamond structure are found in the <110> orientations. Ions entering such a channel reach a much larger R_p than predicted by LSS theory. The ion distribution is no longer Gaussian, but instead has a long tail into the crystal (Gemmel, 1974). An example of ion channeling is shown in Figure 4.20. Beyond a depth of $\sim$50 nm, the profile strongly deviates from a Gaussian distribution.

Ion channeling is important from a technological point of view, especially for dopants implanted into silicon. Dearnaley et al. (1967) investigated this effect by implanting radioactive phosphorus-32 ions into <110> and <111> oriented silicon. Ion concentrations were measured by (1) using an anodic solution to form a well-defined oxide thickness on the surface, (2) dissolving the oxide in HF, and (3) collecting the HF solution with filter paper and measuring its radioactivity. By repeating this process, dopant profiles were obtained. The profiles showed long tails that extended into the crystal, especially for the <110> orientation. As discussed in Section 4.9, channeling can be suppressed by deliberate misorientation of the wafer, preamorphization, or depositing a thin oxide layer on the wafer surface.

11.3 SECONDARY ION MASS SPECTROMETRY

Energetic ion beams are also useful for characterizing trace impurities. In secondary ion mass spectrometry (SIMS), a primary ion beam of energy $\sim$1 to 10 keV hits the sample. In addition to causing damage to the crystal, atoms within a few monolayers of the surface are ejected, or sputtered. SIMS is therefore a destructive characterization technique. Most of the sputtered atoms are neutral, but $\sim$1% are positively or negatively charged (Williams, 1983). These secondary ions, which may be atoms or molecules, are analyzed by a mass spectrometer (Runyan and Shaffner, 1998). O_2^+ primary beams are useful for producing positive secondary ions (e.g., group-III elements), because the electronegative oxygen ion readily accepts an electron. Cs^+ primary beams are used for producing negative secondary ions (e.g., group-V elements) as well as Cs^+–impurity molecules.

In static SIMS, the mass spectrum is recorded for the near-surface region only. In *dynamic SIMS*, more common in semiconductor characterization, the secondary ion yields for selected ions are monitored as a function of time. At the end of the experiment, the depth of the crater formed by the sputtering process is measured. Given a constant sputtering rate, the time can be related to depth, resulting in a plot of the atomic profile with a depth resolution of 5–10 nm (Schroder, 2006). Secondary ion yields for various atoms vary over five to six orders of magnitude (Benninghoven, 1976). To obtain quantitative results, the secondary ion yield must be converted to concentration (cm^{-3}). The best way to do this is to implant a known quantity of the ion into an identical, or nearly identical, host. The ion-implanted standard can then be used as a calibration with an accuracy of a few percent.

The SIMS analysis of the GaAs isotope heterostructure (Figure 8.14) used a primary beam of 3.0- or 5.5-keV Cs$^+$ ions. The detected secondary species were GaCs$^+$ molecules. As discussed in Section 8.6, the knock-on of the Cs$^+$ ions push Ga atoms deeper into the sample, resulting in a broad "back edge." The front edge, in contrast, is sharp, with a resolution of 3.5 nm per decade of concentration change. The instrumental broadening of the front edge is due to surface roughness of the pristine sample plus surface roughness caused by sputtering (Ho and Lewis, 1976). By including the effect of this broadening artifact, Wang et al. (1996) were able to determine the actual broadening of the interface.

A second example of SIMS analysis is shown for interdiffusion of indium and gallium in InGaN quantum wells (Figure 11.7). The depth resolution was limited to 50 nm, due to significant surface roughness. Although the resolution was not sufficient to resolve the individual quantum wells, the quantum-well region is clearly seen in the depth range of 1.0–1.2 μm. In the as-grown sample, the magnesium concentration is constant to a depth of 1 μm. After annealing at 1300–1400°C for 15 min, the magnesium diffuses through the quantum-well region. McCluskey et al. (1998) proposed that the magnesium diffusion enhanced the interdiffusion of In and Ga (Section 8.8).

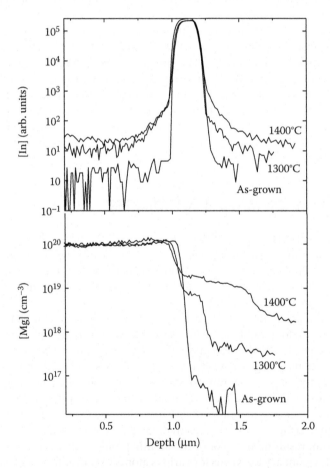

FIGURE 11.7 Secondary ion mass spectrometry (SIMS) profiles of a Mg-doped InGaN heterostructure. The InGaN quantum wells are in the depth range 1.0–1.2 μm. After annealing for 15 min at elevated temperatures, the Mg diffuses into the quantum-well region. (After McCluskey, M.D. et al. 1998. *Appl. Phys. Lett.* 73: 1281–1283.)

11.4 X-RAY EMISSION

As discussed in Section 11.1, RBS measures the backscattering of helium ions to determine the composition and thickness of semiconductor films. A second process that occurs is x-ray emission. When the helium ion strikes an atom, there is a probability that a core electron will be ejected from its shell. An electron in a higher-energy shell will then fall into the hole. To conserve energy, an x-ray photon may be emitted. The energy of this photon corresponds to the specific transition and provides an unambiguous means to identify specific atoms. When helium ions or protons cause the x-ray emission, the technique is referred to as *particle-induced x-ray emission* (PIXE). When electrons are used as projectiles (Section 11.7), then the method for analyzing the emitted x-ray energies is called *energy dispersive x-ray spectroscopy* (EDS or EDX).

The core shells with quantum numbers $n = 1, 2, 3 \ldots$ are denoted $K, L, M \ldots$ (Figure 11.8) An electron in the innermost K shell feels the Coulomb attraction of the positive nuclear charge, partially screened by the other electrons. For $Z > 5$, the binding energy for a K electron can be approximated as

(11.19) $$E_b \approx 13.6\,\text{eV}(Z-2)^2$$

where the factor of −2 accounts for screening. For the outer shells, the screening effect is much greater, and the binding energies are correspondingly reduced. The transition from L to K produces a K_α x-ray, M to K produces a K_β x-ray, M to L produces an L_α x-ray, and so forth. K_α energies span a wide range, from carbon (277 eV) to uranium (98 keV).

Emitted x-rays have a penetration depth on the order of 100 μm, which means that they have no trouble escaping the sample. To detect the x-rays, lithium-drifted silicon detectors with large depletion widths are used. The detectors must be cooled to reduce the leakage current and increase the energy resolution. The number of electron–hole pairs created by the x-ray photon is proportional to the x-ray energy. As in the case of RBS (Section 11.1), a PHA is used to obtain the number of counts as a function of energy.

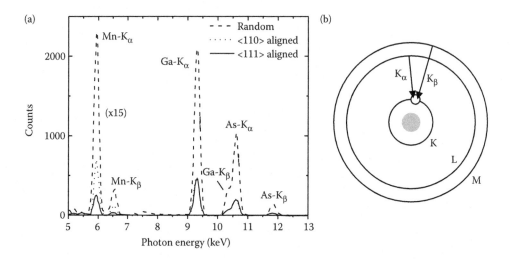

FIGURE 11.8 (a) Particle-induced x-ray emission (PIXE) spectra for $Mn_xGa_{1-x}As$ with $x = 0.09$. X-rays emitted by Mn, Ga, and As atoms are shown. (b) Core-electron transitions that produce the x-ray emission peaks. (After Yu, K.M. et al. 2002. *Phys. Rev. B* 65: 201303.)

288 11.5 X-ray Absorption

The penetration depth for MeV protons or helium ions, or 20–100 keV electrons, is roughly 1–50 μm for typical solids. Hence, PIXE and EDS are useful for probing the composition in a region several μm or tens of μm into the sample. Due to signal-to-noise limitations, the impurity concentration generally must be above 0.1 atomic % for a reliable signal.

A combined RBS and PIXE study was performed on $Mn_xGa_{1-x}As$ with $x = 0.09$ to determine the lattice location of Mn atoms (Yu et al., 2002). The Mn, Ga, and As PIXE spectra were obtained at the same time as the RBS measurements, using a 1.95 MeV $^4He^+$ beam. PIXE spectra are shown in Figure 11.8 for a random wafer alignment and alignments along <110> and <111> directions. Most of the Mn atoms occupy substitutional Ga sites. Due to channeling, the aligned spectra show smaller PIXE (and RBS) signals than the random orientation. The fact that the Mn signal is higher for the <110> alignment than for <111> suggests that a fraction of the Mn atoms reside in interstitial T_d sites. Those sites lie in <110> channels and can therefore be excited by helium ions traveling through the channels. This work has implications for spintronics, where high Curie temperatures require a high concentration of substitutional Mn acceptors.

11.5 X-RAY ABSORPTION

Along with x-ray emission, the core electron levels can be probed by x-ray absorption. In this process, an x-ray photon excites an electron from a core level to the continuum. Such an excited electron is called a *photoelectron*. The cross section for photoionization rises abruptly at the core-electron binding energy E_b. For energies above E_b, the absorption decreases gradually, due to the decreasing wave function overlap between the initial core-electron state and the final photo-electron state. The abrupt absorption edges provide accurate measurements of the core-electron levels and hence the identity of the atom. Monochromatic, tunable x-ray beams from synchrotron sources are typically used.

Above the absorption onset, the absorption profile exhibits oscillations (Figure 11.9). These oscillations, referred to as *extended x-ray absorption fine structure* (EXAFS), arise from interference effects due to the wavelike nature of electrons (Teo, 1986). Outside the atomic core, the wave function of the free electron can be modeled as a spherical wave:

$$(11.20) \qquad \psi(r) = \psi_0 \frac{e^{ikr}}{r}$$

where ψ_0 is the amplitude and k is given by

$$(11.21) \qquad k = p/\hbar = \frac{\sqrt{2m(h\nu - E_b)}}{\hbar}$$

Here, r is the distance from the atom, m is the electron mass, and $h\nu$ is the photon energy.

Now, consider a neighboring atom a distance R away that scatters the wave. The amplitude of the scattered wave is proportional to Equation 11.20. When the scattered wave reaches the first atom, its wave function is

$$(11.22) \qquad \psi_R \sim \left(\psi_0 \frac{e^{ikR}}{R}\right) \frac{e^{ikR+i\varphi}}{R}$$

where φ is a phase shift. The total wave function at the atom is the sum of the outgoing and scattered waves:

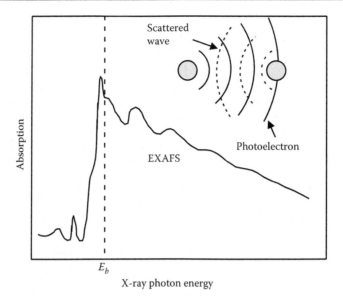

FIGURE 11.9 An x-ray absorption spectrum. Above the core-electron binding energy E_b, extended x-ray absorption fine structure (EXAFS) oscillations are observed, due to the interference between the outgoing and scattered waves. (After Teo, B.K. 1986. *EXAFS: Basic Principles and Data Analysis*. Berlin: Springer-Verlag.)

$$\psi_{tot} \sim \psi_0 + f \psi_0 \frac{e^{2ikR+i\varphi}}{R^2} \tag{11.23}$$

where f is an atomic scattering factor. The electron density at the atom is given by

$$|\psi_{tot}|^2 \sim \left|1 + f \frac{e^{2ikR+i\varphi}}{R^2}\right|^2 \approx 1 + \frac{2f}{R^2}\cos(2kR+\varphi) \tag{11.24}$$

Recall that the absorption strength is proportional to the overlap between the initial and final wave function. When the cosine term in Equation 11.24 is a maximum, then there is constructive interference and the absorption is maximized. Constructive and destructive interference conditions result in the observed sinusoidal oscillation in the absorption. Equation 11.24 is somewhat simplified; a more complete expression can be found in texts such as Teo (1986) or Agarwal (1991).

From Equation 11.24, it can be seen that the period of the EXAFS oscillations, plotted as a function of k, allows one to measure the interatomic distance R. For an atom in a substitutional position, each shell of neighboring atoms will contribute to the EXAFS signal. To extract the R values, a Fourier transform is performed on the absorption spectrum. An example of this analysis is shown for ion-implanted germanium (Figure 11.10). Fourier transforms of EXAFS spectra are shown for as-grown germanium and germanium that was amorphized due to the high implantation dose (Ridgway et al., 2000). Unlike crystalline germanium, amorphous germanium exhibits oscillations only from the nearest neighbors ($R \approx 2$ Å). More distant neighbors, however, are disordered and do not produce EXAFS oscillations.

Along with EXAFS, which is above the absorption edge, near-edge features also provide information about the atom and its environment (Koningsberger and Prins, 1988). The measurement

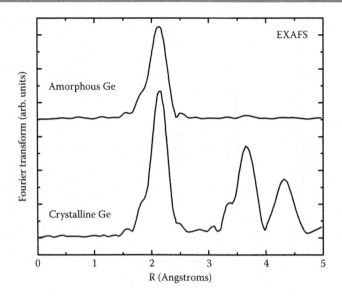

FIGURE 11.10 Fourier transforms of extended x-ray absorption fine structure (EXAFS) data for amorphous and crystalline germanium. Peaks correspond to neighboring atoms that are a distance R from the emitting atom. (After Ridgway, M.C. et al. 2000. *Phys. Rev. B* 61: 12586–12589.)

of these features is referred to as x-ray absorption near edge spectroscopy (XANES). In transition metals, for example, strong XANES peaks (called "white lines") arise from the transition of a $2p$ electron to an unoccupied nd level. These peaks are highly sensitive to the valence state of the atom and its local electronic structure. Other features, above the photoionization threshold, arise from multiple scattering events. For example, the photoelectron may scatter off two neighboring atoms before arriving at the absorbing atom. While the interpretation of XANES spectra is not straightforward, DFT calculations combined with simulation programs have enabled researchers to compare XANES spectra with defect models. For example, XANES was used to show that nitrogen occupies a substitutional oxygen site in ZnO (Fons et al., 2006).

11.6 PHOTOELECTRIC EFFECT

In an x-ray absorption process, a photoelectron is generated. In *x-ray photoelectron spectroscopy* (XPS), the energies of the photoelectrons that escape the sample are measured. Given a monochromatic source of x-rays, the core-electron binding energy can be inferred from

$$E_b = h\nu - E_e \quad (11.25)$$

where E_e is the kinetic energy of the photoelectron in a vacuum. As discussed previously, the binding energy provides a unique "fingerprint" for chemical identification. The exact position of the binding energy depends on the charge state of the element. By using standard reference handbooks (e.g., Muilenberg, 1979), one can determine the chemical identity and charge state (oxidation) of impurities in semiconductors.

XPS uses soft, monochromatic x-rays in the 1–2 keV energy range, produced by Al or Mg K_α lines or by a synchrotron source. The photoelectrons only travel a few nm before scattering inelastically. Such a scattering event changes the characteristic x-ray energy of the photoelectron,

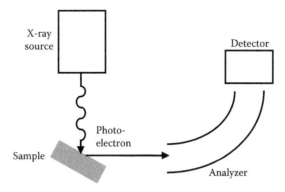

FIGURE 11.11 X-ray photoelectron spectroscopy (XPS) experiment.

causing it to be lost in the background. Therefore, XPS is a surface-sensitive technique. To obtain composition versus depth, or to remove surface contamination, one can sputter the surface with argon ions. The energies of the photoelectrons are selected using an energy analyzer. One type of energy analyzer, shown in Figure 11.11, uses static electric fields to bend electrons into a curve. The curved metal surfaces can be either cylindrical or spherical (Carlson, 1975). Only electrons with the selected kinetic energy reach the detector, which can be an electron multiplier such as a microchannel plate (Section 10.2). The energy analyzer plus detector is an *electron spectrometer*.

An XPS spectrum of ZnO:Cu nanocrystals is shown in Figure 11.12 (Hlaing Oo et al., 2010). The spectrum was obtained with Al K_α x-rays (1.4866 keV), a spherical energy analyzer, and a 16-channel electron multiplier detector. The spectrum shows a Cu 2p peak at 932.5 eV. This binding energy is characteristic of a Cu^{1+} or Cu^0 charge state. The Cu^{2+} charge state, which should occur at a slightly higher energy, was not observed. This is because Cu^{2+} acceptors, observed by Fourier transform infrared (FTIR) spectroscopy (Figure 5.5), were present at a much lower concentration than the Cu^{1+} or Cu^0 impurities. The authors proposed that the XPS peak could arise

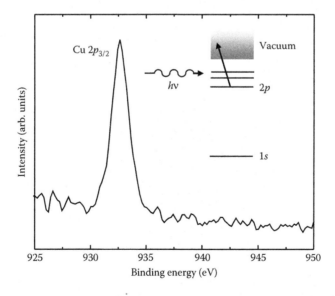

FIGURE 11.12 X-ray photoelectron spectroscopy (XPS) spectrum of Cu in ZnO nanocrystals. The photoelectron comes from a Cu 2p level. (After Hlaing Oo, W.M. et al. 2010. *J. Appl. Phys.* 108: 064301.)

from Cu⁰ atoms on the surface of the nanocrystal. As the nanocrystal grows larger, these atoms are overgrown and become incorporated into the nanocrystal lattice.

A related technique that uses the photoelectric effect is *ultraviolet photoelectron spectroscopy* (UPS). In a UPS experiment, a monochromatic source of photons in the hard-UV range (15–40 eV) is used to generate photoelectrons. Light sources include helium discharge lamps and synchrotron sources. From the energy distribution of the photoelectrons, one can infer the valence-band density of states.

11.7 ELECTRON BEAMS

Beams of electrons are used to characterize sample morphology and composition. Scanning electron microscopy, which uses backscattered electrons to form an image of the sample surface, will be discussed in Chapter 12. In this section, we summarize several interactions that electrons have with the semiconductor host and impurities.

When an electron beam interacts with a sample, numerous scattering events change the direction of motion, resulting in a large sampling volume. An example of a Monte Carlo simulation of electron trajectories is shown in Figure 11.13. The range R of an electron is the distance traveled along a particular trajectory. Because there are many changes in direction, the perpendicular distance from the surface is less than the range. The average range for an electron in a solid can be estimated by the empirical formula

$$R = \frac{1}{\rho} K E_0^n \tag{11.26}$$

where ρ is the mass density of the solid (g/cm³); E_0 is the incident electron energy (keV); K and n are empirical parameters. In the energy range 20–200 keV, one can use $K = 0.0428$ and $n = 1.75$ as a rough approximation for R in units of μm (Leamy, 1982).

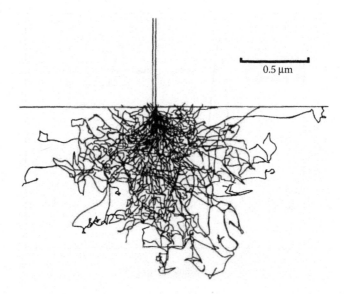

FIGURE 11.13 Monte Carlo simulation of a 20 keV electron beam in iron. (After Goldstein, J.I. et al. 1992. *Scanning Electron Microscopy and X-ray Microanalysis*, 2nd edn. p. 82. New York, NY: Plenum Press. With permission.)

Inelastic scattering arises from a range of physical processes. The electron can excite valence electrons, resulting in collective oscillations called *plasmons*. In this case, the bulk plasma frequency ω_p (Equation 7.19) depends on the concentration N of valence electrons, not just electrons in the conduction band. For silicon, the bulk plasmon results in an energy loss peak at 16.7 eV. Surface plasmons, with a frequency of $\omega_p/\sqrt{2}$, are also excited. Electrons excite lattice vibrations, or phonons, which decay to lower-energy phonons and heat the sample. The deceleration of the primary electrons produces a broad band of radiation, known as *bremsstrahlung* radiation, up to the incident electron energy. The primary electrons also generate numerous electron–hole pairs.

At keV energies, the incident electron can knock out a core electron. When an electron in a higher level fills the hole, a characteristic x-ray may be emitted. This process forms the basis for EDS (Section 11.4). Alternatively, the energy can be carried away by an "Auger electron." For example, an incident electron can knock a K electron out of its shell, and an L electron fills the hole. A second L electron, the Auger electron, is ejected with a characteristic energy (Figure 11.14). This process is denoted a "KLL" transition. Because different elements have different transition energies, *Auger spectroscopy* can be used to identify chemical impurities, using an electron spectrometer as in XPS (Section 11.6). As in the case of a photoelectron, an Auger electron only travels a few nm before being inelastically scattered. Therefore, Auger spectroscopy and XPS are surface sensitive techniques, limited to a depth of 1–2 nm.

An important question is: do energetic electrons create defects? To displace an atom from its lattice site, the bonds must be broken, requiring an energy of 10–30 eV. In contrast to heavy ions, though, an electron is much less massive than the target and imparts a tiny fraction of its energy to the nucleus (Equation 11.2). In practice, MeV electrons are required to produce a significant quantity of defects. To deliberately introduce point defects, *electron irradiation* of samples is performed with Van de Graff or linear electron accelerators. The concentration of defects N (cm^{-3}) produced by electron irradiation is given by

$$(11.27) \qquad N = \eta \phi$$

where η is the defect production rate (cm^{-1}), and ϕ is the electron flux (cm^{-2}).

An early study of defect formation in silicon was performed by Corbett and Watkins (1965). Czochralski-grown silicon samples were irradiated by 0.7–56 MeV electrons at room temperature. The samples contained oxygen impurities that trapped mobile vacancies, forming vacancy–oxygen complexes called A centers (Section 2.4). The EPR signal from A centers was used to monitor the vacancy production rate. Divacancy EPR signals were also monitored. The production

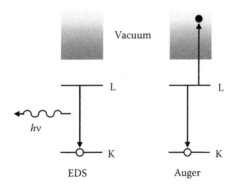

FIGURE 11.14 Energy dispersive x-ray spectrometry (EDS) and Auger processes. In both cases, a core electron fills a hole in a lower level. In EDS, the energy is carried away by an x-ray photon. In Auger spectroscopy, an electron is ejected from the sample.

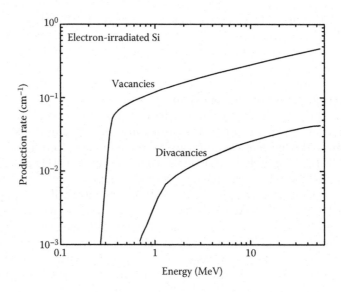

FIGURE 11.15 Production rate of vacancies and divacancies in electron-irradiated silicon at room temperature. (After Corbett, J.W. and G.D. Watkins. 1965. *Phys. Rev.* 138: A555–A560.)

rates are shown for vacancies and divacancies as a function of electron energy in Figure 11.15. The threshold for divacancies, which require the displacement of two silicon atoms, is higher than that for single vacancies.

11.8 POSITRON ANNIHILATION

The anti-electron, or positron, is a positively charged particle that is useful for probing vacancy-type defects in semiconductors (Saarinen et al., 1998; Tuomisto and Makkonen, 2013). When a positron encounters an electron, the particles annihilate via the emission of two 511 keV photons (gamma rays). Energy and momentum are conserved in this process. The lifetime of the positron depends inversely on the electron density. In a solid, typical lifetimes are several hundred ps. If a positron is trapped by an open-volume defect like a vacancy, then the low electron density inside the vacancy results in a longer positron lifetime. This lifetime increase provides a way to identify the types and concentrations of vacancies in a semiconductor.

When an energetic positron enters a sample, it undergoes inelastic scattering processes, similar to an electron (Section 11.7). After a few ps, the positrons thermalize to a Maxwell–Boltzmann (classical gas) distribution. They then diffuse through the sample and can become trapped by defects. In a covalent semiconductor such as silicon, the positron is repelled by the ion cores, and a vacancy acts as a potential well. If the vacancy is neutral or negatively charged, then the positron can become trapped with a binding energy >1 eV. If the vacancy is positively charged, however, then Coulomb repulsion effectively prevents trapping. Positron annihilation is therefore not able to detect positively charged vacancies.

It is possible to tell the difference between negative and neutral vacancies by examining the temperature dependence of the positron trapping rate. Trapping by neutral vacancies is temperature independent. The trapping rate of negative vacancies, on the other hand, increases as the temperature is lowered. One reason for this effect is that a free positron can assume a hydrogen-like orbit around the negative vacancy, and then fall into the potential well. The stability of the hydrogen-like Rydberg state increases as temperature decreases. Therefore, this trapping mechanism is enhanced at low temperatures.

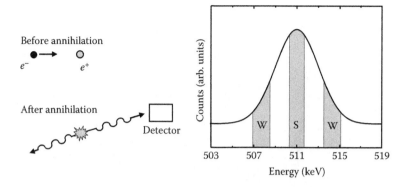

FIGURE 11.16 Positron annihilation spectroscopy. The annihilation of an electron and positron results in the emission of two 511 keV gamma rays. The detected gamma-ray peak is Doppler broadened due to the distribution of electron momenta. The central and wing regions of the peak are denoted S and W, respectively. (After Saarinen, K., P. Hautojärvi, and C. Corbel. 1998. *Semiconductors and Semimetals*, Vol. 51A, ed. M. Stavola, pp. 209–285. San Diego, CA: Academic Press.)

The annihilation rate is proportional to the overlap between the positron and electron wave functions. In general, as the open volume of a defect increases, the annihilation rate decreases (the lifetime increases). In an annihilating electron–positron pair, the momentum of the thermal positron is slow compared to that of the electron (Figure 11.16). From momentum conservation, the net momentum of the emitted photons is approximately equal to the electron momentum. The Doppler shift of one of the photons is given by

(11.28) $$\Delta E_\gamma = \tfrac{1}{2} c p$$

where p is the component of the electron momentum along the direction of the photon emission. Large Doppler shifts, due to annihilation with high-momentum electrons that orbit in the ion cores, result in broadening of the 511 keV line. Valence electrons, in contrast, have low momenta and give negligible Doppler shifts.

For bulk samples, a radioactive isotope is typically used as a positron source. Sodium-22, for example, decays to neon-22 via the emission of a positron and a 1.28-MeV photon, with a half-life of 2.6 years. Positron energies up to 0.54 MeV are emitted. The radioactive isotope is sandwiched between two pieces of identical bulk samples. *Scintillation detectors* are used to detect the 1.28 MeV and 511 keV photons, which serve as the "start" and "stop" signals for determining the lifetime. The scintillator is a material that emits light when an energetic particle impinges upon it; the light is then detected by a PMT. Typical scintillators include NaI, bismuth-germanium oxides (BGOs), and various plastics. The "bulk lifetime" refers to the positron lifetime in an intrinsic sample. In general, the measured positron lifetime will be an average between the bulk lifetime and defect lifetimes.

The Doppler-broadened 511 keV line is measured by a germanium detector. To analyze the peak, the number of counts in the central (S) and wing (W) regions are recorded (Figure 11.16). A positron in an open-volume defect will overlap core electrons less than a positron in the bulk. This results in a narrower line (i.e., the S parameter increases while the W parameter decreases). These parameters provide a way to identify specific defects. For zinc vacancies in ZnO, $S_D/S_B = 1.04$ and $W_D/W_B = 0.87$, where D and B subscripts refer to defect and bulk values, respectively. Gallium vacancies in GaN have the same parameters, to within experimental uncertainty (Tuomisto et al., 2003).

To probe defects at μm depths, slow positrons are produced by sending the positrons emitted by a radioactive isotope through a moderator (Schultz and Lynn, 1988). A linear accelerator can also produce positrons via pair production. The slow positrons are then accelerated to the

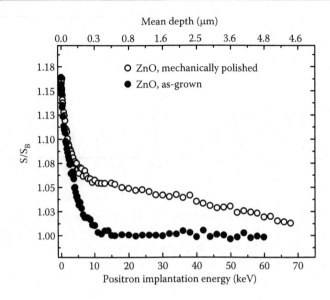

FIGURE 11.17 S parameter (normalized to the bulk) for ZnO as a function of positron implantation energy. The increase in S for polished ZnO indicates that vacancies were introduced several microns into the sample. The mean depth calibration is given by Schultz, P.J. and K.G. Lynn, *Rev. Mod. Phys.*, 60, 701–779, 1988. (After Selim, F.A. et al. 2007. *Phys. Rev. Lett.* 99: 085502.)

desired energy, typically 1–100 keV. Because the mean depth is proportional to the energy, one can determine the vacancy concentration as a function of depth. Such variable-energy beams are useful for studying thin films (Keeble et al., 1993) or near-surface defects. An example is shown for zinc vacancies in ZnO (Figure 11.17) (Selim et al., 2007). At low positron implantation energies (<5 keV), the formation of positronium on the surface results in a narrow 511 keV line (high S parameter). For higher energies, the normalized S parameter (S/S_B) provides a measure of relative vacancy concentration. A sample that was mechanically polished shows evidence of vacancies to a depth of ~6 μm.

11.9 MUONS

In the study of defects in semiconductors, muons are used to gain insight into the behavior of isolated hydrogen (Chow et al., 1998). Positive muons are spin-1/2 particles with a charge $+|e|$, a mass ~1/9 that of a proton, and a mean lifetime of 2.2 μs. A positive muon can capture an electron, forming muonium (Mu), an analogue of hydrogen. Muons are produced at a particle accelerator, by 500 MeV protons that impinge on a carbon or beryllium target. Some of the collisions create pions, which decay into a muon plus a neutrino. From these decays, a spin-polarized beam of positive muons (~4 MeV) can be produced.

The flux of muons is low such that there is only about one muon in the sample at a time, making this technique useful to investigate the properties of isolated hydrogen. When a high-energy muon enters the sample, it loses energy via inelastic collisions (Section 11.7) and thermalizes in a few ns. Initially, nonequilibrium states of Mu are formed. One can then observe these states relax toward equilibrium. Because muons are lighter than protons, they are more mobile, and their zero-point vibrational energies are higher by a few tenths of an eV. The electronic binding properties, however, are essentially identical to those of hydrogen.

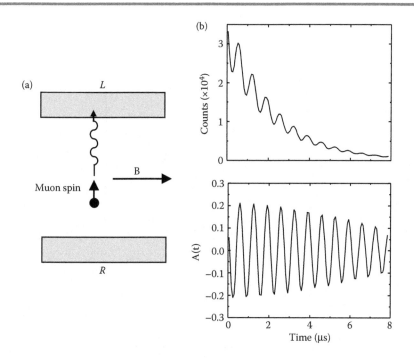

FIGURE 11.18 Muon spin rotation. (a) The spin rotates about the axis defined by the **B**-field. When the muon decays, a positron is emitted preferentially in the direction of the spin. (b) The signal from one of the detectors is shown on the top. The asymmetry parameter A is shown on the bottom. (After Chow, K.H., B. Hitti, and R.F. Kiefl. 1998. In *Semiconductors and Semimetals*, Vol. 51A, ed. M. Stavola, pp. 137–207. San Diego, CA: Academic Press.)

The primary decay mechanism for positive muons is the emission of a positron plus two neutrinos. The positron emission occurs preferentially *in the direction of the spin*, providing researchers with a convenient way to monitor the muon spin as a function of time. Figure 11.18 shows a schematic of an experiment where a magnetic field is applied perpendicular to the initial muon spin. The muon precesses about the magnetic field axis until it decays. Two positron scintillation detectors are situated to the left (L) and right (R) of the muon beam. The signal from one of the detectors is a decaying exponential, with a sinusoidal oscillation due to the spin precession. Assuming the detectors are identical, the asymmetry A is defined

(11.29) $$A(t) = \frac{N_L(t) - N_R(t)}{N_L(t) + N_R(t)}$$

where N_L and N_R are the background-subtracted signals from the L and R detectors, respectively. The plot of A versus t clearly shows the oscillation due to spin precession. To obtain the muon spin rotation (µSR) frequencies, a Fourier transform is performed.

As discussed in Section 8.7, the energetically preferred charge state of hydrogen is positive or negative, depending on the position of the Fermi level. Positive hydrogen occupies a bond-center (BC) location while neutral or negative hydrogen resides in a tetrahedral antibonding (AB) site. In muon-implantation experiments, all three charge states (Mu^0, Mu^+, and Mu^-) may be observed. In GaAs, the paramagnetic Mu^0_{AB} atom sits near three gallium atoms. An isotropic hyperfine interaction between the electron and muon spins results in two µSR frequencies. For the Mu^0_{BC} center, the unpaired electron occupies antibonding orbitals on the neighboring Ga and As atoms, resulting in a

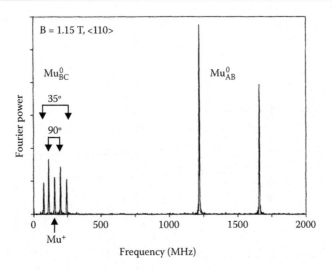

FIGURE 11.19 Muon precession frequencies in GaAs, for a **B**-field aligned along a <110> direction. Mu_{AB}^0 shows two peaks due to an isotropic hyperfine splitting. Mu_{BC}^0 shows two sets of peaks, for complexes oriented 90° and 35° to the **B**-field. (After Chow, K.H., Hitti, B. and Kiefl, R.F., in *Semiconductors and Semimetals* Vol. 51A, pp. 137–207, Academic Press, San Diego, CA, 1998.)

weaker, anisotropic hyperfine splitting. The diamagnetic Mu⁺ and Mu⁻ charge states do not have an unpaired electron. Their μSR frequencies are simply equal to the Larmor frequency, 135.54 MHz/Tesla. Peaks in the μSR Fourier power spectrum correspond to these states (Figure 11.19).

By measuring the μSR peaks as a function of temperature, the donor (0/+) and acceptor (0/−) levels can be determined. To determine the donor level, the Mu_{BC}^0 peaks were monitored (Lichti et al., 2007). As temperature is increased, the rate at which electrons are emitted into the conduction band increases (Equation 9.47). This increase in the Mu_{BC}^0 to Mu_{BC}^+ ionization rate causes the lines to broaden. The broadening of the Mu_{BC}^0 peaks is plotted in Figure 11.20. The increase in linewidth for temperatures above 100 K was modeled by an Arrhenius dependence with an activation energy of 0.17 eV. Hence, the donor level is 0.17 eV below the conduction-band minimum. In the same paper, the acceptor level was determined to be 0.60 eV above the valence-band maximum. The (+/−) level, which lies halfway between these two levels (Figure 8.17), is 0.2 eV above the middle of the band gap, somewhat higher than the value for hydrogen predicted by first-principles calculations (Figure 5.11).

Semiconductors such as InN and ZnO have low conduction-band minima such that the hydrogen (+/−) level lies in the conduction band (Figure 5.11). In that case, the electron falls to the conduction-band minimum. The weakly bound H⁺–electron pair then acts as a shallow hydrogenic donor. In muon-implanted InN, the Mu⁺ and Mu⁰ signals were observed (Davis et al., 2003). Unlike the case in GaAs, the Mu⁰ showed only a small splitting, with a hyperfine constant of 92 kHz. This weak interaction is due to the delocalized nature of the hydrogenic electron wave function, which has a low density at the muon. By plotting the amplitude of the Mu⁺ and Mu⁰ signals as a function of temperature, a donor binding energy of 12 meV was obtained. Qualitatively similar results were found for ZnO (Cox et al., 2001).

11.10 PERTURBED ANGULAR CORRELATION SPECTROSCOPY

Perturbed angular correlation spectroscopy (PACS) is a method to probe the local environment of a radioactive impurity atom (Wichert et al., 1989; Schatz and Weidinger, 1996; Wichert,

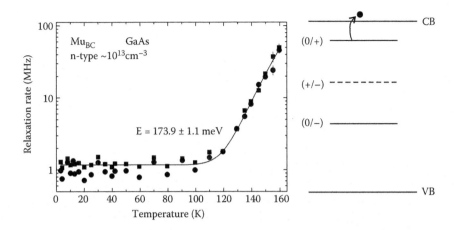

FIGURE 11.20 Plot of the ionization rate (relaxation rate) for the Mu$_{BC}^0$ peak in GaAs. The increase in the rate is due to ionization from the donor (0/+) level to the conduction band. (After Lichti, R.L. et al. 2007. *Phys. Rev. B* 76: 045221. With permission.)

1999). The most commonly used isotope is indium-111, which has a half-life of 2.8 days. It can be obtained commercially, in the form of indium chloride, or by ion implantation at a particle accelerator facility. Indium-111 decays to the excited state of cadmium-111 via electron capture (Figure 11.21). The nucleus decays to an intermediate state via the emission of a 171 keV gamma-ray photon (γ_1). The intermediate state is relatively long lived, with a half-life of 85 ns. The nucleus finally decays to the ground state by emitting a 245 keV photon (γ_2). The γ_1 and γ_2 keV photons are detected by scintillation counters, providing the "start" and "stop" pulse, respectively.

The intermediate-state cadmium nucleus is not perfectly spherical. Instead, it has a positive quadrupole moment, like a cigar or rugby ball. In the presence of an electric field gradient (EFG), the nucleus will precess, similar to a spinning top in a gravitational field. This precession forms the basis for the PACS signal. In the presence of an EFG, the $I = 5/2$ level splits into three levels corresponding to $m_I = \pm 5/2, \pm 3/2$, and $\pm 1/2$. These levels give rise to three precession frequencies ω_1, ω_2, and ω_3, indicated by the transitions in Figure 11.21. For a center with axial (e.g., C_{3v}) symmetry, the ratio of frequencies is $\omega_1:\omega_2:\omega_3 = 1:2:3$. In order for there to be an EFG, the symmetry must be axial or lower. A cadmium-111 atom that resides on a substitutional site with T_d symmetry will not precess and therefore not give rise to a PACS signal.

When the cadmium-111 nucleus decays from its intermediate state to the ground state, the gamma-ray emission is correlated with the nuclear spin direction. It is this angular dependence

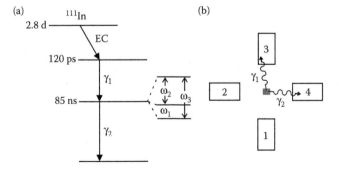

FIGURE 11.21 (a) Nuclear decay diagram for indium-111. (b) A four-detector PACS experiment. (After Skudlik, H. et al. 1992. *Phys. Rev. B* 46: 2172–2182.)

that allows researchers to monitor the spin precession. The gamma-ray emission pattern can be visualized as a lighthouse beam that sweeps across the detectors. Commonly, four scintillation detectors are placed at 90° intervals as shown in Figure 11.21. The coincident count rate N_{ij} is the number of times that detector i produced a pulse, followed by a pulse from detector j a time t later. The experimentally measured spectrum $R(t)$ is defined

$$(11.30) \qquad R(t) = \frac{2}{3}\left[\sqrt{\frac{N_{13}(t)N_{24}(t)}{N_{14}(t)N_{23}(t)}} - 1\right]$$

By performing a Fourier transform on $R(t)$, one can obtain the PACS frequencies.

An interesting application of PACS was the study of acceptor–hydrogen complexes in silicon (Skudlik et al., 1992). Indium-111 ions were implanted into silicon, followed by various methods of hydrogenation. After the formation of indium–hydrogen complexes, the indium decayed to cadmium, producing the PACS signal. The cadmium–hydrogen complexes have C_{3v} symmetry, aligned along <111> directions. Because cadmium is a double acceptor, the cadmium–hydrogen complex is a single acceptor and can have a charge of 0 or −1. In intrinsic silicon, these two charge states result in two distinct complexes. In p-type silicon, only the neutral cadmium–hydrogen complex is observed (Figure 11.22).

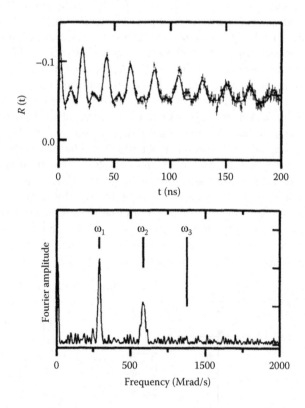

FIGURE 11.22 Perturbed angular correlation spectroscopy (PACS) signal $R(t)$, and Fourier transform, for silicon implanted with indium-111 and hydrogen. The signal arises from the formation of complexes between cadmium-111 and hydrogen. These complexes are aligned along <111> directions and have C_{3v} symmetry. (After Skudlik, H. et al. 1992. *Phys. Rev. B* 46: 2172–2182. With permission.)

11.11 NUCLEAR REACTIONS

Closely related to neutron transmutation doping (Section 4.11), *neutron activation analysis* (NAA) is used to measure the elemental composition of a material. The sample is exposed to thermal neutrons in a nuclear reactor. When a nucleus captures a neutron, it immediately emits a "prompt" gamma ray. If the resulting isotope is unstable, then it will decay with a characteristic half-life, emitting a "delayed" gamma ray in the process. Gamma-ray energies are measured by radiation detectors such as lithium-drifted silicon or ultra-pure germanium. Because the neutrons and gamma rays penetrate through material, NAA is useful for bulk samples.

One advantage of delayed gamma NAA is that the sample can be brought outside the reactor for analysis. The energy of the delayed gamma ray and the half-life of the decay provide an unambiguous identification of the isotope. By comparing the spectrum of emitted gamma rays with a standard exposed to the same neutron flux, the composition of the sample can be determined quantitatively. Prompt gamma NAA can determine the isotopic abundances regardless of whether neutron capture results in an unstable nucleus. In prompt gamma NAA, the gamma rays are detected while the sample is being bombarded by a beam of neutrons.

In *nuclear reaction analysis* (NRA), a high-energy ion beam is used to induce nuclear reactions with host atoms. For example, hydrogen concentration can be determined quantitatively by bombarding a sample with ^{15}N ions (Schatz and Weidinger, 1996). The ^{15}N reacts with a proton to produce ^{12}C plus a helium nucleus (alpha particle). This nuclear reaction, denoted $^{1}H(^{15}N,\alpha)^{12}C$, emits a characteristic 4.4 MeV gamma ray that provides the signature for protons in the sample. The reaction has a sharp resonance for a ^{15}N energy of 6.40 MeV. A ^{15}N ion with an energy above the resonance will penetrate a certain depth into the sample before its energy reaches 6.40 MeV. Therefore, the energy of the ^{15}N ion makes it sensitive to a given depth. By measuring the number of gamma rays versus ^{15}N energy, one can obtain a hydrogen concentration versus depth profile.

PROBLEMS

11.1 A 200 keV proton is scattered by a silicon-28 nucleus at an angle $\theta = 60°$. What is the proton energy after the collision?

11.2 An RBS measurement ($E = 2$ MeV He^+) is performed on a 100 nm thick layer of silicon ($Z = 14, N = 5 \times 10^{28}$ m^{-3}).
a. Calculate the kinematic factor.
b. Calculate the energy loss for a He^+ particle that passes through the layer.
c. Sketch the RBS spectrum, indicating the relevant energies (MeV) on the horizontal axis.

11.3 An RBS measurement (2 MeV He^+) is performed on a ZnS thin film (200 nm). Assume that a He^+ particle in ZnS loses 230 eV per nm. Sketch the RBS spectrum, indicating energies (MeV) on the horizontal axis and relative backscatter yield values on the vertical axis.

11.4 For each property, state the most appropriate characterization technique
Techniques: RBS, EDS, EXAFS, XPS, positron annihilation, μSR, PACS
Properties: Local symmetry of In impurities at low concentrations
Vacancy concentration
Spatially resolved chemical composition of a thin film
Hydrogen donor level
Thin-film thickness
Near-surface (nm) chemical composition
Distance between an impurity/alloy atom (>1% concentration) and its nearest neighbors.

11.5 Using the results of LSS theory (Tables 11.2 and 11.3), estimate the average depth of 100 keV ^{11}B atoms implanted into (a) silicon and (b) germanium.

11.6 A boron-implanted germanium sample is investigated with dynamic SIMS for 10 s. During the first 2 s of the measurement, there is a high boron yield. During the next 8 s, there is very little boron signal. The crater formed by the sputtering process is measured to be 1.3 μm deep. How thick is the boron-doped layer?

11.7 CdTe is doped (alloyed) with Zn atoms, which occupy the substitutional Cd site. The distance between Cd and the Te nearest neighbors is $R = 2.7$ Å.
   a. Estimate the binding energy (keV) for a Zn K electron.
   b. Calculate the first 3 maxima (keV) for EXAFS, assuming no phase shift ($\varphi = 0$).
   c. Sketch the x-ray absorption spectrum, indicating the relevant energies on the horizontal axis.

11.8 An XPS experiment is performed with Al K_α x-rays (1.487 keV). A photoelectron is ejected with a kinetic energy of 554 eV. What was the electron binding energy?

11.9 Estimate the average range for 60 keV electrons in germanium (density 5.3 g/cm^3).

11.10 A 515 keV gamma ray is produced due to the annihilation of an electron and positron inside a semiconductor. The gamma ray travels in the same direction as the electron momentum.
   a. What is the energy of the other gamma ray photon?
   b. Calculate the electron momentum (kg m/s).

REFERENCES

Agarwal, B.K. 1991. *X-ray Spectroscopy: An Introduction*. Berlin: Springer-Verlag.

Benninghoven, A. 1976. Surface analysis by means of ion beams. *Crit. Rev. Solid State Sci.* 6: 291–316.

Bohr, N. 1915. On the decrease of velocity of swiftly moving electrified particles in passing through matter. *Phil. Mag. (Series 6)* 30: 581–612.

Bragg, W.H. and R. Kleeman. 1905. On the α particles of radium, and their loss of range in passing through various atoms and molecules. *Phil. Mag. (Series 6)* 10: 318–340.

Carlson, T.A. 1975. *Photoelectron and Auger Spectroscopy*. New York, NY: Plenum Press.

Carter, G. and W.A. Grant. 1976. *Ion Implantation of Semiconductors*. New York, NY: John Wiley & Sons.

Chow, K.H., B. Hitti, and R.F. Kiefl. 1998. μSR on muonium in semiconductors and its relation to hydrogen. In *Semiconductors and Semimetals*, Vol. 51A, ed. M. Stavola, pp. 137–207. San Diego, CA: Academic Press.

Chu, W.-K., J.W. Mayer, and M.-A. Nicolet. 1978. *Backscattering Spectrometry*. New York, NY: Academic Press.

Corbett, J.W. and G.D. Watkins. 1965. Production of divacancies and vacancies by electron irradiation of silicon. *Phys. Rev.* 138: A555–A560.

Cox, S.F.J., E.A. Davis, S.P. Cottrell et al. 2001. Experimental confirmation of the predicted shallow donor hydrogen state in zinc oxide. *Phys. Rev. Lett.* 86: 2601–2604.

Davis, E.A., S.F.J. Cox, R.L. Lichti, and C.G. Van de Walle. 2003. Shallow donor state of hydrogen in indium nitride. *Appl. Phys. Lett.* 82: 592–594.

Dearnaley, G., J.H. Freeman, G.A. Gard, and M.A. Wilkins. 1967. Implantation profiles of ^{32}P channeled into silicon crystals. *Can. J. Phys.* 46: 587–595.

Eisberg, R. and R. Resnick. 1985. *Quantum Physics of Atoms, Molecules, Solids, Nuclei, and Particles*. New York, NY: John Wiley & Sons.

Feldman, L.C. and J.W. Mayer. 1986. *Fundamentals of Surface and Thin Film Analysis*. New York, NY: North-Holland.

Feldman, L.C., J.W. Mayer, and S.J. Picreaux. 1982. *Materials Analysis by Ion Channeling*. New York, NY: Academic Press.

Fons, P., H. Tampo, A.V. Kolobov et al. 2006. Direct observation of nitrogen location in molecular beam epitaxy grown nitrogen-doped ZnO. *Phys. Rev. Lett.* 96: 045504 (4 pages).

Gemmel, D.S. 1974. Channeling and related effects in the motion of charged particles through crystals. *Rev. Mod. Phys.* 46: 129–227.

Goldstein, J.I., D.E. Newbury, P. Echlin et al. 1992. *Scanning Electron Microscopy and X-ray Microanalysis,* 2nd edn. p. 82. New York, NY: Plenum Press.

Hiraki, A., M.-A. Nicolet, and J.W. Mayer. 1971. Low-temperature migration of silicon in thin layers of gold and platinum. *Appl. Phys. Lett.* 18: 178–181.

Hlaing Oo, W.M., M.D. McCluskey, J. Huso et al. 2010. Incorporation of Cu acceptors in ZnO nanocrystals. *J. Appl. Phys.* 108: 064301 (3 pages).

Ho, P.S. and J.E. Lewis. 1976. Deconvolution method for composition profiling by Auger sputtering technique. *Sur. Sci.* 55: 335–348.

Keeble, D.J., M.T. Umlor, P. Asokakumar, K.G. Lynn, and P.W. Cooke. 1993. Annealing of low-temperature GaAs studied using a variable-energy positron beam. *Appl. Phys. Lett.* 63: 87–89.

Koningsberger, D.C. and R. Prins, eds. 1988. *X-ray Absorption: Principles, Applications, Techniques of EXAFS, SEXAFS and XANES.* New York, NY: John Wiley & Sons.

Leamy, H.J. 1982. Charge collection scanning electron microcopy. *J. Appl. Phys.* 53: R51–R80.

Lichti, R.L., H.N. Bani-Salameh, B.R. Carroll, K.H. Chow, B. Hitti, and S.R. Kreitzman. 2007. Donor and acceptor energies for muonium in GaAs. *Phys. Rev. B* 76: 045221 (5 pages).

Lindhard, J., M. Scharff, and H.E. Schiøtt. 1963. Range concepts and heavy ion ranges. *Mat. Fys. Medd. Dan. Vid. Selsk.* 33(14): 1–42.

Mayer, J.W., L. Eriksson, and J.A. Davies. 1970. *Ion Implantation in Semiconductors.* New York, NY: Academic Press.

McCluskey, M.D., L.T. Romano, B.S. Krusor, N.M. Johnson, T. Suski, and J. Jun. 1998. Interdiffusion of In and Ga in InGaN quantum wells. *Appl. Phys. Lett.* 73: 1281–1283.

Moll, A.J., K.M. Yu, W. Walukiewicz, W.L. Hansen, and E.E. Haller. 1992. Coimplantation and electrical activity of C in GaAs: Stoichiometry and damage effects. *Appl. Phys. Lett.* 60: 2383–2385.

Muilenberg, G.E., ed. 1979. *Handbook of X-ray Photoelectron Spectroscopy.* Eden Prairie, MN: Perkin-Elmer.

Nicolet, M.-A. 1979. Backscattering spectrometry and related analytic techniques. In *Nondestructive Evaluation of Semiconductor Materials and Devices,* ed. J.N. Zemel, pp. 581–630. New York, NY: Plenum Press.

Ridgway, M.C., C.J. Glover, K.M. Yu et al. 2000. Ion-dose-dependent microstructure in amorphous Ge. *Phys. Rev. B* 61: 12586–12589.

Runyan, W.R. and T.J. Shaffner. 1998. *Semiconductor Measurements and Instrumentation.* New York, NY: McGraw-Hill.

Saarinen, K., P. Hautojärvi, and C. Corbel. 1998. Positron annihilation spectroscopy of defects in semiconductors. In *Semiconductors and Semimetals,* Vol. 51A, ed. M. Stavola, pp. 209–285. San Diego, CA: Academic Press.

Schatz, G. and A. Weidinger. 1996. *Nuclear Condensed Matter Physics: Nuclear Methods and Applications.* New York, NY: John Wiley & Sons.

Schroder, D.K. 2006. *Semiconductor Material and Device Characterization,* 3rd ed. New York, NY: John Wiley & Sons.

Schultz, P.J. and K.G. Lynn. 1988. Interaction of positron beams with surfaces, thin films, and interfaces. *Rev. Mod. Phys.* 60: 701–779.

Selim, F.A., M.H. Weber, D. Solodovnikov, and K.G. Lynn. 2007. Nature of native defects in ZnO. *Phys. Rev. Lett.* 99: 085502 (4 pages).

Skudlik, H., M. Deicher, R. Keller et al. 1992. H passivation of shallow acceptors in Si studied by use of the perturbed-$\gamma\gamma$-angular-correlation technique. *Phys. Rev. B* 46: 2172–2182.

Teo, B.K. 1986. *EXAFS: Basic Principles and Data Analysis.* Berlin: Springer-Verlag.

Tuomisto, F. and I. Makkonen. 2013. Defect identification in semiconductors with positron annihilation: Experiment and theory. *Rev. Mod. Phys.* 85, 1583–1631.

Tuomisto, F., V. Ranki, K. Saarinen, and D.C. Look. 2003. Evidence of the Zn vacancy acting as the dominant acceptor in *n*-type ZnO. *Phys. Rev. Lett.* 91: 205502 (4 pages).

Wang, L., L. Hsu, E.E. Haller et al. 1996. Ga self-diffusion in GaAs isotope heterostructures. *Phys. Rev. Lett.* 76: 2342–2345.

Wichert, T. 1999. Perturbed angular correlation studies of defects. In *Semiconductors and Semimetals*, Vol. 51B, ed. M. Stavola, pp. 297–405. San Diego, CA: Academic Press.

Wichert, Th., M. Deicher, G. Grübel, R. Keller, N. Schulz, and H. Skudlik. 1989. Indium-defect complexes in silicon studied by perturbed angular correlation spectroscopy. *Appl. Phys. A* 48: 59–85.

Williams, P. 1983. Secondary ion mass spectrometry. In *Applied Atomic Collision Physics*, Vol. 4, ed. S. Datz, pp. 327–377. New York, NY: Academic Press.

Yu, K.M., W. Walukiewicz, T. Wojtowicz et al. 2002. Effect of the location of Mn sites in ferromagnetic $Ga_{1-x}Mn_xAs$ on its Curie temperature. *Phys. Rev. B* 65: 201303 (4 pages).

Ziegler, J.F., ed. 1984. *Ion Implantation Science and Technology.* New York, NY: Academic Press.

Microscopy and Structural Characterization

12

In this chapter, we discuss characterization techniques that provide spatially resolved information about semiconductor materials, from micron to Angstrom length scales. These methods provide a way to identify extended defects such as dislocations and stacking faults. As the detail continues to reach smaller length scales, our atomic-level understanding of these defects improves. The chapter begins with a discussion of optical microscopy which, when combined with selective etchants, allows researchers to rapidly evaluate defect concentrations and types. We then discuss techniques that involve scanning electron microscopy, including cathodoluminescence and electron beam induced current measurements. After a brief summary of x-ray and electron-beam diffraction, the principles of transmission electron microscopy are outlined. We conclude with scanning-probe techniques that push spatial resolution to the atomic scale and elucidate phenomena in nanoscale semiconductors.

12.1 OPTICAL MICROSCOPY

One of the most venerable pieces of characterization equipment is the optical microscope. A compound microscope consists of two lenses, an objective and an eyepiece. In the reflection geometry, a light source illuminates the sample surface. An imperfection on the surface, represented as an arrow in Figure 12.1, scatters the light. As shown by the ray-tracing diagram, the light emanating from the arrow tip focuses at point *I*. This is a magnified, real image of the object. The eyepiece magnifies this image, and the lens of the human eye collects the rays and projects an image onto the retina. Alternatively, a camera lens projects the image onto the plane of a charge-coupled device (CCD) array or photographic film. The total magnification of the compound microscope is the product of the objective and eyepiece magnifications.

When the light source shines at normal incidence onto the sample, the technique is called *bright field* microscopy. A more useful technique for observing surface irregularities is *dark field* microscopy, in which the light source is incident at an oblique angle. Light that reflects off a smooth horizontal surface (specular reflection) does not travel through the objective and is therefore suppressed, while light scattered from surfaces of pits, bumps, and other morphological

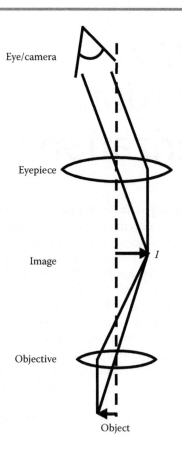

FIGURE 12.1 Compound optical microscope.

features is enhanced (Schroder, 2006). Another method for enhancing defect images is *interference contrast* microscopy, where the reflected or transmitted light is combined with a reference beam. Light that scatters off defects in the sample experiences a change in phase. The phase of the reference beam is adjusted, using a variable phase changer, to bring out the features of interest.

An optical microscope can be adapted to perform spatially resolved photoluminescence (PL) or Raman spectroscopy. Such experiments are useful for inhomogeneous materials, small samples, or device structures. Laser light reflects off a beamsplitter, passes through the objective lens, and focuses on the sample (Figure 12.2). The emitted light travels up through the objective and beamsplitter, and is focused into a spectrometer. Using this apparatus, one can collect a PL spectrum from a small spot. The radius of the diffraction-limited laser spot, or *Airy disk*, is given by (Hecht, 2002).

$$(12.1) \qquad r = \frac{0.61\lambda}{\text{NA}}$$

where λ is the wavelength. NA is the *numerical aperture* of the objective:

$$(12.2) \qquad \text{NA} = n \sin \theta$$

where n is the refractive index of the medium surrounding the lens (typically $n = 1$) and θ is the half-angle of the light cone. The objective can also be described by the *f*-number; for small angles $f/\# = 1/(2\,\text{NA})$.

Chapter 12—Microscopy and Structural Characterization   307

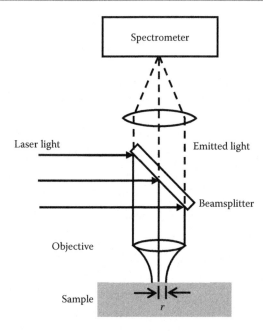

FIGURE 12.2 Photoluminescence (PL) microscope. Laser light (solid lines) reflects off a beamsplitter and is focused by the objective onto the sample surface. The radius of the focused spot is r. Light emitted by the sample (dashed lines) is focused into a spectrometer.

The spatial resolution is defined by the radius of the Airy disk. Depending on the exact resolution criterion, factors slightly different than 0.61 may be used in Equation 12.1, but these variations do not alter the result significantly. From Equation 12.1, an objective lens with NA = 0.3 and a light wavelength $\lambda = 532$ nm provides a resolution of ∼1 μm. Hence, optical microscopy measurements are generally useful for observing feature sizes that are a micron or larger. Two methods that offer improved resolution are electron microscopy (Sections 12.3 through 12.6) and scanning probe microscopy (Section 12.7).

When used in conjunction with chemical etchants, optical microscopy is a powerful technique for characterizing dislocations. *Chemical etching* of semiconductors is not only important in many device technology process steps but also in several analytical preparation techniques (Runyan and Shaffner, 1998). To prepare a semiconductor wafer, the desired crystal geometry is first approximated with mechanical means such as a diamond saw or lapping. Then, chemical etching, or chemomechanical polishing, is used to obtain damage-free surfaces. A large number of etching formulae have been developed for many different purposes. Table 12.1 contains formulations for several semiconductors.

The chemical reactivity of etching compositions is sensitive to strains, impurities (which affect the Fermi level), or other defects near the semiconductor surface. These local effects can lead to preferential etching. The locally accelerated or decelerated etching rate generates a pattern on the crystal surface which is, in many cases, correlated with defects. An example is shown for dopant striations in Te-doped InSb (Figure 4.5). To obtain that image, Witt and Gatos (1966) etched a {211} surface with a 5:3:3:11 mixture of HNO_3:CH_3COOH:HF:H_2O. The samples were observed using interference contrast, adjusted to maximize the contrast from the striations while minimizing the contrast from scratch marks due to mechanical polishing.

Dislocations can be made visible through preferential etching. A preferentially etched {100} surface of a partially dislocated ultrapure germanium crystal is shown in Figure 12.3 (Haller et al., 1977). The etchant was similar to the "No. 1" recipe listed in Table 12.1. The dark large etch

TABLE 12.1 Selected Etchants for Several Semiconductors

Name	Composition	Principal Uses
Germanium		
CP-4	HF:HNO$_3$:CH$_3$COOH:Br$_2$ (5:3:3:0.1)	Etches {111} and {100}, revealing *p-n* junctions and grain boundaries
Hydrogen peroxide	H$_2$O$_2$	Clean, slightly rough surface, good for electrical probe measurements
No. 1	HF:HNO$_3$:10% CuNO$_3$ (1:2:1)	Etches {111} and reveals dislocations; also used for etching silicon
Silver nitrate	HF:HNO$_3$:5% AgNO$_3$ (2:1:2)	Etches {111}, revealing grain boundaries; silver deposits can be removed by cyanide wash
Silicon		
Dash	HF:HNO$_3$:CH$_3$COOH (1:3:8)	Etches all surfaces; forms dip pits, following dislocation lines into the crystal
Secco	HF:4% K$_2$Cr$_2$O$_7$ (2:1)	Etches all surfaces, especially {100}, and reveals dislocations
Sirtl	HF:50% CrO$_3$ (1:1)	Etches {111}
White	HF:HNO$_3$ (1:3)	Chemical polish
GaAs		
AB	HF:0.4% AgNO$_3$ and 50% CrO$_3$ (1:2)	Removes material at 2.5 μm/min and reveals dislocations
NaOH	5% NaOH:30% H$_2$O$_2$ (5:1)	Removes material at 10 to 15 μm/s
KOH	Molten KOH, 350°C	Reveals dislocations on {100} surfaces
InP		
Ferric ion	0.4 M solution of Fe^{+3} in HCl	Etches {111} surfaces; also etches InSb, InAs, and GaAs
Huo	HBr:H$_2$O$_2$:H$_2$O:HCl (10:1:10:10)	Reveals dislocations on {100} and {111} surfaces

Source: After Holt, D.B. and B.G. Yacobi. 2007. *Extended Defects in Semiconductors*. New York, NY: Cambridge University Press; Schroder, D.K. 2006. *Semiconductor Material and Device Characterization*, 3rd ed., New York, NY: John Wiley & Sons; Holmes, P.J. 1962. *The Electrochemistry of Semiconductors*, New York: Academic Press.

Note: Volume ratios are approximate and vary in the literature. For aqueous solutions, the weight percentages of the solutes are indicated.

FIGURE 12.3 Dislocations in ultrapure germanium revealed by preferential etching. (After Haller, E.E. et al. 1977. *Radiation Effects in Semiconductors 1976, Inst. Phys. Conf. Ser. No. 31*, eds. N.B. Urli and J.W. Corbett, pp. 309–318. Bristol, UK: Institute of Physics.)

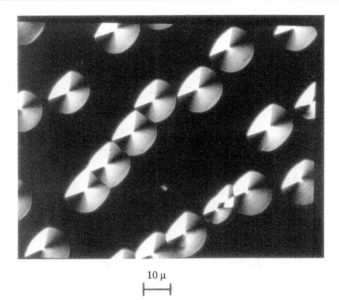

10 μ
⊢——⊣

FIGURE 12.4 Etch pits on a {113} surface of germanium.

pits in the right half of the photo are due to dislocations intersecting the surface. The etching rate near the dislocation core is increased because of the local strain. The structure of the pits is caused by the dependence of etching rate on crystal orientation. On the left-hand side, the dislocation-free section, one sees a large concentration of small, smooth pits without internal structures. It is believed that these pits are correlated with hydrogen or vacancy clusters in the crystal. In the absence of dislocations, the excess hydrogen nucleates and forms hydrogen-filled bubbles. Etch pits on a {113} surface of germanium are shown in Figure 12.4. A special etchant composition (HF:H_2O_2:Cu(NO_3)$_2$, 2:1:1) was developed to obtain preferential etching for this unusual orientation. The pits are formed by four {111} planes, clearly visible in the picture.

As discussed in Section 2.7, dislocations that occur at regular intervals along a grain boundary were first observed in germanium (Vogel et al., 1953). Figure 12.5 shows a dark-field optical micrograph of a {110} surface of germanium. The sample was etched with the CP-4 etchant (Table 12.1). A line of conical etch pits is observed, due to dislocations along a low-angle grain boundary (Figure 2.21).

In general, chemical etching combined with optical microscopy provides a convenient way to estimate the dislocation density in a semiconductor wafer. Etchants that decorate dislocations with metal precipitates provide an additional contrasting agent. Dash (1957) used copper precipitates to image dislocations in silicon, using an infrared (IR) transmission microscope. The IR light transmitted through the silicon but was absorbed by the copper precipitates. A disadvantage of this technique is that copper contamination may ruin the sample for other uses. Along with the elemental semiconductors silicon and germanium, the development of III-V semiconductors such as GaAs and InP relied heavily on preferential etching techniques. For example, the dislocation distribution on a (100) LEC GaAs wafer was measured by Angilello et al. (1975), using a KOH etch.

An optical microscopy technique useful for studying strain fields is *stress birefringence*. A crystal is placed between two polarizers in a cross-polarized geometry (i.e., the axes of the polarizers are orthogonal). Light is transmitted through the polarizers and the sample. For an unstressed cubic crystal, there will be total light extinction and the image will be dark. Internal stresses introduce birefringence, which rotates the polarization of the transmitted light (Iqbal, 1980). This results in a bright spot in the image. Because dislocations introduce

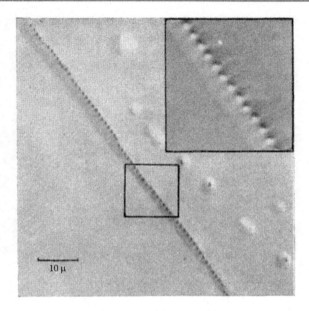

FIGURE 12.5 Line of etch pits due to dislocations along a grain boundary. (After Vogel, F.L. et al. 1953. *Phys. Rev.* 90: 489–490. With permission.)

a strain field, one can identify dislocations by analyzing the stress birefringence image (Holt and Yacobi, 2007).

12.2 SCANNING ELECTRON MICROSCOPY

As discussed in Section 12.1, the spatial resolution of optical microscopy is diffraction limited. Due to the wavelength of light, the minimum spot size is on the order of a μm. Accelerated electrons, on the other hand, have a much shorter wavelength. The de Broglie wavelength for a particle of mass m and kinetic energy E is

(12.3)
$$\lambda = \frac{h}{p} = \frac{h}{\sqrt{2mE}}$$

where h is Planck's constant. From Equation 12.3, an electron with an energy $E = 1$ keV has a wavelength of only 0.4 Å. In typical scanning electron microscopy (SEM) applications, the electron is "particle-like" such that diffraction does not significantly degrade the resolution. One factor that does limit spatial resolution is aberration of the magnetic lenses. In practice, an electron beam in an SEM can be routinely focused down to a spot size of less than 10 nm.

An SEM system consists of the following components (Goldstein et al., 1992) (Figure 12.6):

- An *electron gun* to produce electrons of energy 1 to 40 keV. Typical cathodes are tungsten or lanthanum hexaboride (LaB_6) filaments that are heated by a current. Alternatively, field emission from a sharp tungsten tip can produce electrons without a heating current. The emitted electrons are then accelerated to the desired energy.
- *Magnetic lenses* to focus the electron beam onto the sample.
- A *deflection system* to scan the electron beam across the sample.
- A *detection system* that converts the backscattered and secondary electrons into an electrical signal. Scintillation detectors may be used for this purpose.

Chapter 12—Microscopy and Structural Characterization   **311**

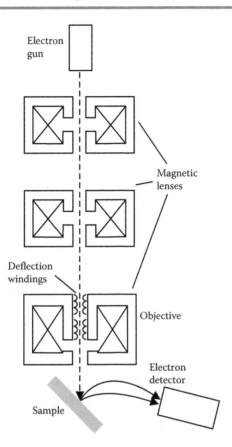

FIGURE 12.6 Scanning electron microscope (SEM). (After Goldstein, J.I. et al. 1992. *Scanning Electron Microscopy and X-ray Microanalysis*, 2nd ed. New York, NY: Plenum Press.)

Unlike the optical microscope shown in Figure 12.1, an SEM does not project an image of the sample onto a CCD array or human retina. Instead, the detector output is recorded point-by-point as the electron beam raster scans the sample. An image is then *reconstructed* from the collected data.

The lenses in an SEM are based on electromagnets. In an electromagnet, wires wound in a coil produce a magnetic flux that is concentrated in iron. The magnetic lens consists of two axially symmetric iron pieces, or "pole-pieces," separated by a gap (Figure 12.7). This gap results in a fringing B field. The electron beam travels through the bore with a speed v_z and experiences a force due to this fringing field:

$$(12.4) \qquad \mathbf{F} = e(v_z \hat{\mathbf{z}} \times \mathbf{B})$$

The direction of this force is perpendicular to **B**, resulting in a rotational velocity $\mathbf{v_{in}}$, directed into the page in Figure 12.7. This rotational velocity interacts with the **B** field to produce a radial force

$$(12.5) \qquad \mathbf{F_r} = e(\mathbf{v_{in}} \times \mathbf{B})$$

that bends the electron toward the z axis. The focal length can be adjusted by varying the current through the coil windings. Typical focal lengths for magnetic lenses are 10 to 40 mm. To

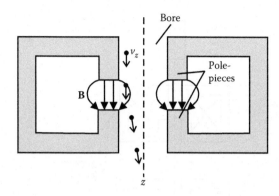

FIGURE 12.7 Cross-sectional diagram of a magnetic lens. The fringing **B**-field causes the electrons to bend toward the z-axis. (After Goldstein, J.I. et al. 1992. *Scanning Electron Microscopy and X-ray Microanalysis*, 2nd ed. New York, NY: Plenum Press.)

minimize spherical aberrations, the NA (Equation 12.2) is kept to a very low value, in the range 10^{-3} to 10^{-2}. Chromatic aberrations, caused by the energy spread in the electron beam, can be reduced by using a field emission electron gun.

In an SEM, the electrons emitted by the electron gun are focused by a series of condenser lenses. The last lens, closest to the sample, is referred to as the objective. Deflection coils in the bore of the objective deflect the electron beam in the x and y directions. By altering the current in these coils, the beam is raster scanned across the sample surface.

When an electron enters the sample, it undergoes numerous elastic and inelastic collisions (Section 11.7). A fraction of the incident electrons undergo large-angle collisions with nuclei and are scattered out of the sample. These are called *backscattered electrons*. In addition, the primary electron can knock low-energy *secondary electrons* out of the sample. A special type of secondary electrons, Auger electrons, was discussed in Section 11.7. To collect the maximum number of electrons, a positive voltage of ~300 V is applied to a wire grid in front of the detector. If one wants to detect backscattered electrons and reject the low-energy secondary electrons, then a repulsive negative voltage (−100 V) is applied. The production of backscattered electrons is proportional to the average atomic number Z of the sample. This effect provides compositional contrast, also called *Z contrast*.

The objective can focus the electron beam to a spot size diameter less than 10 nm, but the effective probe area is significantly larger. The reason for this loss of resolution can be understood from the Monte Carlo simulation in Figure 11.13. When the electrons interact with the sample, they undergo large-angle scattering with negligible energy loss (elastic scattering) and low-angle scattering with significant energy loss (inelastic scattering). The cumulative result of all the scattering events is a broadening of the electron profile inside the sample.

The Auger, secondary, and backscattered electrons originate from different regions of the probed volume. As discussed in Section 11.7, the low-energy Auger electrons come from a layer 1 to 2 nm from the surface. Secondary electrons, which also have energies that are low compared to the primary electron, come from a depth of 5 to 50 nm. Backscattered electrons have the highest energies and come from a depth that corresponds to ~1/2 of the electron range (Equation 11.26). A schematic of these regions is presented in Figure 12.8. The common detection mode for high-resolution SEM is secondary-electron detection. It is also possible to have an EDS detection system integrated into the SEM, allowing for compositional imaging of a sample.

SEM images of copper-doped ZnO and MgZnO samples are shown in Figure 12.9 (Huso et al., 2009). The atomic copper concentrations, measured by EDS, were 0.25% ($Cu_{0.0025}Zn_{0.9975}O$) and 5% ($Cu_{0.05}Mg_{0.30}Zn_{0.65}O$). The samples were made by mixing ZnO, MgO, and Cu_2O powders in

Chapter 12—Microscopy and Structural Characterization **313**

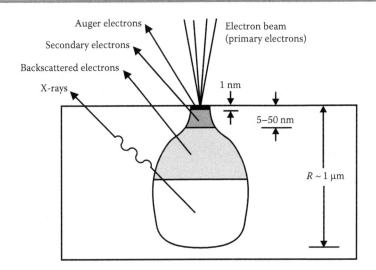

FIGURE 12.8 Region probed by an electron beam. Emitted electrons and x-rays originate from different regions of the probed volume. (After Schroder, D.K. 2006. *Semiconductor Material and Device Characterization*, 3rd ed., New York, NY: John Wiley & Sons.)

a ball mill, pressing them into pellets ~3 mm in diameter, and sintering in the air at 1100°C. During the sintering process, grain growth occurs. It can be seen from Figure 12.9 that the grains in the 0.25% sample are larger than those in the 5% sample. The likely reason for this difference is the lower mobility of atoms in MgZnO as compared to ZnO; sintering at a higher temperature would result in larger grain sizes.

In summary, it is instructive to compare electron-beam microscopy with RBS, another particle-beam method used in semiconductor characterization (Section 11.1). RBS provides information on the elemental composition of the top few thousand Angstroms of a semiconductor. RBS has poor lateral resolution but is quite a good "depth" microscope. High-resolution SEM, on the other hand, can resolve spatial features as small as ~0.2 μm. The large variation in electron range and the large number of possible interactions lead to rather poor depth resolution. The next two sections discuss SEM-based techniques that provide optical and electrical information that complement the morphological information obtained from secondary electrons.

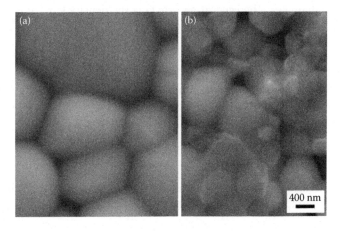

FIGURE 12.9 Secondary-electron scanning electron microscope (SEM) images of copper-doped ZnO (a) and MgZnO (b). (After Huso, J. et al. 2009. *Appl. Phys. Lett.* 94: 061919. With permission.)

12.3 CATHODOLUMINESCENCE

As discussed in Section 12.2, electrons can undergo elastic scattering, which results in high-energy backscattered electrons. Electrons that are not backscattered undergo a variety of inelastic processes. The energy of a primary electron can go toward the creation of

- Secondary electrons (high-resolution SEM)
- Characteristic x-rays (EDS)
- Auger electrons
- Electron–hole pairs
- Lattice phonons.

Cathodoluminescence (CL) experiments measure the light emitted when electron–hole pairs recombine (Yacobi and Holt, 1986). Because CL is integrated with an SEM, it allows one to produce an image with good spatial resolution.

In contrast to PL (Section 10.2), in which the excitation photon generates one electron–hole pair, a single primary electron can generate thousands of electron–hole pairs. The average energy required for a particle to produce a single electron–hole pair is referred to as the ionization energy E_i. Empirically, the ionization energy is given by (Klein, 1968)

$$(12.6) \qquad\qquad E_i = 2.8E_g + M$$

where E_g is the band gap; M depends on the material and is on the order of 1 eV. From Equation 12.6, for $E_g = 2$ eV and $M = 0.4$ eV, a 60 keV electron produces 10,000 electron–hole pairs.

The number of electron–hole pairs generated each second is

$$(12.7) \qquad\qquad N = \frac{V_b I_b (1-\gamma)}{E_i}$$

where V_b and I_b are the electron-beam voltage and current, respectively, and γ is a loss factor due mainly to backscattered electrons. γ is approximately 0.1 for silicon and 0.2 to 0.25 for GaAs. Electrons and holes are produced in a volume V that is determined by the electron range (Equation 11.26). This volume has dimensions on the order of a μm and is much larger than the electron beam diameter. The *generation rate* G is the density (cm^{-3}) of electron–hole pairs generated per second within the generation volume:

$$(12.8) \qquad\qquad G = N/V$$

The relatively large generation volume limits the spatial resolution of any SEM/CL measurement. The resolution of a CL image is further limited by the fact that minority carriers diffuse some distance before recombination occurs. The diffusion length is given by

$$(12.9) \qquad\qquad L = \sqrt{D\tau}$$

where D is the diffusion coefficient and τ is the minority-carrier lifetime. The lifetime is separated into radiative and nonradiative components via

$$(12.10) \qquad\qquad \frac{1}{\tau} = \frac{1}{\tau_{rr}} + \frac{1}{\tau_{nr}}$$

where τ_{rr} and τ_{nr} are the radiative and nonradiative lifetimes, respectively. The values of τ_{rr} and τ_{nr} depend on the concentration and type of defects in the sample. The *internal quantum efficiency* is defined as the fraction of recombination events that are radiative:

(12.11)
$$\eta = \frac{1/\tau_{rr}}{1/\tau} = \frac{1}{1 + \tau_{rr}/\tau_{nr}}$$

A schematic of a CL experiment is shown in Figure 12.10. The focus of the electron beam is located at one focus of an ellipsoidal mirror. An optical fiber is placed at the other focus. Light that is emitted by electrons and holes is collected by the optical fiber and sent to a spectrometer that selects a particular wavelength. As the electron beam raster scans the sample, a two-dimensional (2D) *monochromatic* image of the CL intensity versus position is produced. Alternatively, one can bypass the monochromator and measure the total light intensity integrated over all wavelengths. In that case, the CL image is called *panchromatic*.

Dislocations often produce dark areas in the image. Threading dislocations result in dots on the sample surface, while misfit dislocations, which run parallel to the sample surface, produce straight lines. The contrast comes from nonradiative recombination centers introduced by the dislocations. Dislocations may introduce levels into the gap through dangling bonds or shallow levels caused by strain around the dislocation core. Additionally, and perhaps most importantly, impurities are attracted to the strain around a dislocation, resulting in a Cottrell atmosphere (Section 2.7). These impurities can introduce nonradiative recombination centers, resulting in a dark region in the CL image.

SEM and CL images of GaN, taken at a temperature of 4.6 K, are shown in Figure 12.11 (Paskov et al., 2005). The GaN sample was grown by metalorganic chemical vapor deposition (MOCVD) to a thickness of 1 μm on *r*-plane sapphire. The low temperature provides good spatial resolution, due to the reduced minority carrier diffusion length, as well as sharp peaks in the CL spectrum. The secondary-electron SEM image shows triangular pits on the sample surface. Two monochromatic CL images are shown, for emitted photon wavelengths of 363 nm (3.42 eV) and 376 nm (3.30 eV). These photon energies correspond to excitons bound to stacking faults or impurities, respectively. Both emission wavelengths are enhanced at the triangular pits. The authors proposed that the enhancement of the 3.42 eV emission was due to carrier localization at the pits, whereas the 3.30 eV emission resulted from large concentrations of impurities in and around the pits.

A technique complementary to CL is transport imaging. In CL, the detected luminescence is mapped to the point of excitation. In transport imaging, an optical microscope with a CCD

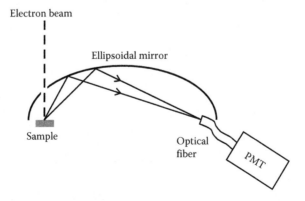

FIGURE 12.10 Cathodoluminescence (CL) experiment. (After Goldstein, J.I. et al. 1992. *Scanning Electron Microscopy and X-ray Microanalysis*, 2nd ed. New York, NY: Plenum Press.)

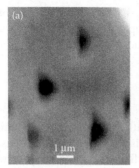 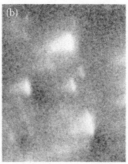 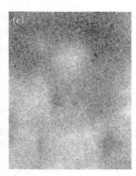

FIGURE 12.11 (a) Secondary-electron scanning electron micrograph (SEM) image of GaN. (b) Monochromatic cathodoluminescence (CL) image for emitted photon wavelength of 363 nm (3.42 eV). (c) Monochromatic CL image for 376 nm (3.30 eV) photons. The enhanced CL emission spots are correlated with triangular pits. (After Paskov, P.P. et al. 2005. *J. Appl. Phys.* 98: 093519 (7 pages). With permission.)

camera, inside the SEM, collects an image of the luminescence. If the SEM is operated in spot mode, with the electron beam fixed at a specific point, then the size and shape of the emission spot allow one to determine the minority carrier diffusion length (Luber et al., 2006). A one-dimensional (1D) line scan provides similar information (Haegel et al., 2009). The SEM can also be operated in the normal raster-scan mode, resulting in a camera exposure that is similar to a CL image, with dark spots corresponding to dislocations and other defects.

12.4 ELECTRON BEAM INDUCED CURRENT MICROSCOPY

Another SEM-based technique is electron beam induced current microscopy (EBIC). In this technique, one measures *current* as the electron beam is raster scanned across the sample surface (Everhart et al., 1964; Varker, 1979). A schematic illustration of an EBIC experiment is shown in Figure 12.12. An energetic electron beam (tens of keV) of an SEM is focused on a semiconductor structure. The electrons and holes, produced during the stopping of the primary electrons, diffuse through the crystal. Some of the minority carriers reach the depletion region of the reverse biased *p-n* junction and are collected. The current that results from this collected charge is plotted as a function of position to produce a 2D image.

If an electron–hole pair is generated *inside* the depletion region, the built-in electric field accelerates the electron toward the *n* side and the hole toward the *p* side. In a closed circuit, this results in current flow given by

$$I_W = |e|G_W WA \tag{12.12}$$

where G_W is the generation rate inside the depletion region, W is the depletion width, and A is the effective area. If an electron–hole pair is generated *outside* the depletion region, then the minority carrier must be within the minority diffusion length L in order to reach the depletion region. This results in a current

$$I_D = |e|G_D LA \tag{12.13}$$

where G_D is the generation rate outside the depletion region. The total current is

$$I = |e|G_W WA + |e|(G_p L_p + G_n L_n)A \tag{12.14}$$

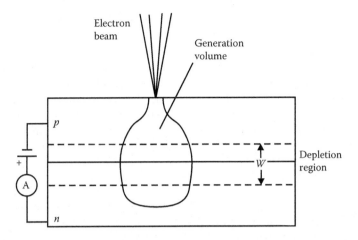

FIGURE 12.12 Electron beam induced current (EBIC) experiment. Minority carriers in the generated volume (electrons on the *p* side and holes on the *n* side) diffuse into the depletion region, resulting in a current. This current is measured as the electron beam is raster scanned across the sample surface. (After Varker, C.J. 1979. *Nondestructive Evaluation of Semiconductor Materials and Devices*, ed. J.N. Zemel, pp. 515–580. New York, NY: Plenum Press.)

where *p* and *n* denote quantities on the *p* and *n* side, respectively. This simplified approach also applies to photovoltaics, in which electrons and holes are generated by photons.

From Equation 12.9, the diffusion length depends on the minority-carrier lifetime that can change by several orders of magnitude within small distances. The lifetime is strongly correlated with defects such as A- and B-swirls in dislocation-free silicon, oxygen precipitates, deep impurities (gold and iron in silicon, copper in germanium), stacking faults, and a host of impurity–defect complexes. By choosing the appropriate geometry of a test device, one can place the EBIC sensitive *p-n* junction at the location of interest. Using this technique, EBIC micrographs have been obtained of dopant striations in silicon (Ravi and Varker, 1974). The contrast arises from changes in the depletion width caused by the varying dopant concentration, which in turn affects the collected current (Equation 12.14). EBIC signals also arise from emitter "pipes" that run through

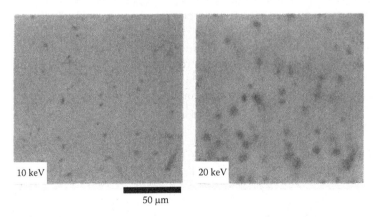

FIGURE 12.13 Electron beam induced current (EBIC) images of $NiSi_2$ precipitates in silicon, which appear as dark regions in the pictures. As the electron energy is increased (from 10 to 20 keV), the generation volume increases and more precipitates are revealed. (After Kittler, M. et al. 1991. *Appl. Phys. Lett.* 58: 911–913. With permission.)

the base layer to the collector of a transistor. These pipes consist of copper precipitates along partially dissociated stacking faults.

An example of an EBIC micrograph is shown in Figure 12.13. In this study, NiSi$_2$ precipitates were formed by diffusion of nickel into FZ silicon at 1050°C, followed by annealing at 500 to 800°C (Kittler et al., 1991). The precipitates are thin platelets, with a typical diameter of 0.8 μm, that lie in silicon {111} planes. Schottky contacts were formed by evaporating gold on the sample surface. Primary electron energies of 10, 20, and 30 keV were used in the EBIC measurement. As the energy increased, the increase in electron range and generation volume revealed more precipitates.

12.5 DIFFRACTION

The previous discussion on electron-beam methods treated electrons as particles, not waves. Additional structural information can be gained by exploiting the wavelike nature of electrons. Electron diffraction was first demonstrated by the famous experiments of Davisson and Germer (1927). Before that, x-ray diffraction (XRD) had been used to determine the structure of solids. In this section, we begin with a summary of XRD and then discuss the electron-beam analogue. Readers interested in the details of XRD can consult numerous comprehensive texts on the subject (e.g., Cullity, 1978).

The basics of XRD can be summarized by the *Bragg diffraction* condition. As discussed in Section 1.2, a crystal has various atomic planes, denoted by their Miller indices (*hkl*). Given a distance *d(hkl)* between planes, x-rays will be reflected when the Bragg diffraction condition is met:

$$2d(hkl)\sin\theta = \lambda \qquad (12.15)$$

where θ is the angle of incidence and reflection, and λ is the x-ray wavelength. Only monochromatic x-rays (single λ) will be considered here. X-rays that are not reflected will be transmitted through the crystal or absorbed. For a cubic cell, the *d* spacings are given by

$$d(hkl) = \frac{a}{\sqrt{h^2 + k^2 + l^2}} \qquad (12.16)$$

where *a* is the lattice constant. With this convention, higher-order reflections give rise to *d* spacings that are a fraction of the actual interplanar distance. For example, (200) and (300) reflections correspond to $d = a/2$ and $a/3$, respectively.

A typical sample geometry is shown in Figure 12.14. The incoming x-ray beam defines the axis. Relative to this axis, a given Bragg diffraction (Equation 12.15) has an angle 2θ. A single-crystal sample must be scanned through various angles, using a mechanical device called a goniometer, to satisfy a Bragg condition. An example of an XRD spectrum is shown in Figure 8.20, where

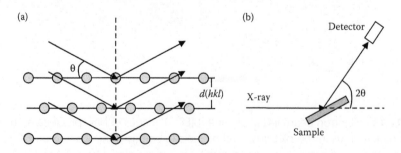

FIGURE 12.14 (a) Bragg diffraction and (b) x-ray diffraction (XRD) experiment.

diffraction peaks from GaN and InGaN are observed. The higher-order InGaN peaks arise from the periodicity of the quantum-well superlattice. A *rocking curve* may be generated by keeping the angle of incidence fixed but varying the angle of reflection. For an ideal crystal, the angle of incidence must equal the angle of reflection, resulting in a sharp rocking curve. Samples that consist of slightly misoriented grains, on the other hand, will have a broad rocking curve. The narrowness of the rocking curve is therefore a figure of merit for overall structural quality.

For a polycrystalline sample, randomly oriented crystallites will diffract at various azimuthal angles about the beam axis (Suryanarayana and Norton, 1998). This results in a "powder diffraction" pattern that consists of a series of concentric rings. By integrating over azimuthal angle, an intensity-versus-2θ spectrum is obtained. For sufficiently small grain sizes, the peaks in the XRD spectrum broaden due to the finite number of atomic planes. For nanocrystals, the broadening of a diffraction peak is given by the Scherrer formula:

$$(12.17) \qquad\qquad \beta \sim \frac{\lambda}{L \cos \theta}$$

where β is the width of the peak (FWHM) in radians, and L is the linear dimension of the nanocrystal. Note that the peak may also be broadened by inhomogeneous strain as well as instrumental limitations.

The wavelike nature of electrons means that, like photons, they also exhibit diffraction. *Low-energy electron diffraction* (LEED) involves electrons with energies <1 keV (Van Hove et al., 1986). Because they barely penetrate into the sample, the electrons are sensitive to the surface atoms. The diffraction pattern is effectively that of a 2D crystal, defined by the periodicity of the atoms on the surface. As discussed in Section 4.6, reflection high-energy electron diffraction (RHEED) occurs when keV electrons are incident at an angle of a few degrees. The small Bragg angle is due to the short wavelength of the electrons (Equation 12.15). This glancing angle makes RHEED sensitive to the surface roughness, enabling MBE growers to achieve submonolayer control.

A beam of electrons transmitting through a thin crystal can diffract off various atomic planes. When these diffracted beams are imaged, one obtains a pattern of spots that depends on the crystal structure and sample orientation (Fultz and Howe, 2001). Bragg diffraction plays an important role in transmission electron microscopy, discussed in Section 12.6. Electron diffraction is also observed in SEM. In that case, the situation is complicated because electrons can scatter inside the material and then diffract. The various scattering events produce "Kikuchi lines" in the SEM image (Randle and Engler, 2000). These lines can be analyzed to determine the crystal structure and orientation, a technique known as *electron backscattering diffraction* (EBSD).

12.6 TRANSMISSION ELECTRON MICROSCOPY

Transmission electron microscopy (TEM) is a powerful technique for investigating the atomic-scale properties of extended defects. The schematic of a TEM is similar to an inverted optical microscope, with electrons and magnetic lenses substituting for photons and glass lenses, respectively (Figure 12.15). First, electrons are emitted by an electron gun and collimated by condenser lenses. The collimated electron beam then transmits through the thin sample and is focused by a series of lenses onto a fluorescent screen or a CCD camera. The lens closest to the sample is referred to as the objective and is followed by the "intermediate" lens. (Sometimes a "projector" lens is placed after the intermediate lens.)

A useful condition for studying defects is *diffraction contrast* imaging (Hirsch et al., 1965). To obtain diffraction contrast, the sample is oriented such that the Bragg diffraction condition (Section 12.5) is satisfied for a single (hkl) plane. In that case, we have two electron beams that

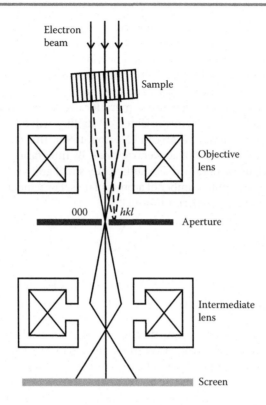

FIGURE 12.15 Transmission electron microscope (TEM). The aperture is placed to transmit the 000 beam (bright-field mode). To take a dark-field image, the aperture would be moved to transmit the *hkl* beam.

exit the sample: the transmitted 000 beam and the diffracted *hkl* beam. An aperture is placed below the sample to transmit either the 000 or the *hkl* beam. When the undiffracted 000 beam is selected, the resulting image is referred to as a "bright-field" image. If a region of the sample diffracts strongly, the intensity of the 000 beam is diminished, resulting in dark contrast.

Regions with high Z scatter electrons strongly and will also give rise to dark contrast. A bright-field image of InGaN quantum wells is shown in Figure 8.21. The high Z of indium makes the quantum wells appear as dark horizontal bands in the micrograph.

When the aperture is positioned such that the diffracted *hkl* beam is transmitted, one obtains a "dark-field" image. For the dark-field imaging condition, a "perfect" crystal gives the brightest image. Defects that strain the crystal reduce the diffraction intensity, resulting in dark regions in the image. An example of a dark-field TEM image is that of the A-swirl in silicon (Figure 2.5). The A-swirl is an extrinsic stacking fault, or an "extra plane" of silicon atoms (Föll and Kolbesen, 1975). The brightness in the image indicates the intensity of the electron beam diffracted from (220) planes. The arrow indicates a <110> direction, which is perpendicular to the diffracting (220) planes. Partial dislocations at the boundary of the A-swirl give rise to a dark loop in the picture. The dark contrast results from a decrease in diffracted intensity caused by the strain around the dislocations.

Because the strain field around a dislocation has a large spatial extent, the region of the diffraction contrast is ~100 times greater than the TEM resolution. To obtain finer resolution, *weak-beam* diffraction contrast is used. In this dark-field technique, the sample is oriented such that a diffracted *hkl* beam is weak. For a perfect crystal, the entire image would be quite dark. Near the core of a dislocation, however, the strain is large enough to satisfy the Bragg diffraction condition. The weak-beam condition therefore shows dislocations as sharp, bright lines. An example

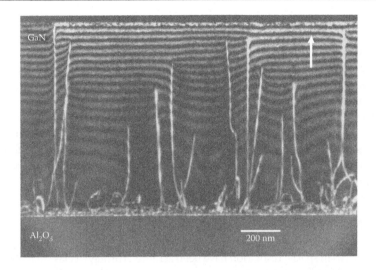

FIGURE 12.16 Transmission electron microscope (TEM) image of GaN grown on sapphire (Al$_2$O$_3$). The arrow indicates the [001] direction, which is perpendicular to the diffracting basal planes. (After Romano, L.T. et al. 2000. *J. Appl. Phys.* 87: 7745–7752. With permission.)

is shown in Figure 12.16 for GaN grown on sapphire (Romano et al., 2000). Numerous misfit dislocations are observed near the highly mismatched GaN/sapphire interface. A few threading dislocations are seen going from the interface to the surface. Because the sample was wedge shaped, the threading dislocations terminate at different heights in the micrograph. The shape of the sample also resulted in interference fringes due to the thickness variation.

In addition to these imaging modes, the user can also examine the diffraction from a specific area of the crystal. To do this, the strength of the intermediate lens is adjusted so that different *hkl* diffracted beams produce focused spots on the fluorescent screen. By consulting standard references (e.g., Fultz and Howe, 2001), these diffraction patterns provide unambiguous information about the crystal structure. The aperture is moved to select a specific area of the sample. This technique, *selected area diffraction* (SAD), is useful for investigating the crystal structures of nanoparticles or inhomogeneous samples.

While the imaging conditions discussed so far provide excellent resolution, it is not at the level of individual atoms. To image columns of atoms, *high-resolution transmission electron microscopy* (HRTEM) is required. In HRTEM, the aperture is widened to allow the 000 and diffracted beams to pass through. When these beams reach the fluorescent screen, they produce an interference pattern. This interference pattern is a real-space image of the columns of atoms in the sample. To see how this works, consider the 000 and a single *hkl* diffracted beam. These beams are focused onto spots in the focal plane of the objective. Like a two-slit experiment, the two sources will produce a sinusoidal interference pattern on the screen, due to the different path lengths traveled by the waves. The sinusoidal pattern is a first-order approximation to the atomic image. This simple treatment only considered a single *hkl* beam; as more diffracted beams are added, the image becomes more refined.

HRTEM provides atomic-level insight into defects and interfaces. Methods for interpreting images, which are not generally straightforward, are reviewed by Schwander et al. (1999). An HRTEM image of a MOS structure is shown in Figure 3.13. The crystalline silicon, with regularly spaced columns of atoms, abruptly ends with a (100) plane. The thin oxide displays a mottled contrast because the atoms are randomly distributed in the amorphous SiO$_2$ glass. The aluminum metallization on top of the thin oxide shows lattice fringes indicating its polycrystalline nature.

As with SEM, an EDS system can be integrated into the TEM, allowing one to obtain the atomic composition of small regions of the sample. In addition, the energy loss of electrons transmitted through the sample can be measured. This technique, *electron energy loss spectroscopy* (EELS), can be used to measure the plasmon frequency (Section 11.7). EELS spectra also show absorption features corresponding to the excitation of core electrons, similar to x-ray absorption spectra (Feldman and Mayer, 1986; Schneider, 2002).

The primary disadvantage of TEM is that the sample must be thinned down to less than a micron (typically <200 nm) to be sufficiently transparent to electrons. The sample preparation is typically destructive and time consuming. TEM is therefore not optimal for routine, high-throughput analysis. When used and interpreted properly, though, TEM images provide useful structural information and, in the case of HRTEM, atomic-level detail.

12.7 SCANNING PROBE MICROSCOPY

Significant advances in spatial resolution have been achieved through a class of techniques known as scanning probe microscopy (SPM). These techniques provide a glimpse into atomic-scale phenomena near the surface of a sample (Friedbacher, 2002) and are especially useful for studying nanocrystals. The first SPM method invented was *scanning tunneling microscopy* (STM) (Binnig et al., 1982). In an STM experiment, a bias is applied between a sharp metal tip and the sample. The tip is brought close enough so that electrons quantum mechanically tunnel through the vacuum gap, producing a current of a few nA. The magnitude of the current decays exponentially with the tip-sample distance. Because of this exponential dependence, STM can resolve vertical surface features well below 1 Å.

The tunneling current is also sensitive to the density of electronic states. For a positive sample bias, electrons tunnel from the tip into the sample (Figure 12.17). The tunneling current is proportional to the local density of *empty* states in the semiconductor. For a negative sample bias, the STM probes the local density of *filled* states. Typically, the STM is operated in constant-current mode, in which the tip is displaced vertically by a piezoelectric translator in order to maintain a constant tunneling current. The signal is proportional to the vertical displacement, and an image is built up by raster scanning the tip across the sample surface.

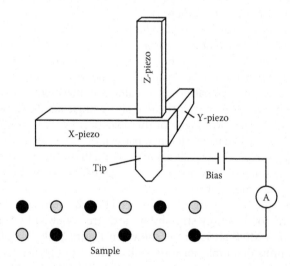

FIGURE 12.17 Scanning tunneling microscope (STM). (Friedbacher, G.: *Surface and Thin Film Analysis.* pp. 276–290. 2002. Copyright Wiley-VCH Verlag GmbH & Co. KGaA. Reproduced with permission.)

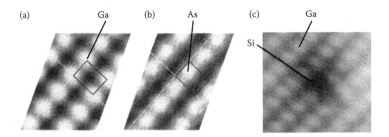

FIGURE 12.18 Scanning tunneling microscope (STM) images of the (110) GaAs surface. Sample biases are (a) 1.5 V, (b) −1.5 V, and (c) 2.5 V. The rectangles in (a) and (b) indicate the surface unit cell. The dark spot in (c) is due to a silicon impurity on the surface. (After Feenstra, R.M. et al. 1987. *Phys. Rev. Lett.* 58: 1192–1195; Zheng, J.F. et al. 1994. *Phys. Rev. Lett.* 72:1490–1493. With permission.)

For III-V semiconductors such as GaAs, the pristine (110) surface is ideal for STM measurements because the dangling bonds do not introduce states into the band gap (Jäger and Weber, 1999). On this surface, each gallium and arsenic atom has one dangling bond. The more electronegative arsenic atom has a filled dangling bond (two electrons), while the gallium dangling bond is empty. The filled arsenic levels lie in the valence band while the empty gallium levels are in the conduction band. As shown in Figure 12.18, the atoms can be selectively imaged by choosing the appropriate sample bias (Feenstra et al., 1987). For a positive bias, the STM signal is sensitive to empty states such that only the gallium atoms are imaged. A negative bias, which probes filled states, shows only arsenic atoms.

STM measurements of GaAs:Si showed silicon donors on the surface and several layers below the surface (Zheng et al., 1994). An image of a silicon donor on the (110) surface, for a positive sample bias, is shown in Figure 12.18. The dark contrast, which looks like a missing atom, was attributed to the localized silicon dangling bond. Because the dangling bond is half-filled, it can trap a tunneling electron and thereby reduce the current. Acceptors, antisites, and vacancies have also been imaged with STM, as reviewed by Jäger and Weber (1999).

The photon analogue to STM is *near-field scanning optical microscopy* (NSOM). In near-field microscopy, a laser spot that is smaller than the diffraction-limited spot size (Equation 12.1) illuminates the sample (Lewis et al., 1984; Pohl et al., 1984). This is accomplished by transmitting light through a small aperture, 50 to 200 nm in diameter. For the far-field condition, in which the distance from the aperture is greater than the aperture diameter, the light spreads out due

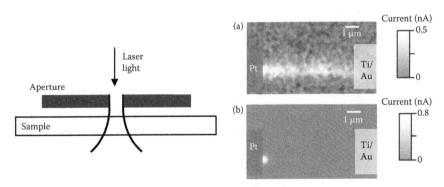

FIGURE 12.19 Near-field scanning optical microscopy (NSOM) experiment. The images show photocurrent data for ZnO nanowires, where Pt forms a Schottky contact. (a) Under forward bias, the photocurrent response is uniform along the nanowire. (b) Under reverse bias, the photocurrent is nonzero only very near the Schottky junction. (After Soudi, A. et al. 2010. *Appl. Phys. Lett.* 96: 253115 (3 pages). With permission.)

to diffraction. In the near field, however, the diameter of the evanescent wave is approximately equal to that of the aperture (Figure 12.19). This near-field condition allows researchers to optically probe nanometer-scale regions of a sample. The aperture is typically defined by the tip of a tapered optical fiber. This NSOM probe is then raster scanned across the sample surface.

NSOM is a preferred method to measure photocurrent in semiconductor nanowires. An example is shown for an *n*-type ZnO nanowire (Figure 12.19). Evaporated Pt and Ti/Au metals form Schottky and Ohmic contacts, respectively. A voltage is applied, the NSOM probe is scanned, and the resulting photocurrent provides the signal. When the Schottky junction is under forward bias, the electric field is constant along the nanowire. This results in a uniform photocurrent, as shown by the photocurrent map in Figure 12.19. When the Schottky junction is reverse biased, the electric field is strongest in the depletion region. In order for a photocurrent to be generated, when an electron–hole pair is generated, the minority hole must diffuse into the depletion region. Because the hole diffusion length is rather short, the photocurrent is large only very close to the Schottky contact. From the photocurrent map, Soudi et al. (2010) showed that the diffusion length decreased as the nanowire diameter decreased. This observation suggests that surface states can trap holes in nanowires with diameters <40 nm.

The final SPM technique to be discussed, *atomic force microscopy* (AFM), measures interatomic forces between a sharp tip and the sample surface (Binnig et al., 1986). The probe tip is mounted on a cantilever, 100 to 200 μm long, which is deflected up and down by a piezoelectric oscillator. The deflection can be measured by reflecting a laser beam off the cantilever. The laser is detected by a two-segment or "split" photodiode (Figure 12.20). As the cantilever bends, the reflected laser beam impinges on one segment more than the other segment; the difference signal is proportional to the cantilever deflection. The sample is scanned in the *x*-*y* plane by piezoelectric translators. The AFM can be operated in "contact mode," where the tip drags along the surface. Because the contact mode may scratch the surface, the "tapping mode" is preferred for soft specimens. In the tapping mode, the cantilever oscillates at its resonance frequency (50 to 500 kHz). The signal is given by the reduction in amplitude, caused by surface features.

An example of an AFM image is shown for germanium nanostructures grown on a silicon (001) surface (Figure 12.21). The silicon surface was first patterned with an array of gold dots. Then, germanium was deposited by MBE at a growth temperature of 600°C. The germanium islands preferentially grew on the silicon surface at regular locations between the gold dots. The islands had a pyramidal shape with {111} side facets and an average height of 60 nm. This kind of metal-mediated growth results in much different island shapes than those grown on bare silicon surfaces (Robinson et al., 2007).

In addition to measuring the surface topography, AFM tips can be modified to measure electrical properties. By coating the tip with *p*-type boron-doped diamond, for example, conductance AFM measurements can be performed. This technique was used to investigate the

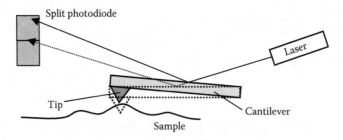

FIGURE 12.20 Atomic force microscopy (AFM) apparatus. As the cantilever is deflected by atoms on the sample surface, the reflected laser beam impinges on a different region of a split (two-segment) photodiode. The photodiode difference signal is proportional to the deflection.

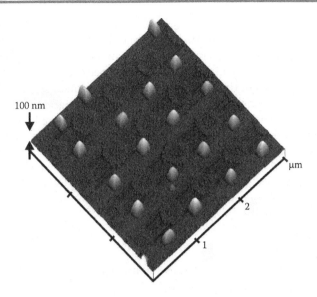

FIGURE 12.21 Atomic force microscopy (AFM) image of germanium islands grown on a silicon (001) surface. (Courtesy of O.D. Dubon.)

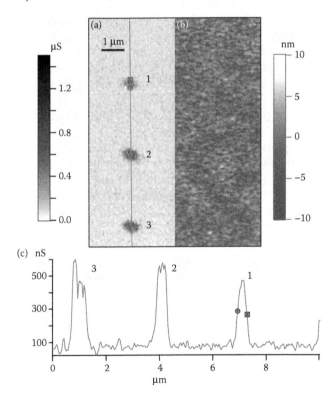

FIGURE 12.22 Atomic force microscopy (AFM) images of a hydrogenated MnGaAs sample that was subsequently exposed to laser pulses. (a) The conductance increased locally in spots 1, 2, and 3, due to laser-induced dissociation of Mn-H complexes. (b) The AFM topographic map shows that the surface is smooth. (c) Plot of the conductance along a line intersecting 1, 2, and 3. (After Farshchi, R. et al. 2009. *J. Appl. Phys.* 106: 103918 (4 pages). With permission.)

326   12.7 Scanning Probe Microscopy

effect of femtosecond laser pulses on $Mn_{0.04}Ga_{0.96}As$ (Farshchi et al., 2009). The MnGaAs sample was hydrogenated by exposure to a hydrogen plasma, resulting in the passivation of manganese acceptors and a drop in the conductivity. Pulses from an ultrafast laser (400 nm wavelength, 100 fs pulse duration, 1 kHz repetition rate) were then focused onto the sample. The laser irradiation caused the dissociation of the Mn-H complexes and an increase in the local conductivity. Figure 12.22 shows AFM maps of the conductance and the topography of the surface. The spots indicate high conductance where the laser was focused onto the sample. (Conductance is defined as I/V and has units of siemens, S.) The topography, on the other hand, shows that the surface is smooth and undamaged by the laser irradiation.

In summary, SPM techniques offer atomic-scale resolution of surface and near-surface features. With the continued interest in semiconductor nanocrystals, they play an important role in studying dopants and defects in these materials.

PROBLEMS

12.1 An objective lens ($NA = 0.7$) focuses 460 nm light on a sample. What is the radius of the diffraction-limited spot?

12.2 Estimate the dislocation density (cm^{-2}) for the sample in Figure 12.4.

12.3 An electron microscope uses 10 keV electrons and has a numerical aperture of 10^{-3}.
   a. Calculate the de Broglie wavelength for the electrons.
   b. Calculate the theoretical resolution for this electron microscope.
   c. Explain why the actual resolution may differ from your answer in part (b).

12.4 A CL experiment is performed on GaAs with a 40 kV, 0.3 μA electron beam.
   a. Estimate the number of electron–hole pairs generated per second.
   b. If the electron–hole pairs are generated in a sphere of radius 0.5 μm, what is the generation rate G?

12.5 An electron beam produces a generation rate G. The minority-carrier lifetime and density are τ and n, respectively.
   a. Write an expression for dn/dt.
   b. From part (a), show that the steady-state minority carrier density is $n = G\tau$.

12.6 This problem illustrates how a magnetic field can "focus" an electron beam. An electron travels at 10^6 m/s in the z direction. It encounters a field
   $\mathbf{B} = (0.01\ \mathrm{T})\hat{\mathbf{x}} + (0.01\ \mathrm{T})\hat{\mathbf{z}}$
   After 1.0 ns, find:
   a. The component of the velocity in the y direction.
   b. The Lorentz force due to the velocity component in part (a).

12.7 An XRD spectrum is taken of a polycrystalline zincblende semiconductor. The x-ray wavelength is 1.54 Å. In the intensity-versus-2θ plot, there is a (111) peak at $2\theta = 29°$ with a FWHM of 0.3°.
   a. Calculate the lattice constant.
   b. Estimate the grain size, assuming the broadening of the peak is due to the finite size.

12.8 A {100} oriented silicon sample ($a = 5.43$ Å) is placed in a TEM, as in Figure 12.15. The electron beam energy is 20 keV. How much should the sample be rotated (degrees) to obtain diffraction from {400} planes?

12.9 In Figure 12.18a, the Ga atoms are bright while the As atoms are dark. In Figure 12.18b, the situation is reversed. Explain.

12.10 A horizontal cantilever in an AFM is 100 μm long. A laser beam reflects off the cantilever at a 20° angle and hits a split photodiode 10 cm away. The AFM tip encounters a bump 10 nm high. What is the displacement of the laser spot at the photodiode?

REFERENCES

Angilello, J., R.M. Potemski, and G.R. Woolhouse. 1975. Etch pits and dislocations in {100} GaAs wafers. *J. Appl. Phys.* 46: 2315–2316.

Binnig, G., H. Rohrer, Ch. Gerber, and E. Weibel. 1982. Surface studies by scanning tunneling microscopy. *Phys. Rev. Lett.* 49: 57–61.

Binnig, G., C.F. Quate, and Ch. Gerber. 1986. Atomic force microscope. *Phys. Rev. Lett.* 56: 930–933.

Cullity, B.D. 1978. *Elements of X-ray Diffraction.* Reading, MA: Addison-Wesley.

Dash, W.C. 1957. The observation of dislocations in silicon. In *Dislocations and Mechanical Properties of Crystals*, eds. J.C. Fisher, W.G. Johnston, R. Thompson, and T. Vreelan, pp. 55–67. New York, NY: Wiley.

Davisson, C. and L.H. Germer. 1927. Diffraction of electrons by a crystal of nickel. *Phys. Rev.* 30: 705–740.

Everhart, T.E., O.C. Wells, and R.K. Matta. 1964. A novel method of semiconductor device measurements. *Proc. IEEE.* 52: 1642–1647.

Farshchi, R., D.J. Hwang, R.V. Chopdekar, P.D. Ashby, C.P. Grigoropoulos, and O.D. Dubon. 2009. Ultrafast pulsed-laser dissociation of Mn-H complexes in GaAs. *J. Appl. Phys.* 106: 103918 (4 pages).

Feenstra, R.M., J.A. Stroscio, J. Tersoff, and A.P. Fein. 1987. Atom-selective imaging of the GaAs(110) surface. *Phys. Rev. Lett.* 58: 1192–1195.

Feldman, L.C. and J.W. Mayer. 1986. *Fundamentals of Surface and Thin Film Analysis.* New York, NY: North-Holland.

Föll, H. and B.O. Kolbesen. 1975. Formation and nature of swirl defects in silicon. *Appl. Phys. A* 8: 319–331.

Friedbacher, G. 2002. Scanning probe microscopy. In *Surface and Thin Film Analysis*, eds. H. Bubert and H. Jenett, pp. 276–290. Weinheim, Germany: Wiley-VCH Verlag.

Fultz, B. and J.M. Howe. 2001. *Transmission Electron Microscopy and Diffractometry of Materials.* Berlin: Springer-Verlag.

Goldstein, J.I., D.E. Newbury, P. Echlin et al. 1992. *Scanning Electron Microscopy and X-ray Microanalysis*, 2nd ed. New York, NY: Plenum Press.

Haegel, N.M, S.E. Williams, C. Frenzen, and C. Scandrett. 2009. Imaging charge transport and dislocation networks in ordered GaInP. *Physica B.* 404: 4963–4966.

Haller, E.E., G.S. Hubbard, W.L. Hansen, and A. Seeger. 1977. Divacancy-hydrogen complexes in dislocation-free high-purity germanium. In *Radiation Effects in Semiconductors 1976, Inst. Phys. Conf. Ser.* No. 31, eds. N.B. Urli and J.W. Corbett, pp. 309–318. Bristol, UK: Institute of Physics.

Hecht, E. 2002. *Optics*, 4th ed. San Francisco, CA: Addison Wesley.

Hirsch, P.B., A. Howie, R.B. Nicholson, D.W. Pashley, and M.J. Whelan. 1965. *Electron Microscopy of Thin Crystals.* Washington, DC: Butterworth.

Holmes, P.J. 1962. *The Electrochemistry of Semiconductors*, 368–377. New York, NY: Academic Press.

Holt, D.B. and B.G. Yacobi. 2007. *Extended Defects in Semiconductors.* New York, NY: Cambridge University Press.

Huso, J., J.L. Morrison, J. Mitchell et al. 2009. Optical transitions and multiphonon Raman scattering of Cu doped ZnO and MgZnO ceramics. *Appl. Phys. Lett.* 94: 061919.

Iqbal, M.Z. 1980. Bi-refringence observations of strain and plastic deformation in GaP. *J. Mater. Res.* 15: 781–784.

Jäger, N.D. and E.R. Weber. 1999. Scanning tunneling microscopy of defects in semiconductors. In *Semiconductors and Semimetals* Vol. 51B, ed. M. Stavola, pp. 261–296, New York: Academic Press.

Kittler, M., J. Lärz, W. Seifert, M. Seibt, and W. Schröter. 1991. Recombination properties of structurally well defined $NiSi_2$ precipitates in silicon. *Appl. Phys. Lett.* 58: 911–913.

Klein, C.A. 1968. Band-gap dependence and related features of radiation ionization energies in semiconductors. *J. Appl. Phys.* 39: 2029–2038.

Lewis, A., M. Isaacson, A. Harootunian, and A. Murray. 1984. Development of a 500 Å spatial resolution light microscope: I. Light is efficiently transmitted through λ/6 diameter apertures. *Ultramicroscopy* 13: 227–231.

Luber, D.R., F.M. Bradley, N.M. Haegel, M.C. Talmadge, M.P. Coleman, and T.D. Boone. 2006. Imaging transport for the determination of minority carrier diffusion length. *Appl. Phys. Lett.* 88: 163509 (3 pages).

Paskov, P.P., R. Schifano, B. Monemar, T. Paskova, S. Figge, and D. Hommel. 2005. Emission properties of a-plane GaN grown by metal-organic chemical-vapor deposition. *J. Appl. Phys.* 98: 093519 (7 pages).

328 References

Pohl, D.W., W. Denk, and M. Lanz. 1984. Optical stethoscopy: Image recording with resolution $\lambda/20$. *Appl. Phys. Lett.* 44: 651–653.

Randle, V. and O. Engler. 2000. *Introduction to Texture Analysis: Macrotexture, Microtexture and Orientation Mapping.* Amsterdam: Gordon and Breach.

Ravi, K.V. and C.J. Varker. 1974. Comments on the distinction between "striations" and "swirls" in silicon. *Appl. Phys. Lett.* 25: 69–71.

Robinson, J.T., F. Ratto, O. Moutanabbir et al. 2007. Gold-catalyzed oxide nanopatterns for the directed assembly of Ge island arrays on Si. *Nano Lett.* 7: 2655–2659.

Romano, L.T., C.G. Van de Walle, J.W. Ager III, W Götz, and R.S. Kern. 2000. Effect of Si doping on strain, cracking, and microstructure in GaN thin films grown by metalorganic chemical vapor deposition. *J. Appl. Phys.* 87: 7745–7752.

Runyan, W.R. and T.J. Shaffner. 1998. *Semiconductor Measurements and Instrumentation.* New York, NY: McGraw-Hill.

Schneider, R. 2002. Electron energy loss spectroscopy (EELS). In *Surface and Thin Film Analysis*, eds. H. Bubert and H. Jenett, pp. 50–70. Weinheim, Germany: Wiley-VCH Verlag.

Schroder, D.K. 2006. *Semiconductor Material and Device Characterization*, 3rd ed. New York, NY: John Wiley & Sons.

Schwander, P., W.-D. Rau, C. Kisielowski, M. Gribelyuk, and A. Ourmazd. 1999. Defect processes in semiconductors studied at the atomic level by transmission electron microscopy. In *Semiconductors and Semimetals* Vol. 51B, ed. M. Stavola, pp. 225–259. San Diego, CA: Academic Press.

Soudi, A., P. Dhakal, and Y. Gu. 2010. Diameter dependence of the minority carrier diffusion length in individual ZnO nanowires. *Appl. Phys. Lett.* 96: 253115 (3 pages).

Suryanarayana, C. and M.G. Norton. 1998. *X-ray Diffraction: A Practical Approach.* New York, NY: Plenum Press.

Van Hove, M.A., W.H. Weinberg, and C.-M. Chan. 1986. *Low-Energy Electron Diffraction.* Berlin: Springer-Verlag.

Varker, C.J. 1979. SEM methods for the characterization of semiconductor materials and devices. In *Nondestructive Evaluation of Semiconductor Materials and Devices*, ed. J.N. Zemel, pp. 515–580. New York, NY: Plenum Press.

Vogel, F.L., W.G. Pfann, H.E. Corey, and E.E. Thomas. 1953. Observations of dislocations in lineage boundaries in germanium. *Phys. Rev.* 90: 489–490.

Witt, A.F. and H.C. Gatos. 1966. Impurity distribution in single crystals. *J. Electrochem. Soc.* 113: 808–813.

Yacobi, B.G. and D.B. Holt. 1986. Cathodoluminescence scanning electron microscopy of semiconductors. *J. Appl. Phys.* 59: R1–R24.

Zheng, J.F., X. Liu, N. Newman, E.R. Weber, D.F. Ogletree, and M. Salmeron. 1994. Scanning tunneling microscopy studies of Si donors (Si_{Ga}) in GaAs. *Phys. Rev. Lett.* 72: 1490–1493.

Appendices

APPENDIX A: COMPLEX EXPONENTIALS

Complex exponentials appear often in physics and are convenient for describing waves. Euler's identity states that

(A.1)
$$e^{ikx} = \cos kx + i \sin kx$$

Note that

(A.2)
$$e^{2\pi i} = e^{4\pi i} = e^{6\pi i} = \cdots = 1$$

From Equation A.1, one can show

(A.3)
$$\cos kx = \frac{1}{2}(e^{ikx} + e^{-ikx})$$
$$\sin kx = \frac{1}{2i}(e^{ikx} - e^{-ikx})$$

The *complex conjugate*, designated by an asterisk, is obtained by replacing i with $-i$. For example,

(A.4)
$$(e^{ikx})^* = e^{-ikx} = \cos kx - i \sin kx$$

The absolute value squared is a function's complex conjugate times the function. For the complex exponential,

(A.5)
$$\left| e^{ikx} \right|^2 = e^{-ikx} e^{ikx} = 1$$

The derivatives of the complex exponential are given by

(A.6)
$$\frac{d}{dx} e^{ikx} = ik e^{ikx}$$
$$\frac{d^2}{dx^2} e^{ikx} = -k^2 e^{ikx}$$

These derivative properties are useful for solving equations of motion. Consider a mass m attached to a spring with a force constant C (Figure A.1). The spring exerts a force $-Cx$, where x is the displacement. There is also an empirical damping factor γ which produces a force $-\gamma mv$, where v is the velocity. Putting these together,

(A.7)
$$F = -Cx - \gamma m \dot{x}$$

329

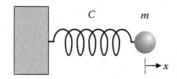

FIGURE A.1 Spring and mass. The displacement from equilibrium is x.

where $\dot{x} \equiv dx/dt = v$. Newton's second law, $F = m\ddot{x}$, yields

(A.8) $$\ddot{x} + \gamma\dot{x} + \omega_0^2 x = 0$$

where $\omega_0^2 = C/m$. Inserting a trial solution

(A.9) $$x = x_0 e^{-i\omega t}$$

into Equation A.8 yields

(A.10) $$-\omega^2 - i\gamma\omega + \omega_0^2 = 0$$

Using the quadratic formula to solve Equation A.10,

(A.11) $$\omega = \frac{1}{2}\left(-i\gamma \pm \sqrt{-\gamma^2 + 4\omega_0^2}\right)$$

If $\gamma \ll \omega_0$, then we can ignore the γ^2 term in the radical. Choosing the positive root,

(A.12) $$\omega \approx -i\gamma/2 + \omega_0$$

Inserting this complex frequency into Equation A.9 yields

(A.13) $$x = x_0 e^{-i\omega_0 t} e^{-\gamma t/2}$$

If we want a real, physical answer, we can just take the real part of Equation A.13,

(A.14) $$\text{Re}(x) = x_0 \cos(\omega_0 t) e^{-\gamma t/2}$$

where we have assumed that x_0 is real.

APPENDIX B: LINEAR ALGEBRA

Consider a system of two linear equations where the right-hand side equals zero,

(B.1) $$\begin{aligned} Ax + By &= 0 \\ Cx + Dy &= 0 \end{aligned}$$

Solving the second equation for y yields

(B.2) $$y = -(C/D)x$$

Substituting this into Equation B.1 gives

(B.3)
$$Ax - (BC/D)x = 0$$

This is solved by

(B.4)
$$AD - BC = 0$$

In matrix form, Equation B.1 is written as

(B.5)
$$\begin{bmatrix} A & B \\ C & D \end{bmatrix} \begin{bmatrix} x \\ y \end{bmatrix} = \begin{bmatrix} 0 \\ 0 \end{bmatrix}$$

For there to be a nontrivial solution (a solution other than $x = y = 0$), the determinant of the matrix of coefficients must equal zero:

(B.6)
$$\begin{vmatrix} A & B \\ C & D \end{vmatrix} = AD - BC = 0$$

which is just a restatement of Equation B.4. However, the determinant in Equation B.6 can be for an arbitrarily large matrix.

APPENDIX C: THE LINEAR CHAIN

A diatomic linear chain is shown in Figure 6.2 (see also Kittel, 2005). Each mass experiences a net force from two identical springs C,

(C.1)
$$F_1 = C(v_s - u_s) + C(v_{s-1} - u_s)$$
$$F_2 = C(u_{s+1} - v_s) + C(u_s - v_s)$$

where F_1 and F_2 are the forces on M_1 and M_2 in unit cell s. Using Newton's second law, we have

(C.2)
$$M_1 \ddot{u}_s = C(v_s + v_{s-1} - 2u_s)$$
$$M_2 \ddot{v}_s = C(u_{s+1} + u_s - 2v_s)$$

To solve these equations, we use complex exponentials,

(C.3)
$$u_s = u e^{i(Ksa - \omega t)}$$
$$v_s = v e^{i(Ksa - \omega t)}$$

where the amplitudes u and v depend on K. They have the following useful properties:

(C.4)
$$\ddot{u}_s = -\omega^2 u e^{i(Ksa - \omega t)}$$
$$\ddot{v}_s = -\omega^2 v e^{i(Ksa - \omega t)}$$
$$u_{s+1} = e^{iKa} u e^{i(Ksa - \omega t)}$$
$$v_{s-1} = e^{-iKa} v e^{i(Ksa - \omega t)}$$

332   Appendices

Plugging the relations in Equation C.4 into Equation C.2 yields

(C.5)
$$-M_1\omega^2 u = C(v + e^{-iKa}v - 2u)$$
$$-M_2\omega^2 v = C(e^{iKa}u + u - 2v)$$

Grouping together terms, we have

(C.6)
$$(M_1\omega^2 - 2C)u + C(1 + e^{-iKa})v = 0$$
$$C(e^{iKa} + 1)u + (M_2\omega^2 - 2C)v = 0$$

To have a nontrivial solution, the determinant (Equation B.6) must equal zero, so

(C.7)
$$(M_1\omega^2 - 2C)(M_2\omega^2 - 2C) - C^2(2 + e^{iKa} + e^{-iKa}) = 0$$

Expanding the terms and using Equation A.3 yields

(C.8)
$$M_1 M_2\omega^4 - 2C(M_1 + M_2)\omega^2 + 2C^2(1 - \cos Ka) = 0$$

Using the quadratic equation to solve for ω^2 yields

(C.9)
$$\omega^2 = \frac{2C(M_1 + M_2) \pm \sqrt{4C^2(M_1 + M_2)^2 - 8C^2 M_1 M_2(1 - \cos Ka)}}{2M_1 M_2}$$

which simplifies to Equation 6.7.

APPENDIX D: THE SCHRÖDINGER EQUATION

Particles such as electrons act like waves. In one dimension, the *wave function* for a particle is denoted $\psi(x)$. The probability for finding the particle between x and $x + dx$ is given by $|\psi(x)|^2 dx$. Since the probability for finding the particle *somewhere* is 1, the probability density $|\psi(x)|^2$ must be normalized:

(D.1)
$$\int_{-\infty}^{\infty} |\psi(x)|^2 \, dx = 1$$

In this book, we generally write $\psi(x)$ as being proportional to some function and lazily ignore the normalization constant.

The Schrödinger equation gives us a way to find the wave function and its corresponding energy E:

(D.2)
$$\frac{-\hbar^2}{2m}\frac{d^2}{dx^2}\psi(x) + V(x)\psi(x) = E\psi(x)$$

where $V(x)$ is the potential energy. As an example, consider a free particle in vacuum, where $V(x) = 0$. Equation D.2 becomes

(D.3)
$$\frac{-\hbar^2}{2m}\frac{d^2}{dx^2}\psi(x) = E\psi(x)$$

Appendices **333**

A solution to Equation D.3 is a complex exponential,

$$(D.4) \qquad \psi(x) = Ae^{ikx}$$

Plugging this solution into Equation D.3 yields

$$(D.5) \qquad E = \frac{\hbar^2 k^2}{2m}$$

This is just the classical kinetic energy of a particle with momentum $p = \hbar k$.

If $V(x)$ is nonzero, then we get different wavefunctions and energies. The simple harmonic oscillator, for example, has $V(x) = \frac{1}{2}kx^2$. In that case, the solutions are given by Equation 6.56. Another popular example is the hydrogen atom, where the electron experiences a $1/r$ potential. The hydrogen wavefunctions are listed in Equation 5.17.

APPENDIX E: DENSITY OF STATES

Consider a free electron confined to a region between $x = 0$ and $x = L$. The wavefunction is given by a complex exponential (Equation D.4). We assume periodic boundary conditions such that $\psi(0) = \psi(L)$. This is satisfied by

$$(E.1) \qquad k = 0, \pm 2\pi/L, \pm 4\pi/L, \ldots$$

This means that there is one k point in an interval $dk = 2\pi/L$. For an arbitrary dk,

$$(E.2) \qquad \text{\# of } k \text{ points (1D)} = \frac{L}{2\pi} dk$$

We now consider an electron in a cube. In 3D, the wavefunction has k_x, k_y, and k_z:

$$(E.3) \qquad \psi = Ae^{i(k_x x + k_y y + k_z z)}$$

The number of k points is given by the product of k_x, k_y, and k_z points:

$$(E.4) \qquad \text{\# of } k \text{ points (3D)} = \frac{L^3}{(2\pi)^3} dk_x dk_y dk_z$$

where L^3 is the crystal volume (V) and $dk_x dk_y dk_z$ is a volume in k space.

Assuming an isotropic effective mass, the electron energy is given by Equation D.5. It only depends on the magnitude of k. A surface of constant k is a sphere, which has an area $4\pi k^2$. The number of points between k and $k + dk$ is

$$(E.5) \qquad \text{\# of } k \text{ points (3D)} = \frac{L^3}{(2\pi)^3} 4\pi k^2 dk$$

To get the number of electron states N, we multiply Equation E.5 by 2, since electrons can be spin-up or spin-down:

$$(E.6) \qquad N = \frac{V}{\pi^2} k^2 dk$$

334   Appendices

From Equation D.5,

$$(E.7) \qquad dE = \frac{\hbar^2 k}{m} dk$$

Substituting the expression for kdk (Equation E.7) into Equation E.6 yields

$$(E.8) \qquad N = \frac{Vk}{\pi^2} \frac{m}{\hbar^2} dE$$

Finally, using Equation D.5 to substitute for k results in

$$(E.9) \qquad N = \frac{V\sqrt{2mE}}{\pi^2} \frac{m}{\hbar^3} dE$$

The density of states $D(E)$ is defined by $N = D(E)dE$. From Equation E.9,

$$(E.10) \qquad D(E) = \frac{V}{2\pi^2} \left(\frac{2m}{\hbar^2} \right)^{3/2} \sqrt{E}$$

To get the density of states for the conduction band (Equation 5.41), we replace m with the electron effective mass m_e. Also, since energy is measured relative to the valence-band maximum, we replace E with $E - E_g$.

For the valence band, we add the states due to the light hole (lh) and heavy hole (hh) bands:

$$(E.11) \qquad D(E) = \frac{V}{2\pi^2} \left(\frac{2m_{lh}}{\hbar^2} \right)^{3/2} \sqrt{-E} + \frac{V}{2\pi^2} \left(\frac{2m_{hh}}{\hbar^2} \right)^{3/2} \sqrt{-E}$$

where E is below the valence band maximum ($E < 0$) and we have assumed parabolic bands that are degenerate at $k = 0$. This can be written as

$$(E.12) \qquad D(E) = \frac{V}{2\pi^2} \left(\frac{2m_h}{\hbar^2} \right)^{3/2} \sqrt{-E}$$

where $m_h = (m_{lh}^{3/2} + m_{hh}^{3/2})^{2/3}$.

APPENDIX F: MATRICES IN QUANTUM MECHANICS

A quantum mechanical system can be represented as a matrix. For a two-level system with energies E_1 and E_2 and corresponding wavefunctions ψ_1 and ψ_2,

$$(F.1) \qquad \begin{bmatrix} E_1 - E & 0 \\ 0 & E_2 - E \end{bmatrix} \begin{bmatrix} \psi_1 \\ \psi_2 \end{bmatrix} = \begin{bmatrix} 0 \\ 0 \end{bmatrix}$$

Setting the determinant equal to zero (Equation B.6) gives

$$(F.2) \qquad (E_1 - E)(E_2 - E) = 0$$

which is solved by $E = E_1$ or $E = E_2$. These are the energy eigenvalues. An N level system would have $E = E_1, E_2, \dots E_N$.

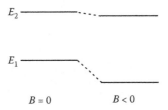

FIGURE F.1 Two level system, where B is a perturbation constant.

In Section 5.4, we labeled electron wavefunctions by their **k** values **k**1, **k**2, ..., **k**N and energies $E_{k1}, E_{k2}, ..., E_{kN}$. The *Green's function* is defined

(F.3) $$G(\mathbf{k},E) = \frac{1}{E - E_{\mathbf{k}} + i\delta}$$

where δ is arbitrarily small. Notice how this function is very large when $E = E_{\mathbf{k}}$. This is called a "pole" of G. Finding the poles of G is equivalent to finding the energy eigenvalues of the system. See Yu and Cardona (1996) and Mattuck (1992) for more details about Green's functions.

A diagonal matrix like Equation F.1 can be used to describe the conduction-band electron energies of an intrinsic semiconductor. Of course, we would need a larger matrix to be accurate, but let's do the two-level case to keep things simple. Suppose a defect is introduced, which adds a term B to all the elements in the matrix:

(F.4) $$\begin{bmatrix} E_1 + B - E & B \\ B & E_2 + B - E \end{bmatrix}$$

Setting the determinant equal to zero (Equation B.6) yields

(F.5) $$(E_1 + B - E)(E_2 + B - E) - B^2 = 0$$

Using the quadratic equation to solve for E, we get

(F.6) $$E = \frac{E_1 + E_2 + 2B \pm \sqrt{(E_1 - E_2)^2 + 4B^2}}{2}$$

If B is small, then the energies in Equation F.6 are only slightly different than E_1 and E_2. An example is shown in Figure F.1 for $B < 0$. The result for a large matrix is shown in Figure 5.10.

APPENDIX G: VAN DER PAUW MEASUREMENT

Consider four contacts (1, 2, 3, 4) on the edge of a semi-infinite thin film of thickness δ (Figure G.1). Their locations are denoted $0, r_2, r_3,$ and r_4. Let current I enter through contact 1 and exit through 2. The change in voltage going from 3 to 4 is

(G.1) $$V_4 - V_3 = \int_{r_3}^{r_4} \left(-\frac{I\rho}{\pi r \delta} + \frac{I\rho}{\pi(r - r_2)\delta} \right) dr$$

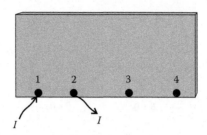

FIGURE G.1 Four contacts on the edge of a thin film of thickness δ. Current is shown flowing into contact 1 and out of contact 2.

where ρ is the resistivity. The first term in Equation G.1 is the voltage drop from current entering $r = 0$ while the second term is from current exiting at $r = r_2$. The solution is

$$V_4 - V_3 = \frac{I\rho}{\pi\delta}\ln\left(\frac{r_3}{r_4} \cdot \frac{r_4 - r_2}{r_3 - r_2}\right) \tag{G.2}$$

This gives the resistance value

$$R_{12,34} \equiv \frac{V_4 - V_3}{I} = \frac{\rho}{\pi\delta}\ln\left(\frac{r_3}{r_4} \cdot \frac{r_4 - r_2}{r_3 - r_2}\right) \tag{G.3}$$

Next, we consider I flowing from 2 to 3. By a similar approach, we get

$$R_{23,41} \equiv \frac{V_1 - V_4}{I} = \frac{\rho}{\pi\delta}\ln\left(\frac{r_3}{r_2} \cdot \frac{r_4 - r_2}{r_4 - r_3}\right) \tag{G.4}$$

From Equations G.3 and G.4, it can be shown that

$$\exp(-\pi R_{12,34}\delta/\rho) + \exp(-\pi R_{23,41}\delta/\rho) = 1 \tag{G.5}$$

From this equation, one can numerically solve for ρ.

In practice, the solution for resistivity (Equation G.5) is expressed as Equation 9.31, where value of f is tuned to give the correct answer. Although this result was derived for a specific geometry, van der Pauw (1958) showed that it holds for four contacts on the periphery of any thin film, regardless of its shape.

REFERENCES

Kittel, C. 2005. *Introduction to Solid State Physics*, 8th ed. New York, NY: John Wiley & Sons.
Mattuck, R.D. 1992. *A Guide to Feynman Diagrams in the Many-Body Problem*, 2nd ed. New York, NY: Dover.
Van der Pauw, L.J. 1958. A method of measuring specific resistivity and Hall effect of discs of arbitrary shape. *Philips Res. Rep.* 13: 1–9.
Yu, P.Y. and M. Cardona. 1996. *Fundamentals of Semiconductors*. Berlin: Springer-Verlag.

Physical Constants

Quantity	Symbol	SI Units	Other Units
Atomic mass unit (amu)	u	1.661×10^{-27} kg	
Avogadro's number	N_A	6.022×10^{23} mol^{-1}	
Boltzmann constant	k_B	1.381×10^{-23} J/K	
Charge of electron/hole	e	$\pm 1.602 \times 10^{-19}$ C	
Free electron mass	m	9.109×10^{-31} kg	$mc^2 = 511$ keV
Permittivity of free space	ε_0	8.854×10^{-12} F/m	
Planck's constant	h	6.626×10^{-34} J s	
	$\hbar = h/2\pi$	1.055×10^{-34} J s	
Speed of light in a vacuum	c	2.998×10^8 m/s	2.998×10^{10} cm/s
Bohr magneton	μ_B	9.274×10^{-24} J/T	14.0 GHz/T
Nuclear magneton	μ_N	5.051×10^{-27} J/T	7.62 MHz/T
Electron volt	eV	1.602×10^{-19} J	
	eV/h	2.418×10^{14} Hz	
	eV/hc	8.065×10^5 m^{-1}	8065 cm^{-1}
	eV/k_B	1.160×10^4 K	

Index

A

AB site, *see* Antibonding site
Absorbance, 133, 248
Absorption, 103, 175, 248–250
    coefficient, 139, 162, 163, 171
    of phonon, 259
    of photons, 21
Acceptor(s), 30, 97
    binding energy, 32, 99
    bound exciton, 178–179
    dopant, 59
    impurity, 19
    level, 118
    wave functions, 105–108
Accumulation, 60
A-centers, 34, 37–38, 293
Acoustical-phonon dispersion, 9
Acoustical branch, 132
Acoustical phonons, 9
AFM, *see* Atomic force microscopy
Airy disk, 306
ALD, *see* Atomic layer deposition
Alloying, 81–83
Aluminum-silicon phase diagram, 81–82
Amorphization, 90
Amphoteric defects, 32
    model, 195–196
Amphoteric impurity, 151
Amplitude, 8
Anharmonicity of potential, 137
Anion vacancies, 34
Annealing, 88–92
Anode, 249
Antibonding site (AB site), 38, 297
Anti-electron, 294
Antimony, 213, 214, 221
Anti-Stokes Raman scattering, 142
Apodization, 257
Applied stress, 262–265
Arc lamp, 248
Area defects, 29
Arrhenius plot, 117, 118, 231
    of emission rate *vs.* inverse temperature, 235
    of free-electron concentration *vs.* inverse
        temperature, 117
    of free-hole concentration *vs.* inverse temperature,
        118
    of Ga self-diffusion in GaAs, 210
    of self-diffusion in silicon, 208
    slope of freeze-out curve, 117
Arsenic, 221
Arsenic antisite, 33

A-swirls, 35, 320
Asymmetric stretch mode, 147
Atomic force microscopy (AFM), 324–325
Atomic layer deposition (ALD), 79
Atomic structure, 29
Atoms, 5, 112
    location in cubic cell, 3
Auger effect, 177–178
Auger electrons, 293, 312
Auger spectroscopy, 293
Autodoping, 78
Avalanche breakdown, 58
Average energy, 109

B

"Back-bonded" orientation, 38
Backscattered electrons, 312
Backscattered light, 253
Ball-and-stick models
    for C-H and Si-H pairs in GaAs, 153
    and energy-level diagram for neutral silicon
        vacancy, 105
Band-gap absorption, 169–173
    of germanium, silicon, and GaAs, 22
Band gaps, 10–11, 18
Band parameters, 16
Band structure, 15
Band-to-band Auger recombination, 177–178
Band-to-band transitions, 171–172
Band theory, 11–16
Base, 2, 58
bcc, *see* Body-centered cubic
BC location, *see* Bond-center location
B-D complexes, *see* Boron–deuterium complexes
Beer–Lambert law, 21, 248
Bending modes, 145
BGOs, *see* Bismuth-germanium oxides
Biexcitons, 180
Binary collisions, 87
Binding energy, 97
Bipolar transistor, 2, 58
Bismuth-germanium oxides (BGOs), 295
Blaze wavelength, 249
Bloch wave functions, 100, 110
Blue diode laser, 63
Blue LED, 62
Body-centered cubic (bcc), 3, 4
Bohr model
    for boron acceptor in silicon, 99
    for phosphorus donor in silicon, 98
Boltzmann tail, 20
Bond-centered hydrogen, 262

340 Index

Bond-centered orientation, 38
Bond-center location (BC location), 297
Born–Oppenheimer approximation, 129
Bond-stretching frequency of diatomic
    quasi-molecule, 136
Boron dopants, 19
Boron-doped silicon, 212
Boron–deuterium complexes (B-D complexes), 146
Boron–hydrogen complexes, 262
    in silicon, 39
Bound excitons, 178–179
Bragg diffraction, 318, 319
    condition, 318, 320
Bragg's rule, 279
Braunsche Röhre, 1
Breakdown, 58
Bremsstrahlung radiation, 293
Bridgman growth, 70
Bridgman technique, 70
Bright field microscopy, 305, 320
Brillouin zone, 8–10
B-swirls, 35
"Bubbler", 77
Bulk crystal growth, 67–70
    dopant incorporation during, 70–74
"Bulk lifetime", 295
Burgers vectors, 43, 44
Burstein–Moss shift, 173

C

Cadmium-111 atom, 299
Cadmium sulfide (CdS), 6, 25
Capacitance–voltage profiling technique (*C-V* profiling
    technique), 231–234
Capture coefficient, 234
Capture cross section, 176
Carbon, 37
    doping, 151
    in silicon, 214
Carrier(s); *see also* Minority carriers
    concentrations as function of temperature, 113–116
    dynamics, 173–178
    emission and capture, 234–235
    free, 23
Cathodoluminescence (CL), 175, 247, 252, 314–316
CBE, *see* Chemical beam epitaxy
CCD array, *see* Charge-coupled device array
CD device, *see* Compact disc device
Cd-O electronic wave function, 183–184
Cell, 3
Central cell correction, 101
Channeling, 87, 88
    RBS technique, 280
Characterization techniques, 221, 247, 305
Charge-coupled device array (CCD array), 61–62, 174,
    305
Charge state, 195–197
Chemical beam epitaxy (CBE), 75
Chemical etching of semiconductors, 307
Chemical "fingerprint", 271
Chemical potential, 197–199
Chemical purification, 71

Chemical vapor deposition (CVD), 75, 77–79
Chemomechanical polishing, 307
CL, *see* Cathodoluminescence
Classical electromagnetic theory, 142
Classical gas distribution, 294
Cleaved surface, 49
Cobalt ions (Co^{2+}), 266
Co-implantation, 91–92
Collective motions, 8
Collector, 2, 58
Compact disc device (CD device), 63
Compatibility tables, 103, 104
Compensation ratio (*k*), 115
Complex, 29
    conjugate, 329
    exponentials, 329–330
Compound optical microscope, 306
Compound semiconductors, 19, 25, 32, 34, 35, 91, 208–209
Conductance, 326
Conduction band, 2, 10
    electron energies, 335
    minima, 15, 16, 106
    minimum, 97, 99, 186
    offsets, 63
Conductivity, 222–223
Contactless techniques, 227
Contact mode, 324
Continuous-wave laser (cw laser), 250
Cooled semiconductor, 259
Copper (Cu), 39
    copper-doped ZnO, 312, 313
    in germanium, 38
Copper oxide (CuO), 1
Correlation
    factor, 207
    part, 112
Cottrell atmosphere, 44–45
Coulomb attraction, 34, 287
Coulomb force, 8, 111
    on electron, 97
Covalent semiconductors, 52, 294
Cross-polarized geometry, 254
Cryogenic Underground Observatory for Rare Events, 124
Crystal
    detectors, 1
    momentum, 9
    orientation, 76
    structures, 8
    vibrations, 130
Crystal growth, 67
    alloying, 81–83
    annealing and dopant activation, 89–92
    bulk crystal growth, 67–70
    CVD, 77–79
    dopant incorporation during bulk crystal
        growth, 70–74
    doping by diffusion, 83–85
    ion implantation, 85–89
    LPE, 75–77
    MBE, 79–81
    neutron transmutation, 92–94
    thin film growth, 74–75
Cubic cell, 3

Cubic crystals, 3–6
Curie temperature, 126
Current density, 222
CVD, *see* Chemical vapor deposition
C-V profiling technique, *see* Capacitance–voltage
  profiling technique
cw laser, *see* Continuous-wave laser
Czochralski crystal puller, 68
Czochralski-grown crystals, 37
  Si crystal, 37
Czochralski-grown silicon samples, 293
Czochralski technique (CZ technique), 68, 69–70
  crystal growth techniques, 70

D

DAC, *see* Diamond-anvil cell
Damage, 87–88
Damping
  bonds, 41
  factor, 139
Dark field microscopy, 305, 320
de Broglie wavelength, 310
Deep levels, 33, 108–113
  defects, 33, 239, 240
  impurities, 33
Deep-level transient spectroscopy (DLTS), 187, 221,
    235–238
  minority carriers and, 238–240
Defect
  classifications, 29
  definitions, 29–30
  dislocations, 42–46
  emission, 250
  energy levels, 30–33
  examples of native defects, 33–35
  examples of nonhydrogenic impurities, 35–38
  formation, 89–90, 193–195
  formation enthalpy, 195, 198
  hydrogen, 38–39
  symmetry, 39–42
  systems, 29
  vibrational modes, 133–138, 140
Deflection system, 310
Deformation
  potentials, 16
  types, 262
Degenerate semiconductor, 125
Degree of iconicity, 25
Delayed gamma NAA, 301
Delocalized wave function, 100, 101
Delta-doped layer, 202
Density functional theory (DFT), 111
Density of states, 113, 333–334
Dephasing time, 141
Depletion
  layer, 49–50
  width, 55–56
Destructive characterization technique, 285
Detection system, 310
DFT, *see* Density functional theory
Diamond-anvil cell (DAC), 263
Diamond crystal structure, 4

Diamond semiconductor, 49
Diamond structures, 9
  first Brillouin zone for, 9–10
Diatomic linear chain, 130, 134, 331
  model, 131
Diatomic model, 136
Dichroism, 262
Dielectric function, 164
Diffraction, 318–319
  Bragg, 318, 319
  contrast imaging, 319
  EBSD, 319
  electron, 318, 319
  LEED, 319
  RHEED, 80, 319
  SAD, 321
  XRD, 216–217, 318
Diffusion, 193, 199–204
  coefficient or diffusivity, 200
  diffusion-driven solid-phase reaction, 53
  doping by, 83–85
  furnace design, 83
  of host and impurity atoms, 214–215
  microscopic mechanisms of, 204–207
Dilute magnetic semiconductors, 126
Dimethyl zinc (DMZn), 78
Dipole-allowed, absorption or emission of light, 168
Dipole-forbidden, 168
Dipole transitions, 168–169, 170
Direct-gap semiconductor, 170
Direct band gap, 15
Directions of cubic crystals, 3–4
Disilicides (MSi_2), 54
Dislocations, 42–46, 308, 315
  core reconstruction, 44
  density, 45
Dispersion relation, 131
Dissociative mechanism, 206, 214
Distribution coefficient, *see* Segregation coefficient
Divacancy, 34
DLTS, *see* Deep-level transient spectroscopy
DMZn, *see* Dimethyl zinc
Donor, 19, 30, 97
  binding energy, 30, 99, 102
  bound exciton, 178–179
  electron, 34
  level, 31
  states, 52
  wave functions, 105–108
Donor–acceptor emission, 178–182
Donor electron wave function, 101
  in one-dimensional crystal, 100
Donor-hydrogen complexes in silicon, 146
Dopant(s), 29, 83, 206–207
  activation, 89–92
  affect band-gap absorption, 172–173
  diffusion, 210–214
  impurities, 32
  incorporation during bulk crystal growth, 70–74
  sources and temperatures, 84
Doped germanium crystals, 69
Doping, 19–20, 67, 277
  alloying, 81–83

342 Index

Doping (*Continued*)
annealing and dopant activation, 89–92
bulk crystal growth, 67–70
conditions, 59
CVD, 77–79
by diffusion, 83–85
dopant incorporation during bulk crystal growth, 70–74
ion implantation, 85–89
LPE, 75–77
MBE, 79–81
neutron transmutation, 92–94
thin film growth, 74–75
Doppler shift, 295
d orbitals, 36
Drift mobility, 240
Drift velocity, 23–24, 119
Drive-in step, 84
Drude model, 163
DX centers, 35, 36
Dynamic SIMS, 285
Dynode, 249

E

E_a, *see* (0/−) level
EBIC, *see* Electron beam induced current microscopy
EBSD, *see* Electron backscattering diffraction
E center, 34
E_d, *see* (0/+) level
Eddy currents, 69, 227
Edge dislocation, 42, 43
EDS, *see* Energy dispersive x-ray spectroscopy
EDX, *see* Energy dispersive x-ray spectroscopy (EDS)
EELS, *see* Electron energy loss spectroscopy
"Effective density of states", 114
Effective mass (EM), 266
theory, 99–100
EFG, *see* Electric field gradient
Einstein relation, 241
EL, *see* Electroluminescence
Elastic, 120
Electrical breakdown, 58
Electrical conductivity, 122, 222
Electrical measurements
capacitance–voltage profiling, 231–234
carrier emission and capture, 234–235
DLTS, 235–238
Hall-effect, 228–231
methods of measuring resistivity, 223–228
minority carrier lifetime, 240–241
minority carriers and DLTS, 238–240
resistivity and conductivity, 222–223
thermoelectric effect, 241–242
Electrical properties of semiconductors, 23
Electric field gradient (EFG), 299
Electroluminescence (EL), 175, 251
Electromagnet, 310
Electromagnetic wave, 139
Electron-beam methods, 318
Electron backscattering diffraction (EBSD), 319
Electron beam induced current microscopy (EBIC), 316–318

Electron energy loss spectroscopy (EELS), 322
Electron-hole
drop, 180–181
pairs, 161
recombination, 62
Electronic
absorption spectra, 257
scheme, 237
stopping power, 283
transport, 23–24
wave function, 182
Electronic properties, 97; *see also* Thermal properties; Optical properties; Vibrational properties
carrier concentrations as function of temperature, 113–116
deep levels, 108–113
donor and acceptor wave functions, 105–108
freeze-out curves, 116–119
hopping and impurity band conduction, 122–125
hydrogenic model, 97–102
scattering processes, 119–122
spintronics, 125–126
wave function symmetry, 102–105
Electron nuclear double resonance (ENDOR), 270–273
Electron paramagnetic resonance (EPR), 34, 265–269
Electron(s), 16–19, 61, 97
affinity, 49
bands in 1D crystal, 14
beams, 292–294
in conduction band, 10–11
density, 14–15
diffraction, 318, 319
effective mass, 17, 18
free, 19, 161, 186
gun, 310
irradiation, 293
secondary, 312
spectrometer, 291
spin, 105
Zeeman splitting, 272
Elemental semiconductors, 25, 30
EM, *see* Effective mass
Emission, 22, 250–252
coefficient, 234, 235
spectrum, 250
Emitted x-rays, 287
Emitter, 2, 58
Emitter push effect, 213
ENDOR, *see* Electron nuclear double resonance
Energetic ion beams, 285
Energy
analyzer plus detector, 291
levels, 29, 30–33
loss, 87
straggling, 279
Energy dispersive x-ray spectroscopy (EDS), 287, 293
Enthalpy, 193, 198, 204
formation, 195, 196, 198–199, 212, 216
migration, 204
self-diffusion activation, 207
Entropy, 193, 204
migration, 204
vibrational, 194

Index **343**

Envelope function, 100
Epitaxial growth techniques, 75
Epitaxy, 74–75
EPR, *see* Electron paramagnetic resonance
Equation of motion, 138
Etchants for semiconductors, 308
Etch pits, 309
Eutectic temperature, 54, 55
Evaporation, 82–83
EXAFS, *see* Extended x-ray absorption fine structure
Exchange-correlation functional, 112
Excitation process, 259
Exciton(s), 171–172
    binding energies, 251
    and donor–acceptor emission, 178–182
    free, 172
    Wannier, 172
Extended defects, 42
Extended x-ray absorption fine structure (EXAFS), 288, 289
Extrinsic defects, 29
Extrinsic semiconductors, 19, 231
Extrinsic stacking fault, 35

F

Fabry–Perot interference fringes, 248, 250
Face-centered cubic (fcc), 3, 4
Fano resonance, 166
Far-infrared (Far-IR), 249
    FTIR spectroscopy, 259
fcc, *see* Face-centered cubic
Feedback current, 227
Fellgett advantage of FTIR spectroscopy, 257
Femto-Farad range (10^{-15} F), 233
Femtosecond (10^{-15} s), 247
Fermi–Dirac distribution function, 20
Fermi distribution function, 113
Fermi energy, 20
Fermi levels, 32, 52–53
Fermi resonance, 146
Fermi stabilization energy, 196
FFT algorithm, *see* Fourier transform algorithm
Fick's first law, 200
Fick's second law, 200
Fingerprints, 290
    IR spectral lines, 188
Finite source, 201
First Brillouin zone, 9–10
    for diamond and zincblende crystals, 10
First harmonic mode, 137
Floating zone technique (FZ technique), 69–70
    crystal growth techniques, 70
Forbidden gap, 132
Force constant, 130
Formation enthalpy, 195, 196, 198–199, 212, 216
Forward bias, 57
    condition, 50
Four-point probe, 223, 224
Fourier transform algorithm (FFT algorithm), 257
Fourier transformation, 100–101
Fourier transform infrared spectroscopy (FTIR spectroscopy), 247, 255–258, 291
    Far-IR, 259

Fellgett advantage, 257
Franck–Condon shift, 185, 186
Frank–Turnbull mechanism, 206
Free carriers, 23
    absorption and reflection, 161–164
    concentration, 222
    recombination, 174
Free electrons, 19, 161, 186
    approximation, 11
    concentration, 116, 234
    laser, 262
Free exciton, 172
Freeze-out curves, 116–119
Frenkel defect, 29
FTIR spectroscopy, *see* Fourier transform infrared
        spectroscopy
"Full-slope" freeze-out regime, 116
"Fully symmetric" representation, 102
"Fundamental" mode, 137
FZ technique, *see* Floating zone technique

G

GaAs, *see* Gallium arsenide
Galena (PbS), 1, 7
Gallium arsenide (GaAs), 3, 25, 42, 80, 106, 297, 298, 309, 323
    deep-level defect in, 33
    defects in, 30
    impurity vibrational modes in, 151–155
Gallium nitride (GaN), 3, 6, 11, 315
Gamma-ray energies, 301
Γ band minimum shifts, 16
Gamma NAA, 301
Γ point, 10, 18
GaN, *see* Gallium nitride
Gap mode, 135
Gas immersion laser deposition, 91
Gate, 59
    oxide, 61
Gaussian distribution, 285
Gaussian doping profile, 88
Gaussian shape, 240
Generation rate, 314, 316
Germanium, 69, 94, 106, 241
    oxygen in, 147–151
Gibbs free energy, 193, 194, 197
"Globar", 248
"Gradient-freeze" technique, 70
Grain boundaries, 45
Great precaution, 239
Green's function approach, 109–111, 335
Ground-state recovery time, 141
Group-IV impurities, 19–20
Group-IV semiconductors silicon and germanium, 30

H

"Half-slope" freezeout regime, 116
Hall bar, 228
Hall coefficient, 228
Hall-effect, 228–231
    measurements, 227

344 Index

Hall voltage, 228
Haynes–Shockley experiment, 240
hcp structure, *see* Hexagonal close packed structure
Heating, 89
Heavy holes, 18
"Helium-like" centers, 101–102
Heteroepitaxy, 74–75
Heterostructures, 62
Hexagonal close packed structure (hcp structure), 6
Hexagonal polytypes, 6
Hexagonal wurtzite structure, 6
High-resolution IR spectroscopy, 151
High-resolution SEM, 313
High-resolution transmission electron microscopy
    (HRTEM), 321
Hillocks, 42
Hole(s), 16–19
    effective masses, 19
    masses, 18
Homoepitaxy, 74–75
Hopping conduction, 122–125
"Hot probe", 241
HRTEM, *see* High-resolution transmission electron
    microscopy
Huang–Rhys factor, 184–185
Hybrid functionals, 112
Hydrogen, 38–39, 112–113, 152
    compensation, 38
    Hyrogen stretch modes, 155
    molecules, 156–157
    passivation, 38
    platelets, 39
    vibrational modes, 155–157
    wave functions, 109
Hydrogenic donors, 30
Hydrogenic model, 97–102
Hydrogenic theory, 108
Hydrogenic wave functions, 184
Hydrostatic pressures, 148–149, 262, 263, 264
Hyperfine splitting, 267, 272

I

Ideal metal-semiconductor junctions, 49–51
Identity element, 40
III-V semiconductors, 323
Implant homogeneity, 86
Improper rotations, 40
Impurity/impurities, 30, 240
    atoms, 29
    band conduction, 122–125
    diffusion, 205
    segregation, 71
    striations, 71, 72
Impurity vibrational modes in GaAs, 151–155
Indirect-gap semiconductors, 21, 23, 106, 169–170
Indirect gaps, 15–16
Indirect mechanisms, 205
Indium-111, 299
Indium gallium nitride (InGaN), 175, 249, 250, 319
    layer, 45, 62, 217
    quantum wells, 62, 216, 250, 286, 320
Indium nitride (InN), 298

Indium phosphide (InP), 309
Indium tin oxide (ITO), 75
Inelastic scattering, 293
Inexhaustible source, 200
Infrared (IR), 3, 25, 106, 129
    absorption, 138–140
    IR-absorption band, 150
    "IR-allowed", transitions, 168
    "IR-forbidden", 168
    measurements, 248
    pump-probe experiments, 262
    spectrum, 161
    spectrum for phosphorus donors in silicon, 106, 107
    transmission microscope, 309
Infrared Astronomical Satellite, 260
InGaN, *see* Indium gallium nitride
InN, *see* Indium nitride
InP, *see* Indium phosphide
Integrated absorption, 139
Integration time, 61
Interactions and lifetimes, 140–142
Interatomic potential, 87
Interdiffusion of indium and gallium in InGaN quantum
    wells, 216
Interfaces and devices, 49
    applications of *p-n* junctions, 58–59
    CCD, 61–62
    depletion width, 55–56
    ideal metal-semiconductor junctions, 49–51
    light-emitting devices, 62–63
    MOS junction, 59–61
    p-n junction, 56–58
    real metal-semiconductor junctions, 51–55
    2DEG, 63–64
Interference contrast microscopy, 306
Interference pattern, 321
Interferogram, 256
"Intermediate" lens, 319
Internal quantum efficiency, 315
Internal stresses, 309
Interstitial(s), 30, 34–35
    host atom, 29
    hydrogen, 112, 156
    oxygen, 37, 264
    in semiconductors, 195
    transition metals, 36
    wind, 213
Interstitialcy mechanism, 35, 206, 214
Intrinsic defects, 29, 33–35, 206–207
Intrinsic point defects, 35
Intrinsic regime, 117
Intrinsic regions, 211
Intrinsic semiconductors, 19, 115
Inverse Fourier transform, 256
Inversion, 42, 60
Ion beam apertures, 86
Ion channeling, 285
Ionic crystal, 8
Ion implantation, 85–89, 277
Ion implantation machine, 85
Ionization energy, 30, 314
Ionized impurities, 121
    scattering, 24

Index   **345**

Ion range, 87, 281–285
Ion scattering processes, 285
IR, *see* Infrared
Isoelectronic impurities, 36, 182–184
Isoelectronic molecule, 183
Isolated vacancies, 34
Isotropic effective mass, 333
ITO, *see* Indium tin oxide

J

Jacquinot advantage of FTIR spectroscopy, 257
Jahn–Teller distortion, 41

K

Kick-out mechanism, 206, 213, 215
"Kikuchi lines", 319
K-space diagram of phonon scattering, 120

L

LA frequencies, *see* Longitudinal acoustical frequencies
"Lamella", 229
Landé factors, 270
Landé g-factor, 266, 267
Laplace transformation, 237
Laplace-transform DLTS, 237
Laser diodes, 25, 62
Lattice
   constant, 3
   relaxation, 33, 184–187
   vacancy, 29
   vibrations, 164–168
   waves, 8
LCAO, *see* Linear combination of atomic orbitals
LDA, *see* Local density approximation
"LDA+U" method, 112
LEC, *see* Liquid encapsulated Czochralski
LEDs, *see* Light-emitting diodes
LEED, *see* Low-energyelectron diffraction
Light-emitting devices, 62–63
Light-emitting diodes (LEDs), 25, 62, 76, 174
Light holes, 18
Light source, 248
Lindhard, Scharff, and Schiøtt theory (LSS theory), 282, 283, 285
Linear algebra, 330–331
Linear chain, 331–332
Linear combination of atomic orbitals (LCAO), 11, 13
Line defects, 29, 42
Liquid encapsulated Czochralski (LEC), 70
Liquid phase epitaxy (LPE), 75–77
Liquid semiconductor, 69
Lithium, 39
   detectors, 287
Lithium-drifted silicon, 301
Local density approximation (LDA), 112
Localon, 147, 264
Local vibrational mode (LVM), 133, 134, 166, 262
Lock-in amplifier, 247, 262
Lock-in detector, 266
LO frequencies, *see* Longitudinal optical frequencies

Longitudinal acoustical frequencies (LA frequencies), 132
Longitudinal mode, 152
Longitudinal optical frequencies (LO frequencies), 132, 165
Longitudinal wave, 132
Lorentz–Drude model, 165
Lorentz force, 228
Lorentzian function, 140–141
Low-energyelectron diffraction (LEED), 319
LPE, *see* Liquid phase epitaxy
LSS theory, *see* Lindhard, Scharff, and Schiøtt theory
Luttinger parameters, 19
LVM, *see* Local vibrational mode

M

M81 galaxy, *see* Messier 81 galaxy
Magnesium diffusion, 286
Magnetic lenses, 310
Magnetic resonance techniques, 265, 273
Magnetron sputtering, 83
Majority carriers, 122, 238
Manganese, 273
Mass, 330
   defect, 133, 135
   mass-action law, 114
   spectrometer, 285
Matrices in quantum mechanics, 334–335
Matthiessen's rule, 122
Maxwell–Boltzmann distribution, 294
MBE, *see* Molecular beam epitaxy
"Mean field" potential, 111–112
Melt, 67
Melting temperature, *see* Eutectic temperature
Mercury arc lamps, 248
Messier 81 galaxy (M81 galaxy), 258
Metal, 50
   Schottky barrier, 55
   tip rectifier, 53
Metal-induced gap states (MIGS), 52
Metal-insulator transition, 125
Metal-oxide-semiconductor (MOS), 3, 38, 59
   junction, 59–61
Metal-semiconductor junctions, 49, 55, 56
Metalorganic chemical vapor deposition (MOCVD), 75, 315
Metalorganic molecular beam epitaxy (MOMBE), *see* Chemical beam epitaxy (CBE)
Metastable state, 186, 268
   supercooled state, 67
MgZnO, 312, 313
Microchannel plate, 251
Microscopic mechanisms of diffusion, 204–207
Microscopy and structural characterization, 305
   CL, 314–316
   diffraction, 318–319
   EBIC, 316–318
   optical microscopy, 305–310
   SEM, 310–313
   SPM, 322–326
   TEM, 319–322
Migration enthalpy, 204

346 Index

Migration entropy, 204
MIGS, *see* Metal-induced gap states
Miller indices, 4, 318
Miller profiling technique, 233
Minimization of dislocation energy, 44
Minority carriers, 122
 and DLTS, 238–240
 lifetime, 240–241
 recombination lifetime, 176
MIPS, *see* Multiband Imaging Photometer
Misfit dislocation, 45, 46
"Missing atoms", 34
MnGaAs, 325–326
Mobility, 23–24
MOCVD, *see* Metalorganic chemical vapor deposition
Molecular beam epitaxy (MBE), 75, 79–81
Momentum relaxation time, 23
Monochromators, 249, 253
 plus detector, 250
Monte Carlo simulation, 312
 of electron trajectories, 292
Morse potential, 137
MOS, *see* Metal-oxide-semiconductor
MOS field effect transistor (MOSFET), 59
MQW, *see* Multiple quantum wells
Multiband Imaging Photometer (MIPS), 258–259
Multichannel detection, 250
Multichannel PHAs, 279
Multiphonon absorption, 165
Multiple quantum wells (MQW), 217–218
Muons, 296–298
Muon spin rotation (μSR), 297
 frequencies, 297, 298
μSR, *see* Muon spin rotation

N

NA, *see* Numerical aperture
NAA, *see* Neutron activation analysis
NaI, 295
Native defects, *see* Intrinsic defects
n-channel MOSFET device, 59
Nd-doped yttrium aluminum garnet laser (Nd:YAG laser), 260, 261
Near-field microscopy, 323
Near-field scanning optical microscopy (NSOM), 323, 324
Nearly free electron approach, 13
Negative-U center, 268
Negative-U defects, 33
Negative Seebeck coefficient, 242
Net donor concentration, 115
Neutral donor–acceptor pair (D^0 A^0), 269
Neutral impurity scattering, 121–122
Neutron activation analysis (NAA), 301
Neutron transmutation, 92–94
Neutron transmutation doping (NTD), 93
Newton's second law, 97
Nitridation, 214
Nitrogen, 37, 251
 Nitrogen-14, 267
NMR transitions, *see* Nuclear magnetic resonance transitions

Nonequilibrium thermal annealing approaches, 90
Nonhydrogenic impurities examples, 35–38
Nonpolar solids, 132
Nonradiative processes, 174–175, 179
Nonstoichiometric materials, 35
Normal-mode coordinate, 136, 184
Normal freezing, 71, 73
"Normalized" diffusion depth, 201
Notch filter, 253
NRA, *see* Nuclear reaction analysis
NSOM, *see* Near-field scanning optical microscopy
NTD, *see* Neutron transmutation doping
N-type contacts on GaAs, 82
n-type germanium, 2
n-type semiconductor, 2, 38, 49, 118–119
n-type ZnO nanowire, 324
Nuclear Coulomb potential, 282
Nuclear magnetic resonance transitions (NMR transitions), 270
Nuclear reaction analysis (NRA), 301
Nuclear reactions, 301
Nuclear spins, 271
Nuclear stopping power, 282
Nuclear Zeeman shift, 271
Nuclear Zeeman splitting, 272
Nucleation sites for single-crystal growth, 68
Numerical aperture (NA), 306
Numerical computation programs, 285

O

ODMR, *see* Optically detected magnetic resonance
Ohmic contacts, 51, 53, 81, 259
Ohm's law, 1, 222
Ohms per square (Ω/Υ), 225
1D harmonic oscillator, 144
"Optical" phonon, 9
Optical branch, 132
Optical chopper, 262
Optical generation, 173–174
Optically detected magnetic resonance (ODMR), 269–270
Optical microscopy, 305–310
Optical parametric oscillator, 262
Optical properties, 21–23, 161; *see also* Electronic properties; Thermal properties; Vibrational properties
 band-gap absorption, 169–173
 carrier dynamics, 173–178
 dipole transitions, 168–169, 170
 exciton and donor–acceptor emission, 178–182
 free-carrier absorption and reflection, 161–164
 isoelectronic impurities, 182–184
 lattice relaxation, 184–187
 lattice vibrations, 164–168
 transition metals, 187–188
Optical spectroscopy
 absorption, 248–250
 applied stress, 262–265
 emission, 250–252
 ENDOR, 270–273
 EPR, 265–269
 FTIR spectroscopy, 255–258

ODMR, 269–270
photoconductivity, 258–260
Raman spectroscopy, 252–255
time-resolved techniques, 260–262
Optical transmission system, 248
Orbital radius, 98
Organometallic vapor phase epitaxy (OMVPE), *see* Metalorganic chemical vapor deposition (MOCVD)
Oscillations, 140
Oscillator strength, 168
Ostwald–Miers range, 67
Out-diffusion of bulk dopants, 201
"Output register", 62
"Overtone" mode, 137
Oxide, 84
materials, 25
Oxygen, 37
atom, 33–34
in silicon and germanium, 147–151
vacancy, 33

P

PACS, *see* Perturbed angular correlation spectroscopy
Panchromatic CL image, 315
Partial dislocation, 44
Partially passivated impurities, 38
Particle-beam methods, 277, 313
electron beams, 292–294
ion range, 281–285
muons, 296–298
nuclear reactions, 301
PACS, 298–300
photoelectric effect, 290–292
positron annihilation, 294–296
RBS, 277–281
SIMS, 285–286
x-ray absorption, 288–290
x-ray emission, 287–288
Particle-induced x-ray emission (PIXE), 287
Particles, 277, 332
Passivate acceptors, 39
Passivation, 34, 38
Pauli exclusion principle, 2, 129
P_b defect, 60
PC, *see* Photoconductivity
Perfect crystals, 49
Persistent photoconductivity, 186
Perturbed angular correlation spectroscopy (PACS), 298–300
PHAs, *see* Pulse height analyzers
Phonon(s), 8–10, 129–133
density of states, 135
in intrinsic semiconductors, 129
phonon-absorption term, 171–172
replicas, 178
scattering, 24
scattering in k space, 120
"Phosphors", 25
Phosphorus, 221
Phosphorus–hydrogen complex in silicon, 39, 144–145
Photoconductivity (PC), 258–260, 263

decay measurements, 241
Photoelectric effect, 290–292
Photoelectron, 288
spectroscopy, 247
Photographic film, 305
Photoluminescence (PL), 175, 250, 252, 258, 306, 307
emission process, 269
Photoluminescence excitation spectroscopy (PLE spectroscopy), 251, 252
Photomultiplier tube (PMT), 249
Photons, 58, 61, 238, 247, 259
cause vertical transitions, 21
energy, 21
Photothermal ionization spectroscopy (PTIS), 259–260
Photovoltaic detectors, 58
Physical constants, 337
Physical vapor deposition, 82–83
"Physics of dirt" phenomena, 1
"Pile-down" impurities, 202
"Pile-up" of impurities, 202
PIXE, *see* Particle-induced x-ray emission
PL, *see* Photoluminescence
Planck's constant, 310
Planes, 4
Plasma frequency, 163
Plasma reflection, 163
Plasmon–LO coupling model, 166
Plasmons, 293
PLD, *see* Pulsed laser deposition
PLE spectroscopy, *see* Photoluminescence excitation spectroscopy
PMMA, *see* Poly(methylmethacrylate)
PMT, *see* Photomultiplier tube
p-n diodes, 58
p-n junction, 56–58
applications, 58–59
Point contact transistor, 2
Point defects, 29, 39–40, 42
Poisson equation, 55
Polarizers, 253
"Pole" of G, 335
Poly(methylmethacrylate) (PMMA), 264
Polytypes, 6
Positive Seebeck coefficient, 242
Positive voltage, 57
Positron, 294
annihilation, 34, 294–296
Post-growth doping methods, 67
"Powder diffraction" pattern, 319
Preamorphization of silicon, 88
Predeposition step, 84, 204
"Projector" lens, 319
Prompt gamma NAA, 301
"Pseudolocalized" mode, 149
Pseudomorphic, InGaN layer, 45
Pseudomorphic strain, 45
Pseudopotentials, 111
PTIS, *see* Photothermal ionization spectroscopy
p-type semiconductors, 2, 57, 117, 175, 195
Pulsed laser annealing, 90
Pulsed laser deposition (PLD), 75
Pulse height analyzers (PHAs), 279
Pump-probe experiment, 261–262

348 Index

Pure semiconductors, 1–2, 19
Purification, 71
Pyrite (FeS), 1

Q

Quality factor (Q), 266
Quantum-mechanical derivation, 142
Quantum-mechanical system, 334
Quantum-mechanical tunneling, 169
Quantum mechanics
    matrices in, 334–335
    rules, 107
Quantum wells, 62, 75
    InGaN, 62, 216, 250, 286, 320
    intermixing, 214–218

R

Radiation
    detectors, 301
    radiation-enhanced diffusion, 89
Radiative recombination, 174–175
Radioactive isotope, 295
Radio frequency (RF), 68, 227, 271
Raman scattering, 142–144
Raman spectroscopy, 252–255, 306
Raman spectrum, 129
Raman tensor, 253
Rapid thermal annealing (RTA), 91, 167, 168
Ray-tracing diagram, 305
RBS, *see* Rutherford backscattering spectrometry
Real metal-semiconductor junctions, 51–55
Reduced zone scheme, 14
Reflectance, 162
"Reflecting" surface, 201
Reflection high-energy electron diffraction (RHEED), 80, 319
    oscillations, 80
Reflections, 40
    effects, 21
Refractive index, 139
"Relative" dielectric function, 162
Resistivity, 222–223, 229
    methods of measuring resistivity, 223–228
Resonant mode, 135, 149
Resonant Raman scattering, 254–255
Reststrahlen band, 132, 153, 165
Reverse-biased diode, 233
Reverse bias, 50–51, 58
RF, *see* Radio frequency
RHEED, *see* Reflection high-energy electron diffraction
Rocking curve, 319
Rocksalt (NaCl) structure, 7
Rotations, 40
RTA, *see* Rapid thermal annealing
Rutherford backscattering, 277
Rutherford backscattering spectrometry (RBS), 277–281
Rutherford scattering differential cross section, 278
Rutile structure, 7
    crystal structure, 8

S

SAD, *see* Selected area diffraction
Saturation
    regime, 116
    velocity, 24
Scanning-probe techniques, 305
Scanning electron microscopy (SEM), 292, 305, 310–313
Scanning probe microscopy (SPM), 322–326
Scanning tunneling microscopy (STM), 322
Scattering mechanisms, 97
Scattering processes, 119–122
Scherrer formula, 319
Schottky barrier, 50, 51, 53, 57
    heights, 54, 55
    junction, 231
Schottky defect, 29
Schottky diode, 266
Schottky junctions, *see* Metal-semiconductor junctions
Schrödinger equation, 332–333
Scintillation detectors, 295, 310
Screw dislocation, 43
Secondary electrons, 312
Secondary ion mass spectrometry (SIMS), 207, 277, 285–286
"Second harmonic" mode, 137
Second harmonic generation, 260
Seebeck coefficient, 242
Segregation coefficient, 70, 202
Selected area diffraction (SAD), 321
Selenium, 1, 32
Self-compensation, 195
Self-diffusion, 205, 207–210
    activation enthalpy, 207
SEM, *see* Scanning electron microscopy
Semiconductor(s), 1, 50, 97, 161, 298
    band gap, 10–11
    band theory, 11–16
    characterization, 277
    cubic crystals, 3–6
    doped with transition metals, 126
    doping, 19–20
    electronic transport, 23–24
    electrons and holes, 16–19
    examples, 24–25
    historical overview, 1–3
    nanocrystals, 326
    optical properties, 21–23
    other crystals, 6–8
    phonons and brillouin zone, 8–10
    properties, 221
    properties of important semiconductors, 12–13
    work function, 49
Semi-insulating GaAs (SI GaAs), 20, 70
Shallow donors, *see* Acceptor(s)
Shallow dopants, 258
Shallow implants, 88
Shallow levels, 30, 33
Shallow-to-deep transition, 109
Sheet resistance, 225
Shock compression, 264
Shockley–Queisser limit, 58

Shockley–Read–Hall mechanism (SRH mechanism), 176, 177
$^{28}Si_{Ga}$-H stretch mode, 154
SI GaAs, *see* Semi-insulating GaAs
Silicides, 53
Silicide-silicon technology, 54
Silicon, 2, 5, 34, 37, 41, 42, 69, 94, 106, 152–153, 211, 241, 294
    donor, 323
    oxidation, 213–214
    oxygen in, 147–151
    vacancy, 34, 104
    VPE, 77
Silicon carbide (SiC), 3
Simple cubic (sc), 3, 4
SIMS, *see* Secondary ion mass spectrometry
Single-crystal silicon, 68
Sinusoidal pattern, 321
60° dislocation, 43, 44
Slip direction, 42
Sodium-22, 295
Solar cells, 58
Solid phase epitaxy (SPE), 90
Solid solubility, 70
Sound waves, 9
Spatial resolution, 307, 322
SPE, *see* Solid phase epitaxy
Spectrometer, *see* Monochromator plus detector
Sphalerite structure, *see* Zincblende structure
Spin-on process, 85
Spintronics devices, 125–126
Spitzer telescope, 260
Split interstitial, 35
Split-off band, 19
SPM, *see* Scanning probe microscopy
Spreading resistance probe, 226–227
Spring, 330
Sputtering, 75, 83
    of aperture material, 86
SRH mechanism, *see* Shockley–Read–Hall mechanism
Stacking faults, 35
Stacking sequence, 6, 7
Static SIMS, 285
Sticking coefficient, 80
STM, *see* Scanning tunneling microscopy
Stokes process, 252
Stokes Raman scattering, 142
Stopping cross section, 279
Streak camera, 261
Stress birefringence, 309
Stretch mode, 145, 147, 149
Strontium titanate (SrTiO$_3$), 25
Substitutional
    copper in germanium, 32
    impurity, 40
    site, 19
    zinc acceptor, 215
Supercells, 111
Superlattices, 75
Surface plasmons, 293
Symmetric stretch mode, 147, 149
Symmetry, 41

T

TA frequencies, *see* Transverse acoustical frequencies
"Tapping mode", 324
TD, *see* Thermal donors
TEES, *see* Thermoelectric effect spectroscopy
TEGa, *see* Triethyl gallium
TEM, *see* Transmission electron microscopy
Temperature, carrier concentrations as function of, 113–116
Temperature gradient, 241
Tetragonal unit cell, 7
Tetrahedral bonding, 5
Thermal activation, 89
Thermal donors (TD), 37
Thermal energy, 119
Thermal instability, 58
Thermally stimulated current (TSC), 242
Thermal properties, 193; *see also* Electronic properties; Optical properties; Vibrational properties
    charge state, 195–197
    chemical potential, 197–199
    defect formation, 193–195
    diffusion, 199–204
    dopant diffusion, 210–214
    microscopic mechanisms of diffusion, 204–207
    quantum-well intermixing, 214–218
    self-diffusion, 207–210
Thermoelectric effect, 241–242
Thermoelectric effect spectroscopy (TEES), 242
Thermoluminescence, 252
Thermopower measurement, 242
Thermoprobe, 241
Thin film growth, 74–75
"Thirteen nines" pure material, 69
Threading dislocation, 46
Three-dimension (3D)
    crystals, 9
    harmonic oscillator, 144
Three-particle process, 22
Tight-binding, 11
Time-averaged power absorption, 139
Time-constant selection methods, 237
Time-correlated single photon counting experiments, 260
Time-resolved optical technique, 261
Time-resolved PL
    emission, 260–261
    of GaN, 261
Time-resolved techniques, 260–262
Time dependence of concentration, 200
Tin (Sn), 24
Tin dioxide (SnO$_2$), 7
Titanium, 7
Titanium dioxide (TiO$_2$), 7, 25
TMAl, *see* Trimethyl aluminum
TO frequencies, *see* Transverse optical frequencies
"Total energy" of electron, 33, 98, 111, 112, 129
Total impurity wave function, 106
Transient bleaching signal, 262
Transient enhanced diffusion, 214
Transition metals, 36, 187–188
    impurity, 36

350 Index

Transmission electron micrograph, 35
   of defects, 90
   of GaAs, 81
Transmission electron microscopy (TEM), 60, 319–322
Transverse acoustical frequencies (TA frequencies), 132
Transverse gap mode, 146–147
Transverse mode, 147, 149, 152
   energies, 150
Transverse optical frequencies (TO frequencies), 132, 165
Transverse waves, 132
Trap-assisted Auger recombination, 178
Triethyl gallium (TEGa), 78, 79
Trimethyl aluminum (TMAl), 78
TSC, *see* Thermally stimulated current
Tungsten filament, 248, 322
Tunneling, 58
Two-dimension (2D)
   defects, 35
   monochromatic image, 315
   potential, 149
Two-dimensional electron gas (2DEG), 63–64
Two-step diffusion process, 204

U

Ultrafast Ti:sapphire amplifiers, 260
Ultra-pure germanium, 123–124, 301
Ultraviolet (UV), 248
Ultraviolet photoelectron spectroscopy (UPS), 292
Uniaxial strain, 262, 264, 265
Uniaxial stress, 262–263
Unperturbed wave functions, 102
UPS, *see* Ultraviolet photoelectron spectroscopy
Urbach's rule, 173
UV, *see* Ultraviolet
UV/visible spectroscopy, 250

V

Vacancy–antisite pair ($V_{As} + As_{Ga}$), 196
Vacancy-assisted diffusion, 206
Vacancy-assisted process, 209, 213
Vacancy–hydrogen complexes, 155
Vacancy mechanism, 205
Vacuum tube, 1
Valence band (VB), 2, 10, 13, 50
   maxima, 18
   offsets, 63
   splitting, 19
Valence electrons, 295
Valley-orbit splitting, 106
van der Pauw geometry, 229, 230
van Der Pauw measurement, 335–336
Vapor phase epitaxy (VPE), 75
   growth, 77
   reactor, 78
Variable-temperature Hall-effect measurements, 231
"Variable range", electrons, 123
$V_{As} + As_{Ga}$, *see* Vacancy–antisite pair
VB, *see* Valence band
Vibrational "signatures", 147
Vibrational entropy, 194
Vibrational frequencies, 132

Vibrational modes, 147
Vibrational properties; *see also* Electronic properties;
       Optical properties; Thermal properties
   defect vibrational modes, 133–138
   hydrogen vibrational modes, 155–157
   impurity vibrational modes in GaAs, 151–155
   infrared absorption, 138–140
   interactions and lifetimes, 140–142
   oxygen in silicon and germanium, 147–151
   phonons, 129–133
   Raman scattering, 142–144
   wave functions and symmetry, 144–147
Vibrational quantum creation, 143–144
Vibrational quantum destruction, 143–144
VPE, *see* Vapor phase epitaxy

W

Wag modes, 145, 152
Wannier excitons, 172
Wave function, 11, 332, 333
   symmetry, 102–105, 144–147
Weak-beam diffraction contrast, 320
"White lines", 290
Work function ($e\Phi_M$), 49
Wurtzite crystals, 6

X

XANES, *see* X-ray absorption near edge spectroscopy
X-band minimum shifts, 16
X center, 37
XPS, *see* X-ray photoelectron spectroscopy
X-ray
   absorption, 288–290
   emission, 287–288
   photon, 287
X-ray absorption near edge spectroscopy (XANES), 290
X-ray diffraction (XRD), 216–217, 318
X-ray photoelectron spectroscopy (XPS), 290, 291
XRD, *see* X-ray diffraction

Z

Z contrast, 312
Zeeman splitting, 266
Zero-phonon line, 178
(0/−) acceptor levels, 36
(0/−) level, 32
(0/+) level, 31–32
Zinc, 19–20
   diffusion, 207, 215
Zincblende, 5, 9
   crystal structures, 4
   first Brillouin zone for, 9–10
   semiconductor, 49
   semiconductors, 102
Zinc oxide (ZnO), 6, 298
Zone-edge TO phonon, 165
Zone melting, 74
Zone purification, 74
Zone refining, 2